Amateur Radio
on the Move

...from Your Car, Boat, Airplane, Motorcycle or Backpack!

ARRL The national association for AMATEUR RADIO

Authors

Roger Burch, WF4N; Mike Gruber, W1MG; Terence Rybak, W8TR; and Mark Steffka, WW8MS: **Automobile Mobile**

Steve Waterman, K4CJK: **Maritime Mobile**

Dave Martin, W6KOW: **Aeronautical Mobile**

Al Brogdon, W1AB: **Motorcycle Mobile**, **RV Mobile**

John Bee, N1GNV: **HF Unplugged—Backpack Mobile**

Production

Michelle Bloom, WB1ENT

Sue Fagan— Cover Design

Jodi Morin, KA1JPA

Joe Shea

David Pingree, N1NAS

Michael Daniels

Published by **ARRL**, *The National Association for Amateur Radio* • Newington, CT 06111-1494

Contents

Foreword

We at ARRL are pleased to publish *Amateur Radio on the Move*. This book covers all forms of mobiling in Amateur Radio. It has how-to chapters describing mobiling in your car, your boat, your airplane, your motorcycle, your RV or even from your backpack while hiking in the wilderness.

Each chapter has been written by experts, people with lots of hands-on, practical experience with the joys, as well as the problems and pitfalls, of mobile operation.

- Roger Burch, WF4N; Mike Gruber, W1MG; Terence Rybak, W8TR; and Mark Steffka, WW8MS, combine their considerable talents to fully describe mobiling from your automobile.
- Steve Waterman, K4CJK, brings his decades of experience in the field of maritime communications to his chapter on maritime mobiling.
- Dave Martin, W6KOW, really brings the excitement of aeronautical mobiling to life in Chapter 3.
- Al Brogdon, W1AB, details his long love affair with mobiling from his motorcycle and from his RV in his two chapters.
- And last but not least, John Bee, N1GNV, wrote "HF Unplugged," the chapter on backpack mobile.

Newcomers and old-timers will both benefit from the vast store of practical information in *Amateur Radio on the Move*. There's something for everyone who has ever entertained an idea about getting on the air while moving.

As always, we want to hear back from you. Help us make this book even better. We've provided a handy form at the back of this book that you can use to share your ideas with us.

David Sumner, K1ZZ
Executive Vice President
Newington, Connecticut
March 2005

About The ARRL

The seed for Amateur Radio was planted in the 1890s, when Guglielmo Marconi began his experiments in wireless telegraphy. Soon he was joined by dozens, then hundreds, of others who were enthusiastic about sending and receiving messages through the air—some with a commercial interest, but others solely out of a love for this new communications medium. The United States government began licensing Amateur Radio operators in 1912.

By 1914, there were thousands of Amateur Radio operators—hams—in the United States. Hiram Percy Maxim, a leading Hartford, Connecticut, inventor and industrialist saw the need for an organization to band together this fledgling group of radio experimenters. In May 1914 he founded the American Radio Relay League (ARRL) to meet that need.

Today ARRL, with approximately 170,000 members, is the largest organization of radio amateurs in the United States. The League is a not-for-profit organization that:

- promotes interest in Amateur Radio communications and experimentation
- represents US radio amateurs in legislative matters, and
- maintains fraternalism and a high standard of conduct among Amateur Radio operators.

At League headquarters in the Hartford suburb of Newington, the staff helps serve the needs of members. ARRL is also International Secretariat for the International Amateur Radio Union, which is made up of similar societies in 150 countries around the world.

ARRL publishes the monthly journal *QST*, as well as newsletters and many publications covering all aspects of Amateur Radio. Its headquarters station, W1AW, transmits bulletins of interest to radio amateurs and Morse code practice sessions. The League also coordinates an extensive field organization, which includes volunteers who provide technical information for radio amateurs and public-service activities. ARRL also represents US amateurs with the Federal Communications Commission and other government agencies in the US and abroad.

Membership in ARRL means much more than receiving *QST* each month. In addition to the services already described, ARRL offers membership services on a personal level, such as the ARRL Volunteer Examiner Coordinator Program and a QSL bureau.

Full ARRL membership (available only to licensed radio amateurs) gives you a voice in how the affairs of the organization are governed. League policy is set by a Board of Directors (one from each of 15 Divisions). Each year, one-third of the ARRL Board of Directors stands for election by the full members they represent. The day-to-day operation of ARRL HQ is managed by an Executive Vice President and a Chief Financial Officer.

No matter what aspect of Amateur Radio attracts you, ARRL membership is relevant and important. There would be no Amateur Radio as we know it today were it not for the ARRL. We would be happy to welcome you as a member! (An Amateur Radio license is not required for Associate Membership.) For more information about ARRL and answers to any questions you may have about Amateur Radio, write or call:

ARRL—The national association for Amateur Radio
225 Main Street
Newington CT 06111-1494
(860) 594-0200
Fax: 860-594-0259
E-mail: **hq@arrl.org**
Internet: **www.arrl.org/**

Prospective new amateurs call (toll free):
800-32-NEW HAM (800-326-3942)
You can contact us also via e-mail at **newham@arrl.org**

CHAPTER 1

Mobile in Your Automobile

Most of this chapter on automobile mobile is updated from *Your Mobile Companion*, published by ARRL and written by Roger Burch, WF4N. ARRL thanks Roger for his kindness in allowing major portions of his book to be included here.

Although Roger became interested in ham radio in the mid 1960s, he finally took a break from SWLing and building gadgets to get his Novice license in 1982. He upgraded to Extra class six months later. He has worked for 30+ years in the electric power generation and automotive industries. His first love is mobile CW, but declares that his workbench is his favorite QTH! Keeping in line with his interest in mobile amateur radio, Roger also loves to build high-powered "muscle" cars.

Mike Gruber, W1MG, has also provided updates and additions to this chapter. Mike has held both a commercial and amateur radio license for over 30 years. He has a BSEE and Electrical Engineering Technology degree and is the EMC engineer in the ARRL Laboratory. Mike also did Product Reviews in the ARRL Laboratory for seven years. He operates 80 through 2 meters from both his home and mobile stations.

INTRODUCTION

Does this exchange sound familiar? *"Andy, it certainly has been a pleasure. I hate to rush off on you but I'm in the parking lot. We have our monthly office meeting this morning and I don't want to be late. Thanks for riding along with me; it sure makes the trip go faster, 73. W1EPG, clear."*

Fig 1-1—Photo of the serious engine modifications for W6ZV's 75-meter mobile installation. An additional generator, carbon-pile voltage regulator and reverse-current relay combined with hash filters were all needed for this state-of-the-art installation. This picture was published in Jan 1952 *QST*.

Fig 1-2—W6ZV's vehicle, as it looked as he tooled around 6-land in the early 1950s. The original caption read: "The high-power mobile antenna of W6ZV is a potent putter-outer and a real attention-getter." From January 1952 *QST*.

"That's perfect timing, Paul. I just pulled into the parking lot myself. I expect to get out at the usual time so may see you tonight. Have yourself a splendid day, 73. W1FG clear."

Tune in on any given day to any of the several thousand repeaters across the country and you're likely to hear a conversation a lot like this one. Besides the fact they are both hams, what else did they have in common? That's right—they were both operating mobile. There's nothing unusual about hams combining their favorite hobby with our national love affair with the automobile. When you consider the average American motorist drives more than 12,000 miles per year, it's only logical those motorists who are hams would take along a radio of some sort. Or in other words, *go mobile*.

In this chapter, we'll examine the nuts and bolts of mobile operation from your car. We'll look at who operates mobile and why. We'll take you from selecting what bands to operate to helping select the radio equipment you'll use. We'll show you how to install your equipment (and how not to). We'll help you with antenna selection and installation, as well as the routing of cables. We'll look at automotive electronic systems and how you can help your radio equipment coexist peacefully with them. And we'll be looking at lots of ways to operate mobile. But before we get down to who, what, when, where and how, let's look at a little bit of the history of mobile ham radio.

BACK IN THE GOOD OLD DAYS

Perhaps you may think the practice of driving around with a radio in the car and talking to other hams has only recently become popular. In fact, hams have been taking their radios to the streets in our hobby for a long time.

The accompanying photos in **Fig 1-1** and **1-2** show a mobile station from the early 1950s. Although some VHF mobile operation was taking place at the time, it was, for the most part, limited to metropolitan areas. The absence of repeaters on the VHF bands dictated HF operation for communication over any appreciable distance.

A January 1952 *QST* article included photos describing in detail how W6ZV put together a 1000-watt mobile installation for 75 meters, complete with a dual 10-foot long whip antenna system, engine mounted 28-volt, 150-amp generator and trunk full of kW amplifier. The ham of 50 years ago may have sported a hundred pounds of radio gear stuffed under the dash

of his Hudson Hornet, and another couple of hundred pounds of gear in the trunk. Yes, those tubes, transformers and dynamotors were pretty heavy—and quite bulky, too!

WHY ME?

Surely by now you are convinced that the Biblical observation "there is no new thing under the sun" certainly applies when it comes to mobile hamming. But have you ever given serious thought to why we operate our radios from cars, trucks, vans, boats, planes, motorcycles, bicycles busses, trains and a host of other forms of transportation? Okay, I can hear some of you seasoned "Road Warriors" out there chuckling to yourselves, "'because it's convenient!" or "it's fun!" or "everyone's doing it!" Well... all those arguments actually are quite valid. Mobile operation is convenient and it's fun, too. And it does seem just about every ham *is* doing it. But perhaps there *is* more to this concept of operating mobile than meets the eye.

ROLLIN' DOWN THE HIGHWAY

Operating mobile can add a whole new dimension to *keeping in touch*. It's one of the most popular reasons hams operate

Fig 1-3—They like their mobile antennas *big* and rugged down in Mississippi. K5IJX's Jeep Cherokee with a "Texas Bug Catcher" mobile installation. *(Photo courtesy of K5IJX.)*

mobile. It doesn't matter whether we're going around the corner or around the world—across town or across the continent. Nor does it matter whether we are staying in contact with a spouse, child, parent, friend or whoever happens to show up on frequency; having a radio along when we are on the go is our umbilical cord to the rest of the world. Granted, modern technology (cellular phones, for example) has made it possible for non-hams as well as hams to stay in touch. Although a cell phone can be quicker and more convenient, especially if you need to talk to someone who isn't a ham, if you like to ragchew for hours on end as you travel across the country, please don't send me the phone bill.

Many a traveling ham has been kept awake, informed or entertained by a fellow ham at the other end of the circuit. As those who do it will quickly attest, it's one of the best reasons in the world to operate mobile. But there are many other good reasons, too.

Fig 1-3 shows a modern mobile station belonging to K5IJX in Mississippi. They do like their big mobile antennas down in the deep south.

COMPETITIVE MOBILING: THE HEAT IS ON

Are you the competitive ham-radio sort? Does a friendly game of one-upmanship strike your fancy? If so, there is a good possibility you may have already tried your hand at contesting. There's a ham-radio contest of some sort just about every weekend throughout the year on one or more of the amateur bands. On some weekends, you may find several contests in progress, each one pitting hundreds or thousands of hams against each other, all vying for the coveted Numero Uno position in the score box. Since contests generate so much interest, each month *QST* devotes an entire column to upcoming contest rules and other operating information.

So why not try mobile contesting? If you've contested, but not while mobile, you're probably asking what possible reason there could be to operate mobile during a contest. After all, isn't it pretty demanding to just stay in the thick of a contest, much less trying to contest and drive at the same time? And besides, you don't get any extra points for operating mobile, do you?

Not necessarily. Actually, there *are* some contests that do allow extra points for contacts made while mobile. Check the contest column of your favorite magazine to find those that reward mobile operators with extra points or multipliers. They usually include, but aren't limited to, statewide contests.

Another reason for mobile contesting, and a very popular one at that, is that some contests effectively encourage mobile operation by allowing you to count multiple contacts with the same station as long as one of you changes location for each contact. Generally, on the HF bands this will require you to operate from a different county in a given state for each same-station contact to count. On the VHF and higher bands, a system of *grid squares* is used to determine your location, with each grid square you operate from allowing you to count same-station contacts the same as contacts with new stations.

It doesn't take a lot of imagination to see the implication here: The mobile ham has the advantage of no longer being the hunter; he or she is now the hunted. Each time you enter a new county or grid square, you again become "fresh meat"—just what every contester wants! You only *thought* it was necessary to mount a DXpedition to be on the receiving end of a pileup.

DX—A LITTLE EXTRA SPICE

Oh, and speaking of DX, mobile operation works the same way when you are chasing DX on the HF bands from your home station. While mobile whip antennas and barefoot rigs can't easily compete with kilowatt stations and their beam antennas when it comes to chasing DX, it is surprising how many DX stations will go

Mobile DX

DX from the car? It's not as hard as you might think. I've worked many DX stations, mostly on 15 meters late in the afternoon when things start to open up. Japan, England, Australia, New Zealand and others are there for the working if you have a little patience. DX on 40 meters is available, especially in the evening. However, I have regularly worked Mexican, Canadian and Central American stations during the day. All you need is an opportunistic spirit and a will to go for it.—*Fred Anderson, K5LX*

the extra mile to work a ham who is signing mobile. The mobile operator seems to enjoy a sort of underdog status in the heart of many DX operators— something you can certainly use to your advantage when you are trying to work a rare one.

Fig 1-4 shows how KH6DX/M6 has outfitted his pickup with a full kW mobile station on 160 meters. How well does this station work? Don has confirmed more than 100 countries worked on 160 mobile!

FILLING THE NEED

So far, we have looked at the casual or entertaining reasons for operating mobile. There is certainly nothing wrong with operating mobile strictly for the pleasure of doing so. After all, that's why most hams do it.

Now let's shift gears and look at one

very compelling reason every Amateur Radio operator should have some means of mobile operating capability. It's called *public service*. Granted, this term can encompass many activities, including message handling through the National Traffic System and providing tactical communication at parades, marathons and other public events.

Perhaps the most important aspect of public service communication, however, is furnishing communication during times of emergency or disaster. That's when mobile operators really shine. Sometimes traveling hundreds of miles to make their communications services available to an area stricken by disaster, they are the unsung heroes to many folks who have been left helpless, hurting and without communication. So if you're the type of person who likes to help others (it seems to come with the ham ticket), you'll find the ability to operate mobile is an excellent way to funnel your helping-hands energies.

For example, if you live anywhere near "Tornado Alley," you can be a valuable asset to local authorities by volunteering as a weather spotter—someone who goes to a location with a good vantage point and watches for the approach of severe weather. Or you may be dispatched to an area devastated by a tornado, where you might be the first person to provide damage reports.

On the other hand, you may live in an area prone to earthquakes. Again, your ability to get out and be on the scene of a disaster soon after it occurs can be of immeasurable worth to the victims—who may find you to be their only link to the

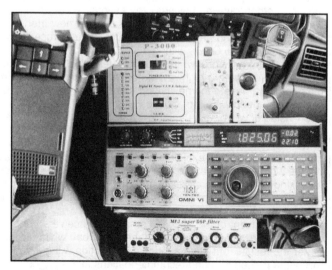

Fig 1-4—Don Stribling, KH6DX/6, has worked more than 100 countries—on 160 meters from his truck! He can run as much as 900 W output power. Don's amazing feat of working DXCC was a first from a mobile station on Topband.

outside world. Fact is, no matter where you live, the possibility of a disaster occurring naturally or otherwise, always exists. If it does occur, you'll never regret being able to provide essential communications services to those who need them. Having a well-equipped mobile station is the first step to being prepared to meet this need.

But what if you don't live on the San Andreas Fault or in Dorothy and Toto's neighborhood? Does that make it unlikely your mobile communications skills will ever be called in to action? Hardly! Each year, hundreds of mobile hams provide emergency communications by alerting authorities to the scene of an accident or potential hazard. In fact, it's the most common form of emergency communications hams provide. You might be the person who calls a service truck for a mom and her kids who have spent the past hour shivering in a freezing car, broken down on the side of the road. Or maybe you'll be the first to report a road hazard to the state police and save some traveler from receiving serious damage to their vehicle. Or perhaps you'll suddenly find someone's life dependent on you and your radio. While fortunately not a common occurrence, it has happened.

Get Involved

Most areas have a volunteer known as the *Emergency Coordinator*, or EC. The EC is part of the ARRL *Field Organization*—individuals who serve the community and their fellow hams by organizing local hams to be responsive in times of emergencies. You can contact your EC by contacting your local Section Manager, or SM. Contact information, including your SM's email address and phone number, appear each month in *QST*. You can also find this information on the ARRL Web site at **www.arrl.org/sections/?sect=**.

Years ago, only hams had the radios that let police, fire, Red Cross and other emergency services talk and coordinate activities with each other. Back then, a modern town or county may have had one hotline frequency and a few radios that could be switched to this frequency, but generally the radios and manpower to cross organization boundaries came from hams. Today, it appears that everyone carries a hand-held radio. Trunked systems on 800 MHz are replacing the VHF/UHF systems previously used, and coordination is the name of the game with trunked systems. So why do hams still play an important role?

Emergencies are, by definition, unexpected. All sorts of surprises occur and trained people are needed—those who can support themselves with equipment, spares and the know-how to respond quickly. Need a communications point at a school?—send a ham. Power out at a location and you need to talk to someone?—send a ham. Buildings in the way and normal radio can't be heard in the area?—see if the hams with a portable repeater can help.

Flexibility, quick response and trained people—this is the reputation of hams in many communities. You, your mobile and your hand-held radio are a significant part of the service.

PUBLIC SERVICE IS EXPECTED

Okay, so you're ready to head out on the highway and do some public service work. Good. You'll have lots of company; there's nothing hams like better than to get out and show off their communications skills. But are you aware hams are actually *expected* to provide public service communications? I can hear you now: *"Nope, looked over my ticket and the FCC Rule Book, and they don't say anything at all about me being required to provide*

Meet Fred, K5LX Mobile

Ham radio to me is enjoyment of communicating with others. It's also most of my daily activity as I drive 65,000 miles a year in my job as a pharmaceutical sales representative in West Texas and New Mexico—typically operating 40-meter SSB. In my 41 years on the air, HF mobile has been consistently the most interesting, challenging and rewarding aspect of amateur radio I've ever encountered. Conditions and location change daily, even moment by moment, which sometimes can really test my operating skills. Equipment, antennas and installation must also be optimized for top performance and reliability. Once conquered, little compares with 70 mph on the open road, cruise control on, headset on, 40 meters in the display and rag chewing with friends, new and old.

Mobile operation consists of so many fun and rewarding aspects it's hard to mention them all. Whether heading out early talking to other traveling business representatives or chasing the evening sun into El Paso, TX, ham radio is my constant companion, introducing me to hundreds of great new friends, many of whom I have gotten to know personally. There are frequent dinners and breakfasts with hams in Las Cruces, NM—Bill, W5UMQ, and Jack, N5PK—not to mention stopping by for short eyeball QSOs with C.B., W5FLA, Dave, W5DBC, and George WB5RWF, in west Texas. One of the greatest challenges is tracking down my buddies Jeff, N5UJJ, and Carl, KC5RWN, who are mobile all the time like me. We try to cross paths out on the highway whenever we can. There are many others I am privileged to talk to every day. A good many of them have become lifelong friends. Ham radio friendships never QSB, and are always 5 by 9 plus.

If all this sounds like fun, mobile HF is for you, oh—and I haven't even mentioned the pleasure of just watching the wide-open vistas of west Texas and New Mexico pass by as I drive down the highway to my next account. I've also had the pleasure of visiting radio club meetings in other towns and making new friends. As you might expect by looking at my antenna, I also get lots of questions from non-hams. These are great opportunities for me to introduce Amateur Radio to the general public as I explain that, although I probably can talk to MARS, most of my time is spent visiting with other people who, like me, have a love of the radio art called amateur or "ham" radio. Mobile operation can also demonstrate what kind of people hams really are. There are times when only we as hams can provide communications and others can't. Our support in times of emergency saves time, money, and most of all, lives. Hams give of their time and expertise to help others, and they do it gladly!

With the advent of the new and smaller transceivers, mobile operation has become easier than ever, affording every radio enthusiast the chance to see the open road while talking to others. Of course, the better the antenna, the better the better the results, but even a modest installation will get you on the air. Be aware. If you try it, you'll be hooked. If you follow the procedures and recommendations in the pages of this book, many years, and miles of enjoyment await you in mobile ham radio. Get out there! See you on 7.195 MHz.

public service communications!"

Well, you're absolutely right. We aren't required. But it is expected. "*By whom?*" you ask? By the FCC. The FCC has always expected Amateur Radio operators to provide public service communications. And realistically, we hams owe the very existence of our hobby to our long and notable history of providing communications when no one else could. Our ability to do so contributes to justifying our frequency allocations.

While some may look at Amateur Rspadio as just a hobby, the agencies that oversee the allocation and usage of our radio frequency spectrum see it much differently. They see Amateur Radio as a source of skilled communicators who are trained and prepared to step forward when needed to fill the gaps in the emergency communications systems, be it local or nationwide. And this is a good thing for us. In just the slice of radio extending from the 3.5-MHz band to the upper end of our 440-MHz band, we have more than 40 MHz of radio spectrum—much of it prime RF real estate. And it's high priced real estate at that! We are now watching as the privilege to use radio frequencies is being sold to the highest bidder, sometimes for prices effectively approaching $800 million per megahertz. And since megahertz aren't being made any anymore, it's a safe bet prices will go up. As prices go up, so will the competition for any and all available spectrum.

Today, more than ever, hams must be

Repeaters: Lifeline for the Mobile Ham

Nearly all hams have used repeaters. Why are they so popular? There are many possible answers, but the thing you will like best about repeaters is that they greatly increase the range of your radio.

Usually installed at a high location—on a tall building or perhaps on a mountain or hilltop—the repeater can hear and be heard for a much greater distance than stations operating point to point. What does that mean to you, the mobile operator? By relaying your signal the repeater can make it possible for you to talk to someone who wouldn't normally be able to hear you. You might simply be too far away, or there might be an obstruction preventing your signal from reaching him. Either way, the repeater steps in to pass the signals back and forth.

With such range-multiplying capability, repeaters must surely be super-complicated, right? Not necessarily.

What Do Repeaters Actually Do?

In its simplest form, a repeater has but one function: To receive a signal and retransmit it.

Picture it like this. You take one of your kid's toy space-ranger radios, tape the push-to-talk switch down and set it in front of your stereo system's speakers. You then take the other space-ranger radio out to the garage where you can now be kept entertained—thanks to your homemade repeater. That's pretty much how an Amateur Radio repeater works.

In a basic repeater, the receiver has its audio output fed into a transmitter, which simultaneously retransmits whatever the receiver is hearing. (Technically, it's more correct to say the repeater is simultaneously transmitting your signal, but the name "repeater" sounds better than "simultaneous-transmitter"!) Since the repeater-transmitter doesn't have a microphone, we can't tape the push-to-talk button down. Besides, we don't want the transmitter to be on all the time—just when there's someone using the repeater. So how does the transmitter know when to transmit? Again, in the most basic repeater, a carrier-sensing circuit is used to key the transmitter any time it detects a signal on the receiver. That done, your melodious voice arriving at the receiver is being squirted back into the airwaves by another transmitter.

But wait. If the transmitter and receiver are both operating at the same time, isn't the repeater hearing its own signal? It would be, if it weren't for some basic facts of repeater life. One of those facts is something called *offset*. This simply means the repeater has its receiver tuned to a different frequency than the transmitter. Let's say, for example, a repeater transmits on 146.64 MHz. This means its receiver is tuned to

146.04 MHz, an offset of 600 kHz (standard for 2-meter repeaters). You'll find different offsets for different bands: 440 MHz uses 5 MHz, on 222 MHz it's 1.6 MHz; 6 meters uses 1 MHz and on 10 meters it's 100 kHz.

Making a Contact

Let's suppose you want to use our theoretical 146.04/.64 repeater to give your friend a call. First, you would set your radio for 146.64 MHz. Then, unless your radio does so automatically, select a negative offset. When you key your radio, it will switch automatically to a transmit frequency of 146.04 MHz, where the repeater will receive your signal and retransmit it on 146.64 MHz. When you've finished talking and un-key, your radio will switch back to its receive frequency of 146.64 MHz. You will now be able to hear the repeater transmitter—and your friend, who's answering you.

Having the receiver and transmitter frequencies offset is necessary to keep a repeater from talking to itself, but it isn't enough. Even with separate transmit and receive antennas, early repeater operators found the signal from the transmitter could easily desensitize, or *desense* the receiver, making it impossible for all but the strongest signals to make it into the machine. Unsolvable dilemma? It might be if not for a device called a *duplexer*. Connected three ways between the repeater's receiver, transmitter and antenna, a duplexer is a filter that's so selective, one of its two sections passes a signal to the receiver virtually unaffected, while completely blocking the signal from the transmitter. At the same time, the other section allows almost all the power from the transmitter to pass through, while stripping off any spurious energy that might be present on the receiver frequency. The duplexer is so efficient at what it does that the transmitter and receiver can share a common antenna. That's something you would never want to try without a duplexer.

Okay, take one transmitter, one receiver and one duplexer, tune carefully and you have... yes—a repeater. Not a very exciting repeater but it's still a repeater. To spice up their machines a bit, many repeater owners add another ingredient called a *controller*. Essentially a self-contained computer, a controller takes full command of the repeater. It keys the transmitter, provides the tone known as a *courtesy beep*, connects the repeater to the phone line when you want to make an autopatch call, transmits the repeater call sign at proper intervals—and it may even talk.

Although sophisticated controllers aren't required for repeaters, everyone agrees they make repeater use much more enjoyable.

ready to prove our frequency allocations are justified and being put to good use for more than just our amusement. Naturally, one excellent way to do it is to have a well-equipped mobile station, ready to hit the road whenever it's needed. In the meantime, of course, it'll make for some very enjoyable hamming!

LET'S GO!

Okay, so you're brimming with enthusiasm. You've got the old battle-wagon cleaned up, tuned up, gassed up and shod with new rubber, ready to become a superstation on wheels. So where do you go from here? There's a lot to consider when it comes to mobile operation: Where to mount radios and antennas, what bands to operate and a host of other concerns. Let's now continue our journey into the world of mobile hamming with a closer look at these and several other aspects of Amateur Radio on wheels.

WHAT BANDS?

Before you purchase or install a mobile radio, you have to decide what bands or bands you will be operating. Sound easy? Well… perhaps. If you limit yourself to using commercially available equipment, it is still possible to operate mobile on a total of 16 different Amateur Radio bands spanning the range from 1.8 to 1300 MHz.

With such a vast expanse of radio spectrum at your disposal, your choice of bands may seem overwhelming. It need not be. With a careful evaluation of what each band has to offer, you can be sure to find your niche in mobile hamdom.

Before getting into a detailed examination of the various bands, let's look in a more general sense at what makes a band attractive to a mobile operator.

BAND RESTRICTIONS

On HF bands, there are restrictions on the 60 and 30-meter bands. All bands above the 1¼-meter band are amateur secondary and as such, amateurs can't cause harmful interference to the primary users. For the specific band sharing agreements, see Sections 97.303 and 97.307 (see **www.arrl.org/FandES/field/ regulations/news/part97/**).

It's also important to know that you are secondary in many bands in the event you hear non-amateur operation. The reason amateurs have so many allocations is because we are good sharing partners with other government and non-Government services. As for mobile operation, you should also know that in some military

bases and near certain observatories, there may be additional limitations. These are mentioned in Chapter 4 of the *ARRL FCC Rule Book*. Some of the restrictions are not mentioned in FCC Part 97 rules, but are footnotes to Section 2.106 that apply to all services, not just amateurs. All of these are mentioned in the *ARRL FCC Rule Book*.

Is Someone There?

Here's an easy one: What is the purpose of being a ham radio operator? To talk to other hams, of course. So how does that affect your choice of band(s) for mobile operation? To start with, you are going to want to choose a band with plenty of activity. You will find this choice to be even more crucial for the mobile VHF or UHF operator, whose range is typically less than a ham operating from home on the same band. Keep in mind, too, that sparsely populated bands are not limited to the spectrum above 900 MHz. Many things affect the popularity of an amateur band. Local convention, the number of hams in a given area—even the sunspot cycle—are all factors in *who* operates *where*.

Of course if you are already active on one or more bands from your home station, this might be a persuasive argument for assembling a similar setup for mobile operation. After all, you will already be familiar with the characteristics of these bands as well as knowing many of the hams who frequent them. In addition to not having to plow new ground, operating mobile on the same bands as you do from home can allow the use of the same rig for both locations. You will have the chore of transferring the equipment to and from your vehicle.

But what if you are not yet active on any of the amateur bands? Furthermore, you would like to make sure your debut as a mobile operator will not play to an empty house. Check with your Elmer (the ham who introduced you to Amateur Radio) or the members of your local ham radio club to find out what bands *they* are using for mobile operation.

Be careful, however, not to rely strictly on the opinions of others. Hams are a diverse group. You may find what appeals to other hams may not interest you at all. So once you have gotten some valuable input from your ham friends, do a little detective work on your own.

If you are considering the bands above 30 MHz, it is likely you will want to center much of your mobile operation around one or more repeaters. If this is the case, the latest copy *of The ARRL Repeater Directory* will be a valuable asset as you

map out repeaters of interest in the areas where you will be mobiling. The *Directory* provides you with the locations, call signs and frequencies of all the repeaters in the US and Canada. It also furnishes pertinent information on each repeater, such as whether an autopatch is available or if it is necessary for your rig to transmit a subaudible tone to access the machine.

Indispensable to the traveling ham, *The ARRL Repeater Directory* also comes in electronic form for use with a personal computer. With the *TravelPlus for Repeaters* CD-ROM, you have the power of *The ARRL Repeater Directory* on your computer! Use this map-based software to locate repeaters along US and Canadian travel routes. You can also prepare custom lists of repeaters from its large database. If you have a GPS, TravelPlus will pinpoint repeaters for you in real-time. For a convenient reference, *The ARRL Net Directory*'s data base is added to *TravelPlus for Repeaters* CD-ROM."

MAKING CONTACT WITHOUT REPEATERS

Keep in mind that not all VHF/UHF activity occurs on repeaters. In areas where repeaters are often busy, many hams use simplex frequencies whenever possible. Going simplex helps relieve the congestion on busy repeaters (and makes the repeater available to other hams who need the communication range a repeater provides).

Since many hams hang out on non-repeater frequencies and since the presence of a repeater in your area does not guarantee it sees a lot of use, there is no substitute for doing some *listening*. You probably will not have radios for every band you will be considering, so one of the best ways to scout for activity is with a programmable scanner. If you own or can borrow a scanner, you can use the search feature to quickly survey a band from edge to edge. Once you have easily pinpointed those frequencies popular with the hams in your area, you will be able to decide which frequencies are of interest to you when you are mobile.

AND THE WINNER IS…

Okay, suppose you have graduated from the Sherlock Holmes School of Frequency Sleuths, you have done some intensive detective work and you have found that plenty of hams in your area are on practically every band. So what do you do? Equip a mobile station that can operate from dc to daylight? If you have the means to do so you will certainly be the envy of

US Amateur Bands

ARRL *The national association for* **AMATEUR RADIO**

160 METERS

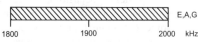

E,A,G

1800   1900   2000   kHz

Amateur stations operating at 1900-2000 kHz must not cause harmful interference to the radiolocation service and are afforded no protection from radiolocation operations.

80 METERS

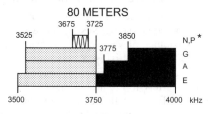

3675 3725
3525   3850   N,P *
      3775
                    G
                    A
                    E
3500   3750   4000   kHz

60 METERS

General, Advanced, and Amateur Extra licensees may use the following five channels on a secondary basis with a maximum effective radiated power of 50 W PEP relative to a half wave dipole. Only upper sideband suppressed carrier voice transmissions may be used. The frequencies are 5330.5, 5346.5, 5366.5, 5371.5 and 5403.5 kHz. The occupied bandwidth is limited to 2.8 kHz centered on 5332, 5348, 5368, 5373, and 5405 kHz respectively.

40 METERS

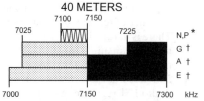

7100 7150
7025   7225   N,P *
                    G †
                    A †
                    E †
7000   7150   7300   kHz

† Phone and Image modes are permitted between 7075 and 7100 kHz for FCC licensed stations in ITU Regions 1 and 3 and by FCC licensed stations in ITU Region 2 West of 130 degrees West longitude or South of 20 degrees North latitude. See Sections 97.305(c) and 97.307(f)(11). Novice and Technician Plus licensees outside ITU Region 2 may use CW only between 7050 and 7075 kHz. See Section 97.301(e). These exemptions do not apply to stations in the continental US.

30 METERS

E,A,G

10,100   10,150   kHz

Maximum power on 30 meters is 200 watts PEP output.
Amateurs must avoid interference to the fixed service outside the US.

20 METERS

14,025   14,150   14,225
          14,175
                    G
                    A
                    E
14,000   14,150   14,350   kHz

17 METERS

E,A,G

18,068   18,110   18,168   kHz

15 METERS

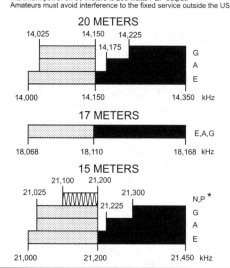

21,100   21,200
21,025   21,300   N,P *
          21,225
                    G
                    A
                    E
21,000   21,200   21,450   kHz

June 1, 2003

Novice, Advanced and Technician Plus Allocations

New Novice, Advanced and Technician Plus licenses are no longer being issued, but *existing* Novice, Technician Plus and Advanced class licenses are unchanged. Amateurs can continue to renew these licenses. Technicians who pass the 5 wpm Morse code exam *after* that date have Technician Plus privileges, although their license says Technician. They must retain the 5 wpm Certificate of Successful Completion of Examination (CSCE) as proof. The CSCE is valid indefinitely for operating authorization, but is valid only for 365 days for upgrade credit.

12 METERS

E,A,G

24,890   24,930   24,990   kHz

10 METERS

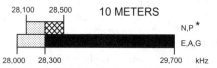

28,100 28,500
                    N,P *
                    E,A,G
28,000 28,300   29,700   kHz

Novices and Technician Plus Licensees are limited to 200 watts PEP output on 10 meters.

6 METERS

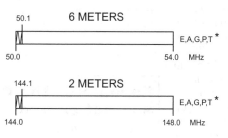

50.1
                    E,A,G,P,T *
50.0   54.0   MHz

2 METERS

144.1
                    E,A,G,P,T *
144.0   148.0   MHz

1.25 METERS ***

E,A,G,P,T,N *

222.0   225.0   MHz

Novices are limited to 25 watts PEP output from 222 to 225 MHz.

70 CENTIMETERS **

E,A,G,P,T *

420.0   450.0   MHz

33 CENTIMETERS **

E,A,G,P,T *

902.0   928.0   MHz

23 CENTIMETERS **

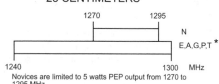

1270   1295
                    N
                    E,A,G,P,T *
1240   1300   MHz

Novices are limited to 5 watts PEP output from 1270 to 1295 MHz.

US AMATEUR POWER LIMITS

At all times, transmitter power should be kept down to that necessary to carry out the desired communications. Power is rated in watts PEP output. Unless otherwise stated, the maximum power output is 1500 W. Power for all license classes is limited to 200 W in the 10,100-10,150 kHz band and in all Novice subbands below 28,100 kHz. Novices and Technicians are restricted to 200 W in the 28,100-28,500 kHz subbands. In addition, Novices are restricted to 25 W in the 222-225 MHz band and 5 W in the 1270-1295 MHz subband.

KEY

= CW, RTTY and data

= CW, RTTY, data, MCW, test, phone and image

= CW, phone and image

= CW and SSB phone

= CW, RTTY, data, phone, and image

= CW only

E = AMATEUR EXTRA
A = ADVANCED
G = GENERAL
P = TECHNICIAN PLUS
T = TECHNICIAN
N = NOVICE

* Technicians who have passed the 5 wpm Morse code exam are indicated as "P".

** Geographical and power restrictions apply to all bands with frequencies above 420 MHz.
See *The ARRL FCC Rule Book* for more information about your area.

*** 219-220 MHz allocated to amateurs on a secondary basis for fixed digital message forwarding systems only and can be operated by all licensees except Novices.

All licensees except Novices are authorized all modes on the following frequencies:

2300-2310 MHz
2390-2450 MHz
3300-3500 MHz
5650-5925 MHz
10.0-10.5 GHz
24.0-24.25 GHz
47.0-47.2 GHz
75.5-76.0, 77.0-81.0 GHz
119.98-120.02 GHz
142-149 GHz
241-250 GHz
All above 300 GHz

ARRL *We're At Your Service*

ARRL Headquarters	860-594-0200 (Fax 860-594-0259)	hq@arrl.org
Publication Orders	Toll-Free 1-888-277-5289 (860-594-0355)	orders@arrl.org
Membership/Circulation Desk	Toll-Free 1-888-277-5289 (860-594-0338)	membership@arrl.org
Getting Started in Amateur Radio	Toll-Free 1-800-326-3942 (860-594-0355)	newham@arrl.org
Exams	860-594-0300	vec@arrl.org
ARRL on the World Wide Web	www.arrl.org/	

all hams in your neighborhood. You will also find yourself much in demand anytime there is a need in your area for emergency communication!

For most hams, however, the constraints of budget and limited vehicle space will dictate a more conservative approach. Realistically, your choices will likely be governed by the distance you want to be able to communicate and by the type of operating you plan to do.

ABOVE 30 MHZ

The VHF and UHF bands are far and away the most popular for mobile hams. It is easy to see why. The equipment is compact and easy to mount, as are the antennas. In addition, there are lots of other hams to talk to, no matter where you may travel. What is really nice, though, is that communication on these bands is predominantly local and very reliable.

Want to check in with the local "Early Risers Group" every morning as you head to work? If they hang out on one of the VHF or UHF repeaters, you do not have to worry about whether HF band conditions will be good enough for you and the group to hear each other. If there are other hams in your family, you can catch up on the day's last-minute family plans that you forgot to discuss before you left for work.

Offering reliable local communications and a large pool of available operators makes the VHF and UHF bands a natural for emergency use as well. While a major disaster will sometimes require long-range communication, the overwhelming majority of emergency communications are local and take place on the VHF/UHF bands. Mobile hams are ideally suited for local emergency communications. Let's have a look at some of the appealing characteristics of bands over 30 MHz.

6 Meters

Sandwiched between the 30- to 50-MHz business band and the broadcast television channel 2, the 6-meter band extends from 50 to 54 MHz. Once considered a "no ham's land" because of the propensity of 6-meter transmitters to cause interference to neighborhood TV sets, 6 meters has seen resurgence in activity in recent years. Why?

Cable TV

Probably the greatest blessing ever to be bestowed on the 6-meter band, cable TV provides a closed system of shielded transmission lines and remotely located antennas and amplifier systems. There is no doubt cable has helped eliminate the TVI curse from the prospective 6-meter

operator. This has encouraged many new hams, as well as some old-timers, to visit this formerly forsaken band.

Lots of Space to Roam

Another reason is that 6 meters is not usually crowded. Of course, this does not mean the band is desolate. Many metropolitan areas, as well as some rural locations, have 6-meter repeaters. And as the other bands grow much more crowded, you are sure to see an increase in activity on 6.

CQ DX on 6 Meters

If you like to work DX, you will be glad to hear 6 meters experiences more frequent periods of exceptional propagation—and over longer distances—than any other VHF or UHF band. The reliable range during normal conditions on 6 meters might be 30 to 50 miles for an FM mobile working a base station or a repeater. With the push provided by band openings, your 6-meter mobile signal will be transported many times that distance. Does the idea of driving around town as

---

Repeaters: The Good and Not so Good

Repeaters are found on all bands from 10 meters on up, but their real popularity is on the 2-meter and 440-MHz bands. The quiet in the car (due to the *squelch* control), increased driving safety with channelized operation and the fact that you can stay in touch with your local group are all plus features. Most stations within range of the repeater are Q5—armchair copy. But if you never operated on this mode, you'll need to prepare for some challenges.

On long trips, every 40 or 50 miles, depending on the number of hills and mountains, you will have to switch to a new repeater frequency. During the day, many repeaters have very little activity. Don't be surprised to find repeaters occupied by only a handful of hams and these only during rush hours.

Other repeaters, in metropolitan areas, are constantly busy—perhaps too busy for you to have a leisurely chat. In addition, one station accidentally (or otherwise) keying the repeater can keep everyone else from using it.

Despite these challenges, repeaters are the most popular means of ham communication today. So jump right in—it's the best way to keep in touch with friends and family and to make new friends.

---

you talk to other hams who may be 1500 miles or more away sound exciting? The 6-meter band is the only band above 30 MHz where it is going to happen.

Repeaters, occasional long distance propagation, a growing number of users, and some really neat mobile gear—all good reasons to try 6 meters as *your* mobile band.

2 Meters—Where the Action Is!

Two meters is one of the most popular amateur bands. Nestled into a 4-MHz wide chunk of VHF spectrum spanning 144 to 148 MHz, it is a bustling beehive of activity, offering a smorgasbord-like variety to the mobile ham.

Expect reliable mobile-to-base simplex range on this band to be approximately 20 to 40 miles under normal conditions, although periods of enhanced propagation may increase the range to several hundred miles.

Although 2-meter mobile simplex operation is common, one of the most appealing aspects of the band is the presence of repeaters—thousands of them. There is hardly a place in the US where the mobile ham might travel and not be within range of a 2-meter repeater. In fact, 2 meters was the birthplace of the Amateur Radio FM repeater. As 2-meter FM grew ever more popular in the 1960s, hams discovered that a strategically located repeater could give their mobile rigs the same range as a well-equipped fixed station.

As 2-meter activity has evolved over the years, so too has the complexity of repeater systems—fueled by the ingenuity of the hams who construct them. With a feature known as *repeater linking*, you are no longer limited to a range of perhaps 100 miles when working through a 2-meter repeater.

Suppose you are driving from Evansville, Indiana, to Chicago, and you would like to stay in touch with your ham spouse. No problem. By using a system of linked repeaters, the signal from your 2-meter rig will be relayed back to a repeater within range of your home. Your spouse will hear you as though you were next door, even though you are hundreds of miles away.

Of course, linking is not limited to just connecting one 2-meter repeater to another. Sometimes you will find a 2-meter repeater with a *crossband* link, enabling you to have your signal retransmitted on another band. If the crosslinked band is 10 meters, with good band conditions you may find yourself talking to someone halfway around the world.

Autopatches are an indispensable

feature to the mobile ham—and you will find them available on many 2-meter repeaters. Simply an electronic interface between the repeater and a telephone line, autopatches make it possible for you to use your radio to call someone on the telephone. It works much the same way as a cell phone but is typically limited to simplex operation.

Stuck in traffic as you head home from work? A quick call using the repeater autopatch will allow you to inform your family they'd better start dinner without you.

Lot's of repeaters, autopatches, plenty of activity—it is easy to see why 2 meters is the band of choice for thousands of mobile hams. So come join the fun.

222 MHz (1.25 meters)

Want to escape the VHF rat race? Then 222 just may be your ticket to Serenityville. Situated just above the top of the top of the VHF broadcast television band, it resides in a 3-MHz slot extending from 222 to 225 MHz.

Increased crowding in the other VHF/UHF bands has led to increased interest in 222 MHz, which offers range coverage comparable to the 2-meter band. Although you will not find a bewildering maze of repeaters on 222 MHz, there are still many to choose from, with more being brought on-line each year.

If you are a Novice, that's especially good news for you, since 222 MHz is the only VHF band available for your use. The good news gets even better with the knowledge that some 222-MHz repeaters are linked to repeaters on other bands. This gives you the ability to talk on frequencies where you do not normally have operating privileges. Depending on what band is linked to the 222-MHz repeater you are using, it may be possible for you to join the conversations on the local 2-meter repeater. Or, you may even be able to work some DX on 10-meter FM.

As with 2 meters, you will find a good assortment of 222-MHz mobile gear available for your choosing. Antennas, by the way, are noticeably shorter than 2-meter antennas with comparable gain.

440 MHz (70 cm)

The first UHF band in our upward exploration of places to operate mobile, the 440-MHz (70-cm) band covers an astounding 30-MHz-wide territory stretching from 420 to 450 MHz. To the mobile operator, the 440- to 450-MHz slice of the 70-cm band is generally the one of greatest interest. This is the portion of the band where you will find FM repeaters—not as many as on 2 meters, but the gap is closing.

Although the range on 440 is somewhat

Safety First

Driving and operating can be a problem, if you let the concentration needed for driving be overshadowed by the attention paid to the radio. Mobile CW is fun, and many hams enjoy it regularly. But if you are not all that comfortable with your CW ability, or drive where the road and traffic require all your concentration, then perhaps you should stick to voice operation.

less than on the lower frequency bands, linked repeaters can extend the range of your 440-MHz mobile rig far beyond the normal limits.

Get the Picture on 440?

Just because most hams are using 440 MHz to talk to other hams does not mean this is all you can do on this versatile band. You see, 440 is the lowest frequency band where you can operate fast-scan amateur television (ATV). Using the same type of format as the television in your living room, ATV allows you to transmit live, full-motion video—in color if you like. Want to show your ham friends how bad the traffic on the freeway really is? Mount a camera and ATV transmitter in your car and let them see for themselves. Or use a similar setup on your boat to show your envious friends just how nice the water is as you skim across the lake. The NASCAR guys are not the only ones who can have their very own Car-Cam!

Of course, mobile ATV can have a serious side, too. The ham with mobile ATV capability can be a real asset after a disaster strikes by transmitting live video from an affected area, providing valuable information to damage assessment officials. You may also find several ATV repeaters on the 440-MHz band in your area.

With plenty of activity, lots of repeaters, ATV and a great selection of mobile gear, it is no wonder many hams prefer 440 MHz.

900 MHz (33 cm)

This band is commonly referred to the 33-centimeter band and stretches from 902 to 928 MHz. Repeaters are located between 906 and 921 MHz, with simplex FM between 927 and 928 MHz. Repeater range and quantity are similar to 1200 MHz. This band has limitations in the states of Colorado, New Mexico, Texas and Wyoming. For details, see the *ARRL FCC Rule Book*.

1200 MHz (23 cm)

The last in our survey of the VHF/UHF

bands, the 1200-MHz band—or as it is commonly called, the 23-centimeter band—stretches from 1240 to 1300 MHz. Although the band is rather large, it is only the 1270 to 1295 MHz portion that is of interest to the mobile ham. This is where you'll find FM repeaters. Granted, you will not find a tremendous number of them but there are some, mostly in metropolitan areas. While range on the 23-centimeter band is relatively short, crossband linking of some 23-cm repeaters provides enhanced coverage.

Novices will find 23 cm appealing as it is the only UHF band where they have voice privileges. But before you buy equipment, make sure there is enough activity on the band in your area to justify your purchase.

THE HF BANDS

It would appear that with all the VHF/UHF bands have to offer, they must be the only way to go. Wrong! Don't rule out the frequencies below 30 MHz until you have read about how much fun it is to work DX—or the next state—from your vehicle.

The HF part of the spectrum consists of ten diverse bands spanning from 1.8 to 29.7 MHz (even though technically the 1.8 to 2.0-MHz band is a Medium Frequency band). They are commonly referred to as the "160- through 10-meter bands." Even though the total amount of operating spectrum offered by the HF bands is slightly less than the width of the 2-meter band alone, don't let this fool you. The ten HF bands provide a potpourri of propagation, as well as a tantalizing variety of operating opportunities to the mobile ham.

Most hams get pretty excited when enhanced propagation makes it possible for them to use their mobile rigs to talk to someone a few hundred miles away on VHF. On HF you can do it every day.

The Other Digital Mode: CW

Another facet of HF you may find particularly appealing is mobile CW. Although it sounds more challenging than juggling BBs while wearing boxing gloves, many mobile operators are discovering mobile CW is both loads of fun, and much easier that they thought it would be. Most hams use an electronic keyer—instead of the old telegraph-type straight key—since the keyer requires much less effort.

CW is inherently a more efficient mode of communication than voice. You will find mobile CW signals will be heard farther and better than an SSB transmission using the same power level. CW is so good at getting through that most hams who like the challenge of running QRP (transmitter power output of 5 W or less) use it almost exclusively. Some hams even run CW QRP when mobile. That's like bagging a grizzly bear with a pellet gun and a lot more fun. You can leave the kilowatt amplifier at home when you go mobile HF on CW.

Has your appetite for mobile HF been whetted? Are you ready to work some DX? Maybe even give mobile CW a try? Good. Let's have a little closer look at what each of the HF bands has to offer.

10 Meters

You say you really enjoy the FM repeaters on VHF/UHF but you've grown a little tired of talking to the same people all the time? Then check out the repeaters on 10 meters. These FM machines work just like the ones you are used to, with one exception. They can frequently be heard for distances of several thousand miles. Don't expect that on your local 2-meter repeater!

The only HF band where standard narrowband FM is allowed is on the top end of the 10-meter band, above 29.0 MHz. Repeater inputs and outputs above 29.5 MHz are specified by the FCC.

On 10 meters you can also use SSB or CW, in accordance with the privileges offered by your class of license. Although worldwide communication is often possible on 10 meters, propagation is strongly affected by the 11-year sunspot cycle. The best and most frequent DX comes in the years at or near the cycle's peak. During the low part of the cycle the band is often, but not always, dead. Short-range communication via ground wave is possible anytime, however.

12 Meters

First made available to hams in 1985, the 12-meter band extends from 24.890 to 24.990 MHz. With its close proximity to the 10-meter band, 12 meters has similar propagation characteristics.

The mobile operator will find this relatively uncrowded band an ideal place for making voice or CW contacts. As you might expect, the sunspot cycle affects this band almost as much as it does 10 meters. It is great during the cycle maximum and very quiet during minimums.

15 Meters

The third largest of the nine HF bands plus one MF band, the 15-meter band offers the mobile ham lots of elbow room between 21.000 and 21.450 MHz. Offering world-wide propagation, it's a favorite of many hams who like to pursue DX in a more relaxed and roomier atmosphere. Daytime openings are frequent, even during periods of poor propagation. It is considered to be the best long-range DX band available to Novice licensees, who have CW privileges on 15 meters.

17 Meters

One of the newer HF bands, the 17-meter band offers 100 kHz of operating space, at 18.068 to 18.168 MHz. With worldwide communications potential, this small but still uncluttered band is a favorite of many mobile hams.

20 Meters

Would you like to pass the time on your morning commute by chatting with hams all over the US? Or perhaps talk to hams in Europe or Asia as you make your afternoon drive home? Then 20 meters is the band for you.

Residing in the space between 14.000 and 14.350 MHz, 20 is the most popular of all HF bands. Under most conditions, 20 meters is open to some area of the world pretty much around the clock. Many mobile hams have used 20 meters to work 100, 200 or even 300 or more foreign countries. Especially in the evening and on weekends, however, you will be competing with well-equipped high-power stations using multielement Yagi antennas. Be prepared for a great deal of very strong QRM.

Of course, you don't have to be a DXer to like 20 meters. If you frequently travel more than a few hundred miles from home, on vacation or business, you'll find the 20-meter band to be the perfect connection to the hams in your hometown.

In addition to general types of operation, you'll also find various contests and nets on 20 meters, many of them catering to the mobile operator. So whether you plan to chase DX or just work hams in the US, whether you favor CW or phone, be sure to include 20 meters in your HF mobile portfolio.

30 Meters

Weighing in with only 50 kHz of space, 30 meters is the second smallest of the HF bands, running from 10.100 to 10.150 MHz. However, don't let the small size of this band fool you.

You'll find 30 meters an excellent band for mobile operation, especially if you like CW, since voice operation isn't allowed on this band. Couple that with the fact that contests aren't conducted here either, and 30 meters has the makings of a nice place to look for mobile contacts. Propagation characteristics are similar to those on the more crowded 40-meter band.

40 Meters

Beginning at 7.000 MHz and ending at 7.300 MHz, the 40-meter band is capable of providing a variety of communications opportunities to the mobile ham. Want to talk to hams in the next county? 40 will work. Want to chase rare DX? It can be found on 40 too.

Just as with the other HF bands, propagation varies on 40 meters from day to night, as well as sometimes from day to day. That's one of the things that make this band so interesting and fun for the mobile operator. Of course this doesn't mean 40 is unreliable. It can provide dependable and reliable propagation during both day and night time conditions.

Although you won't find 40 meters exploding with DX the way 20 meters does, it won't disappoint you. In the early mornings and late evenings you will find 40 meters to have long-range propagation, providing worldwide communication. When this happens, there's no telling who might show up on the band. You may find Japanese or perhaps Russian stations with signals sounding like locals.

Daytime operation is very popular on 40 meters, with several nets specifically catering to mobile stations. You can call in and then ask anyone interested to go off frequency with you to have a chat. In the evening and at night, commercial broadcasters from Europe make this band a wild and annoying cacophony of high-power carriers and noise in many areas of the country.

With a great mix of long and short range capability, it's no wonder 40 meters is the band of choice for many mobile operators, especially during the day when the broadcast stations can't be heard.

60 Meters

The newest and smallest amateur band, 60 meters is unique in several respects. It consists of five USB-only discrete 2.8-kHz-wide channels. Amateurs have secondary access on this band, meaning they cannot cause inference to and must accept interference from the Primary Government users. Channel centers range from 5.332 to 5.405 MHz. The NTIA says that hams planning to operate on 60 meters "must assure that their signal is transmitted on the channel center frequency." This means that amateurs should set their carrier frequency 1.5 kHz *lower* than the channel center frequency. Despite these restrictions, 60 meters has gained some popularity with mobilers. Commercial antennas, for example, are available for 60 meters.

80 Meters

The second largest of our HF bands, 80 meters offers 500 kHz of space, at 3.500 to 4.000 MHz. While foreign DX is possible on this band, the mobile US ham will find it to be primarily a good place for ragchewing with North American stations. Conditions on 80 meters are greatly affected both by time of day and time of year. Coast-to-coast contacts are easily possible in the late evening hours, but ionospheric absorption so greatly attenuates signals during the day you may find it impossible to even talk farther than across town. You can also expect activity on 80 meters to dwindle somewhat during the spring and summer months—those seasons plague 80 with annoying levels of thunderstorm crashes and static.

During the day, 80 meters is often very quiet. Daytime propagation is less than about 100 miles. During the evening, local QSOs, high-power stations, traffic nets and chatting groups can make it difficult to stay in a QSO. With a well-equipped fixed station on the other end you can stay in contact as you drive for several hundred miles.

Despite a few drawbacks, 80 meters is still home to many mobile hams. Various voice and CW nets, as well as a plethora of other activities, make 80 meters attractive to the ham on the go.

160 Meters

Located just above the AM broadcast band, 160 meters occupies the slot from 1.800 to 2.000 MHz. Propagation under average conditions can best be described as short range. Affected by daytime absorption and static in a similar manner as 80 meters—though more severely—160 meters is considered to be most useful during the evening hours of the winter months, when longer-range propagation occurs. If you are looking for a challenge, give 160 meters a try.

So Many HF Bands... So Little Time

There you have it: a brief survey of the HF bands. With so many good bands to choose from, deciding which bands *not* to operate many be your hardest decision!

SELECTING EQUIPMENT

A safe, convenient and aesthetically pleasing mobile installation will often appear challenging, if not even impossible, at first glance. And the current crop of cars doesn't exactly offer lots of good mounting options for our radios. To make matters worse, once we've somehow figured out how to squeeze a radio or two into our vehicles, we sometimes find that the automakers haven't paid a lot of attention to designing cars that are good electronic neighbors with our equipment. And with many new autos costing more than some of us paid for our homes a few years ago, it's pretty hard to get up the nerve to set a mag-mount antenna on the roof, much less to even think about drilling holes.

Don't let all this discourage you. It's really not all that bad. When it comes to choosing our mobile radio equipment we hams have never had it so good. The current generation of compact, feature-packed mobile ham gear is vastly superior to anything available even just a few years ago. In fact, if this book were being written in the not-so-distant past, this chapter would have consisted largely of schematics and construction information showing you how to build your own mobile gear. Sure, there's something very satisfying, and perhaps a bit romantic, about operating equipment that you've built from scratch and a good number of hams still do it. But it's unlikely today that you'll find many hams who mourn the passing of the days when you were forced to build your own mobile gear.

So there's nothing left to do besides run out, buy a radio and install it. Right? Wrong!

If you're the impulsive type, you can easily find yourself in trouble if you leap before you look. Once you've made up your mind to buy a new piece of gear, it can be easy to get seduced by the pretty pictures and the slick sales talk. You'll be happier in the long run if you use those sales techniques as a jumping-off point for your research. Unless you figure out exactly what features you want in a mobile rig, you could be stuck with far more radio than will serve your particular needs.

Before you pull out that credit card, check the Product Reviews. (The ones in *QST* are based on the results of sophisticated, independent lab tests.) Read all the manufacturers' literature you can get your hands on. Ask around. Perhaps a member of your club just bought one of the rigs you are considering. See how he or she likes it before you invest in the same model. Aside from the monetary considerations, your choice of mobile ham gear can directly affect your ability to operate your vehicle safely. We will explore these details and other factors that should go into any purchasing decision.

PLAYING IT SAFE

If you are going to operate your mobile radio while you are driving, then your attention is going to be diverted at least partially away from your driving. It doesn't take a genius to figure out that a difficult-to-operate radio for mobile use can be a prescription for disaster. If you operate such an unruly animal at your home station, at most it might cause occasional frustration or maybe a missed contact every now and then. Run a mobile rig with that same sort of disposition and you could wind up missing much more. Always remember the most important thing when it comes to mobile operation is safety, safety, safety!

While we're on the subject of safety, it's worth noting that in recent years most of the radio manufacturers have become more conscious of the need to address ergonomics, especially with mobile equipment. Ergonomics is just a fancy way of saying that the radio should be designed to be easy and convenient for a human to operate.

That sounds simple enough, but it's not always so easy to accomplish. For instance,

---

Portable, Mobile and Signing Your Call

The FCC requires us to identify ourselves by transmitting our call signs at the end of each contact, and every 10 minutes during each contact. This must be done in English. This is all that's required. As a courtesy to your fellow hams, however, it is customary to provide a little more information.

Generally, you would say you are mobile if you are using a station capable of operating while in motion. Car, truck, bicycle or on foot—it makes no difference. A station that is not at your license location could be called portable, especially if you will remain at this location for an extended period. Many hams add the call area to their identification. For example, while on vacation in Arizona, I like to sign *N1II portable 7*. While driving around my hometown in Connecticut, I usually use *N1II mobile 1*.

Al Brogdon, W1AB (ex-K3KMO), operates mobile CW from both his car and his motorcycle. He says many mobiles on CW use /M and other use / M3 (telling the other station they are mobile in the third call area). Still others operate mobile CW without any addition to their call sign.

The *FCC Rule Book*, published by the ARRL, contains the full set of FCC requirements. Now matter how you classify your station, and no matter how you sign your call, you must meet the minimum FCC requirements. Anything more is up to you.—*Paul Danzer, N1II.*

---

back in the days when a 2-meter mobile rig had a front panel with only three knobs—the on/off-volume control, squelch control and channel selector—it was pretty hard to get confused while operating a rig. About the only thing that would invoke frustration on the part of the operator was if the channel indicator light burned out! Contrast that with today's multi-featured 2-meter rigs with upwards of a dozen front panel controls, some with dual or even triple functions, and it's easy to see why ease of operation is often one of the most important considerations when shopping for a mobile rig.

Keep in mind too when you are shopping for a mobile rig that while it's always good to ask for your friends' opinions and recommendations about the radios they've used, it's possible that your tastes may be quite different. That's why it's imperative to become thoroughly familiar with any rig you may be considering for mobile use. The best way to do that, of course, is to actually sit down and operate the rig, simulating as closely as possible the conditions under which you would be operating it if it were installed in your own vehicle. Does it seem intuitive to operate?

For instance, if it's a VHF or UHF FM rig, are the more commonly operated controls located on the microphone? Does a simple command such as switching from a memory channel to the VFO require you to look at the rig (and away from the road)? If it does, then you may want to consider a different radio. While you may find that there's no rig made that you can operate without ever taking your eyes off the road, your first priority must be to spend a minimum amount of time looking at the radio and a maximum amount of time with your eyes on the road.

Unless you plan to never operate mobile at night, another essential consideration is whether or not the controls you will most often use are illuminated. As you shop around you may find rigs that have

The NOAA Weather Channels

My mobile 2-meter transceiver has a pair of up and down buttons on the microphone and wideband receive coverage. I programmed the local NOAA weather channel into the highest channel number and my local repeaters in descending order of use starting with memory channel 1. When I surf through my memory channels, I use the weather channel as an audio "marker" to indicate when I'm starting back at channel 1. The weather channel is always on so I never need to take my eyes off the road. It doesn't take long to remember the order of my most frequented repeaters, either.

I've also added weather channels for other locations, such as frequent vacation destinations. Whenever I hear the local weather channel, I know the next set of channels is programmed for the local repeaters. You can, of course, adopt this same idea to suit your individual needs or requirements.—*Mike Gruber, W1MG.*

addressed this consideration rather well; other haven't addressed it at all.

Keep in mind that having to turn on the dome light to see how to access a particular function of your rig doesn't make for safe motoring. An additional way to determine how convenient a rig will be for you to operate mobile is to examine the operator's manual carefully. (Many manufacturers now provide operator's manuals as a download from their web sites.) Does a simple operation such as turning on CTCSS encode have such a steep learning curve that you must read the instructions several times before you can

successfully enable that option? If so, you know that operating that rig mobile isn't as easy as you'd like. Check it out.

Granted, with so many features being packed into such small packages, some functions must be activated by multiple presses of a particular control—or sometimes by the simultaneous use of more than one control. Obviously, this is a price we pay for wanting radios that do so much while occupying so little space.

But what if you don't need all that complexity in a mobile rig? Well, you may want to consider one of the barebones rigs that some manufacturers are now producing. Those rigs concentrate on the basic operating necessities. This not only results in a rig with a clean, uncluttered control panel and a relatively smooth learning curve, it means you'll probably save money on the purchase price. No sense paying for features in a radio you'll never use.

Regardless of whether you select a radio with Cadillac-like sophistication or Volkswagen-style simplicity, be sure the features you are likely to use most often don't require you to divert your attention from your driving. Always remember that using your radio should *never* compromise the safe operation of your vehicle.

Okay, we've driven home the need for safety considerations. Now let's move onto some other criteria you may want to use in making your selection of a rig for mobile use.

VHF/UHF RIGS

If this is your first shot at mobile hamming, VHF/UHF will probably be your initial choice. So let's have a look at some of the features you may want to consider when shopping for mobile VHF/UHF gear. **Fig 1-5** shows a basic 2-meter-only mobile radio from Yaesu. **Fig 1-6** shows a basic 220-MHz-only mobile radio from Kenwood.

Fig 1-5—This FT-2800M is a single-band 2-meter transceiver. Its maximum transmit power is 65 watts. It has 221 memories and 10 NOAA weather channels. Receive coverage extends from 137 to 174 MHz.

Fig 1-6—This Kenwood TM-331A is a 220-MHz FM transceiver. Maximum power output is 25 watts, and it weighs 2.7 pounds. Take advantage of 1.25 meters. There's plenty of room for any application and potential for good use on this band!

Remoting Your Mobile Rig

There's no question that the bane of present day installation of mobile ham gear is the diminished space that's available in the modern automobile. What's surprising is that even though VHF/UHF rigs have steadily shrunk in size, we find fewer and fewer available places to mount them in our vehicles! What to do? Go remote, of course.

And you know what? That's not a new idea at all. You see, back in the earliest days of mobile ham stations, many hams—and amateur radio manufacturers—found that by providing a separate control box for a mobile rig, the bulky heat-producing main chassis could be mounted in a remote location, most often the trunk. Needless to say, that made for much less crowding in the passenger compartment, which translated into happy hams and even happier family members. Nor was that lesson lost on the manufacturers of commercial communications equipment. Mobile business-band radio equipment in the early 1970s often consisted of a remote-mounted main chassis with an under-dash control head.

In the world of Amateur Radio, however, an interesting change had begun to take place by the 1960s. With the arrival of the transistor, it suddenly became fashionable and practical to make mobile gear self-contained and designed for under-dash mounting. Not that the radios were necessarily all that small; by today's standards, they weren't. It just happened that as Amateur Radio gear began to get more compact, automobile interiors were also becoming more expansive, a convenient combination of events.

But as we all know, half of that happy equation didn't last. With the gas crisis of the 1970s and government-mandated fuel economy standards, we suddenly found the interior space of our vehicles shrinking faster than a cheap sweater in a red-hot clothes dryer. Even with the current generation of mobile rigs that are smaller than ever before, the arrival of unsupportive plastic dashboards, full-length center consoles and downsized interiors have made it difficult to find a place to mount ham gear in many vehicles.

Fortunately, the Amateur Radio equipment manufacturers have come to our rescue with remote-mountable radios for mobile use. As the name implies, on a rig with remote-mountable capability, the main radio chassis can be mounted in a remote location, perhaps under a seat or even in the trunk. Then the detachable control panel, usually rather small and easily mounted, can be positioned in a convenient location that's easily visible and accessible by the operator. A cable (usually part of a remote-mounting kit available at additional cost from the manufacturer) then connects the main chassis of the radio to the detached control panel. This sort of installation has obvious advantages. No longer do you have to worry that your under-dash-mounted rig may someday end up on the floor, with a large chuck of plastic dashboard still attached. Nor do you have to worry about cutting up a portion of your car's interior to make a place to install your radio, thus dooming the resale value when you go to trade.

Salesman: *"What's the big hole in the dash for?"*

Ham: *"Hole? What hole? Oh, that's an optional air conditioning duct I ordered on the car when it was new. It's really a rare and valuable option, you know."*

Salesman: *"Sure buddy, and I'll just bet that Mario Andretti is your uncle too!"*

Seriously though, in some vehicles you may find that a remotable rig is the only practical way you can even operate mobile. There can also be a fringe benefit to having a remotable rig that you might not have considered. You see, if you install the remote control panel so that it is easily removed when you leave the vehicle and then complement it with an easily removed antenna, you have a very effective theft deterrent. If you do a great deal of traveling, or if you often leave your vehicle in an unattended parking lot, remote mounting of your radio gear is a good idea. After all, there's not much worse than having your new rig become someone else's new rig—and that's not even including the possibility of damage that could occur to your vehicle! Concealing your radio installation removes it from the view of a would-be thief.

As you can see, the advantages to remotable gear are numerous. So if you've come to the conclusion that your vehicles isn't very hospitable toward a radio installation, perhaps a remotable rig is the answer. See **Fig 1-7**.

Two Rigs In One

Two meters is hot, with thousands of repeaters in operation in the United States. But it's not the only burner on the stove. More and more, the 440-MHz band is gaining enthusiastic advocates.

The rapidly expanding population of 440-MHz repeaters, coupled with increased availability of equipment, make this band a popular preserve for mobile operators. No matter where you might travel throughout the country, you're likely to find either of these two most popular bands bustling with activity. So how do you choose which band is best for you? The good news is you don't have to choose. You can operate both bands from one rig by using a *dual bander*.

As the name suggests, a dual-band radio is one that has the ability to operate on two different bands. While various combinations of band coverage are offered (and some rigs even offer coverage of more than two bands), the most common dual-band offering is the combination of 2 meters and 440 MHz. With a dual-band radio, a whole new world of mobile operation opens up to you. Not only do you have access to nearly twice as many repeaters as you would with just a single band rig, you also gain the

Fig 1-7—The main unit (under the remote head) of this ICOM IC-2720H measures only about 1.6 × 5.5 × 7.4 inches, and the remote head measures 2 × 5.5 × 1 inches—about the size of a large candy bar! The radio covers both 2 meters and 440. Maximum power output is 50 W on 2 meters and 35 on 440. A total of 212 memory channels provide plenty of programming capability.

capability to operate *duplex*.

Here's how duplex works. Normally, when we talk on a radio, regardless of whether we are using FM, SSB or even CW, we can hear the person we are talking to only when we stop talking and listen. That's because we transmit and receive on the same frequency. This is what is known as *simplex* operation. On the other hand, with a dual-band radio, we are no longer limited by that restriction. Because a dual bander can simultaneously receive on one band while transmitting on the other, you are able to hear the person you are talking to even as you speak. That's duplex operation, which allows a more natural style of conversation. It's as if you were using a telephone instead of a radio. That's a particularly nice feature when you are making an autopatch through a cross-band repeater. The person you are talking to on the autopatch doesn't have to wait for you to unkey your radio before he or she can reply or interrupt.

Of course duplex operation isn't limited to autopatch operation. It doesn't even require you to use a repeater. But regardless of whether you use a repeater or talk direct, the ham on the other end will also have to be using a rig with duplex capability.

HAVE YOUR OWN REPEATER

The ability of your mobile rig to transmit on one band while simultaneously receiving on another makes available another very attractive option known as *cross-band repeat*. Available on practically all dual-band radios, it adds a slight twist to the normal form of duplex operation. Although the rig is still receiving on one band and transmitting on the other, instead of the transmitted audio coming from the microphone, it is now coming from the receiver. To put it another way, whatever is being received on one band is simultaneously being retransmitted on the other. Think of it—you have your very own mobile repeater.

Consider a case in which you are in a rather remote area that's ruggedly landscaped with deep valleys and rugged hills. As you might suspect, this is the kind of terrain that renders an HT virtually useless, even to work a local repeater. The remoteness of such an area also puts a significant premium on communications—especially emergency communications. Should a situation arise in which you require emergency assistance, but unable to reach someone on the air with your HT, cross-band repeat may be an option.

Simply install a higher power dual-band

rig with cross-band repeat capability in your vehicle. Now, when you venture deep into the wilderness, park your vehicle on a hill overlooking the area where you will be and set up your mobile rig to cross-band repeat. With 50 W and a good antenna, your mobile rig should be able to get to and from local repeaters with a full quieting signal. Assuming you're never more than a few miles from your vehicle, your dual-band HT should work your mobile rig with very little effort. You can now enjoy the great outdoors with a great deal more peace of mind.

That is but one example of how handy a cross-band repeat mobile rig can be. Some hams use their cross-band repeating radios along with their dual band HTs to stay in touch with the gang on the local repeater while they browse deep inside the shopping malls—where it's sometimes hard to hit any repeater, unless of course it's sitting in the parking lot. Other hams dispense with the need to have a second VHF/UHF rig and antenna for their homes by leaving their mobile rigs in the cross-band repeat mode and using only an HT from the house. Now that's something that becomes even more attractive when the mobile rig is equipped for *remote control*. That simply means that you can use your HT, or any other rig with a DTMF pad, to remotely control some of the functions of your mobile rig.

For instance, say that you have your mobile rig set for cross-band repeat and you are using it to retransmit your HT's 440-MHz signal to 146.52 MHz, where you are chatting with a friend. As he travels farther away, he starts to lose your signal and asks you to move to the local repeater. If your mobile rig has remote control capability, all it takes is just a few key presses and, presto, you've switched over to the repeater. You are now continuing your conversation without having missed a word. And you didn't have to run out to your car to do it. That's why it's called remote control.

By the way, as handy as cross-band

repeat and remote control are, keep in mind that when you use those functions, there are several FCC rules, and sometimes other restrictions, that must be observed.

A Little Caution on Cross-Band Repeating

If you do use your rig as a repeater, keep in mind it was probably designed for a low duty cycle. Many rigs will overheat when they are forced to transmit for extended periods, as may happen if you allow your rig to retransmit the output of the local repeater. You might be able to select a low power output setting or only pick this repeater function during periods of low use of the local repeater. Remember, you are responsible for the output of your rig. Always make sure you can shut it off, if it is transmitting something it shouldn't.

ALL GOOD THINGS...

Hopefully this isn't beginning to sound like a sales pitch for all the new mobile rigs. With so many good features available—dual bands, remote mounts, repeat, extended receive, remote control, dual in-band-receive and well, you get the idea—it's sometimes hard to determine just exactly what capabilities you'll need and want in your mobile installation. The preceding wish list of desirable features from a mobile operator's standpoint should get you started on your search for the right VHF/UHF mobile rig for you. As always, thoroughly gather your information and consider it carefully before you buy. And by all means, ask lots of questions. After all, it's your money.

WE NOW QSY TO HF

Were all you prospective HF mobile operators thinking you were going to be left out? Not a chance. Let's take a look now at some possible criteria to use when selecting an HF rig for mobile operation.

Picking an HF Rig

How to pick a VHF/UHF rig still hold for HF rigs. The starting point is the *QST* reviews. Be sure to pick an HF mobile rig with enough flexibility for your use. Single-band rigs are fine, but recognize their limitations. A 10-meter-only rig during the low part of the sunspot cycle will not provide many QSOs, nor will a 20-meter SSB-only rig do very well if you plan to operate only on Sunday afternoons, when the band is full of kilowatt stations.

HF Features

- Size compatible with your car
- Bands covered
- Built-in antenna tuner
- Ease of control

- Control lock?
- Sufficient audio to overcome road noise
- Effective noise blanker
- Mobile accessories

It's a Nice Rig But Where Am I going to Put It?

What's one of the most difficult aspects of performing an HF mobile installation? If you guessed that it's finding a satisfactory location to mount the rig, you are right. Think it's hard nowadays to find a place in your car for your 2-meter rig? Pity the poor HF operator who's dealing with a rig that is considerably larger. Fortunately, the situation isn't really as bad as it seems. Just as with VHF/UHF gear, manufacturers are beginning to make greater strides in the direction of more easily mounted HF mobile rigs.

Shrinking the Rig

One very logical approach to a mobile installation is to make the rig physically smaller. As electronic technology has advanced, giving us surface mount and other space-saving components, it has made possible the packaging of more and more features into an ever-smaller box. This has resulted in a new generation of HF rigs that, while still somewhat larger than their VHF/UHF siblings, are much more space conscious than ever before.

At least one manufacturer has taken downsizing of HF mobile gear a step farther by leaving off the bells and whistles and offering only the most essential functions necessary for an HF mobile rig. The result is a package only a smidgen larger than a 2-meter rig, yet offering complete coverage of the HF bands.

If you can live without all the unnecessary frills, as well as making do with 5 W or less of RF output, you may want to consider one of the QRP rigs being offered by several manufacturers. Depending on your operating needs, your choice of QRP rigs ranges from the ultra-small single-band CW-only rigs to multi-band rigs offering the AM, CW, SSB and FM modes as well as many of the features found in their larger brothers. See **Fig 1-8**, a photo of the Ten-Tec Argonaut V QRP transceiver.

But what if you don't want a stripped-down rig with limited features? And QRP isn't the kind of challenge you're looking for when operating mobile; but you simply don't have enough room to mount a conventional HF rig in your vehicle? Well, you may want to consider an HF rig that has remote control capability. As with the VHF/UHF rigs that offer the same option, you can place the main radio chassis in a convenient spot that's out of the way, then mount the removable control panel in a location that's easily and safely accessed. See **Fig 1-9**, which features the Yaesu FT-857D and its front-panel remoting capability.

Fig 1-8—This Ten-Tec Argonaut V Model 516 is covers 160 through 10 meters with a 20-W maximum output. The 20 W suggests QRP with an edge, but it's a great power level for those who enjoy the slow lane, and it's demonstrably plenty of power to work DX as well as casual contacts. As a bonus, reduced power can mean reduced RFI issues.

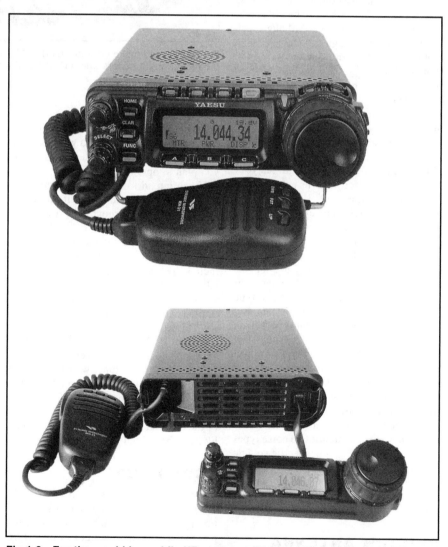

Fig 1-9—For the would-be mobile HF operator with limited vehicle space, the Yaesu FT-857D is a dream come true. In addition to its small size, its removable control panel makes it remote-mountable. With the main chassis tucked away in a safe location, the control panel can be mounted in an easily accessible spot, making almost any vehicle suitable for an HF installation. This rig covers all amateur bands from 160 through 70 centimeters.

As you can see, finding a place in your vehicle to mount an HF rig has never been easier, if you choose the right equipment. **Fig 1-10** is a photo of the very popular ICOM IC-706MKIIG transceiver that covers amateur bands from 160 meters to 70 cm (although not 220 MHz).

Which Bell? Which Whistle?

Okay, now that we've solved the "*I can't find a rig that will fit in my car*" problem, what else is there to look for when choosing an HF mobile rig? Well that depends on what sort of operating you plan to do. But regardless of your planned mobile operation, there are a few attributes that you'll always find useful.

No Static, Please

One is a good noise blanker. Designed to suppress the various forms of electrical interference that we commonly refer to as static, a noise suppressor can be a lifesaver for the mobile operator. Although you might be lucky enough to have a vehicle that generates little or no static of its own, though it's unlikely, the vehicles you share the road with might be not so well mannered.

Keep in mind also that automobiles aren't the only source of electrical noise. Power and transmission lines are another hideous source of noise, and they run alongside the roads we travel. Since your vertically polarized mobile antenna is a very efficient gatherer of all that electrical garbage, an *effective* noise blanker can be a really valuable asset to your mobile rig. Note here the emphasis on the term effective. Noise blankers vary greatly in their performance and ability to suppress various noise types. Some blankers, for example, can best eliminate impulse type noises from your car's ignition system. But they may have no discernable affect on power-line noise. Other noise blankers are pretty much useless for any type of noise. In some cases, the cure can actually be worse than the disease.

If you are shopping for an HF mobile rig, a real-world evaluation of the noise blanker performance is a wise move. Be sure to pay special attention to noise types that may be unique to your specific vehicle or expect to encounter in your operating environment. Remember that just because it says "NB" on the panel doesn't necessarily make it so.

BUILT-IN ANTENNA TUNERS

Once available only as outboard accessories that were often bigger than the rigs they were used with, antenna tuners

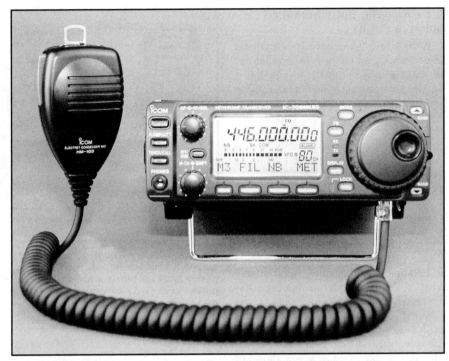

Fig 1-10—Can't make up your mind? This ICOM IC-706MKIIG covers 160 to 10 meters, plus 6 meters, 2 meters and 440 MHz, all modes—in a very compact package.

are now available as a built-in option for many mobile-sized rigs. Do you need one for your mobile rig? That depends on how you plan to operate.

If you intend to tune your antenna and rig to 3950 kHz and leave it there, an antenna tuner will probably be a waste of money, especially if the rig you decide to use only provides a tuner as an option. On the other hand, if your operating plans include frequent excursions between band edges (especially on 40 and 80 meters, where most mobile antennas have very limited bandwidth), and you don't want to have to stop and adjust your antenna each time you QSY, then you will probably benefit from a built-in tuner.

Beware too that a tuner won't allow you to use an antenna that's not made for the band you're operating. Even if the tuner should achieve a match, the performance of the antenna on the wrong band is most likely to be dismal at best!

LOCK IT UP

Another feature that's really desirable in a mobile rig is a VFO lock to disable the tuning knob. Once enabled, this control prevents accidental changes in operating frequency, even on bumpy roads. Want to become notorious with your fellow hams? Then let your rig do a vibration-induced QSY boogie all over the band as you transmit.

DAH DI DAH DIT DAH DAH DI DAH

Are you considering joining the rapidly growing ranks of mobile CW operators? If so, you may be thinking that your needs in a rig are much different than for other modes. Fortunately, they're not. As a rule, the preceding list of features will be an asset to the CW as well as the phone operator. There are, however, a few things you will want to look for if you are planning to become a *brass pounder* on wheels.

Keyers

It seems that hardly anyone pounds brass on a hand key anymore, especially when mobile. Instead, we relegate all that "dit-dah" work to electronic keyers. And even though it sometimes appears that CW capability has been added to some HF rigs as an afterthought, most rigs offer an internal keyer as either standard or optional equipment. Since the last thing most mobile installations need is added cabling and another box to mount, it makes sense to use a rig with a built-in keyer. While you're at it, don't forget to check out how the keyer speed is controlled. Unless you will be setting the speed once and then forgetting it thereafter, which is very unlikely, it will pay you great dividends to see how easily the speed can be set while you are mobile.

On some rigs, you may find this control to be very touchy to adjust, making it extremely difficult to get the keyer speed you want. Other rigs have very easily adjusted controls, and some even give a display panel read-out of keyer speed. Which type of control would you prefer to use?

Filters

Anyone who's done much operating in the crowded CW bands will testify to the value of a good CW filter. But to the mobile CW operator, a filter isn't a luxury, it's a necessity. And not just because of the need to eliminate adjacent frequency interference, or *QRM*. A narrow-bandwidth CW filter is also very effective in reducing noise.

Reduced noise and QRM when you operate mobile CW can greatly reduce operator fatigue. And fatigue is something you don't need when you are driving. So get a mobile rig with a CW filter—one with a 500-Hz bandwidth is a good compromise between ease of tuning and effective performance. Your ears will thank you.

Whew! We've looked at a lot of good features to consider when choosing a rig for mobile operation, but before we leave the topic of equipment selection, let's look briefly at choosing a vehicle.

DEALS ON WHEELS?

What does choosing a vehicle have to do with selecting equipment for mobile operation? Perhaps you heard a story about some guy who was out several hundred dollars to replace the electronic control unit in his vehicle after it died from exposure to RF from his HF rig? Well, that and other horror stories of incompatibility between ham gear and automotive electronic systems have circulated in recent years. And while it can sometimes be hard to separate fact from fiction, one thing is for sure: There's no guarantee that your ham gear is going to coexist peacefully with all the electronic widgets found in the modern automobile. Gone are the days when you could solve the worst case of interference in a mobile installation with resistor spark plugs, suppression wires and perhaps a few bypass capacitors. It's a new day, with plenty of new challenges.

So what should you do? The subject of RFI, both to and from radio equipment, will be dealt with in-depth in the last section of this chapter. But if you are planning to trade vehicles soon, or if you are planning a mobile installation in a vehicle that you've never operated radio gear from before, you may want to do a

---

How About Those Hybrids?

Hybrid vehicles are becoming increasingly popular. Using both an electric and gasoline power, they incorporate some fairly sophisticated electronics as well as high-power control circuitry. Not surprisingly, development of hybrid vehicles could have an impact on mobile Amateur Radio installations. This is due to the extensive electronic systems required to operate a hybrid vehicle. In addition, high currents and possibly high voltages may also be present. Switching of either can generate unwanted RFI. And since hybrid vehicles still utilize a gasoline engine, they still have all the traditional ignition and other systems that can cause problems in conventional vehicles.

So how are mobile hams with hybrids doing? Based on a recent survey, it doesn't seem they are any less compatible with VHF/UHF FM equipment than conventional vehicles. HF equipment however may present some additional challenges for the would-be mobile ham. Due to the small number of respondents and vehicles sampled in our survey, the jury is still out on this new technology. Some comments from the survey participants:

- Frederick Johnson, KD6DKH, 2003 Toyota Prius: "*I work 2 meters and 440 in my Prius and will be installing an ICOM 706MIIG in it.*"
- Kyle Brewer, KG4YUQ, 2004 Honda Civic Hybrid: "*I'm using a dual-band 2-m/70cm rig. Never had any problems with interference either to or from the vehicle.*"
- Betty Hamilton, WA4MKF, 2003 Toyota Prius: "*My husband, W5HFT, and I have operated 2-meter and 450 HTs from our hybrid. Communications have been both vehicle-to-vehicle and vehicle-to-repeater.*"
- Glenn Valenta, KC0SHS, 2001 Toyota Prius: "*I've done it with a 2-m/70cm rig in the trunk tied directly to the battery without any problems.*"
- Dick Bokern, WB9PUJ, 2003 Toyota Prius: "*A radio installation in the Prius at best is a challenge. The total car is a 'Mixmaster' of electrical and data control signatures along with the many digital sensors.*" Dick required additional filtering for his installation.

---

little detective work first.

One form of RFI that's easy to check for originates from the vehicle electronics and interferes with your radio. Since several of the electronic devices in a modern auto, from the AM/FM radio/clock to the engine-control computer, are generators of radio-frequency energy, it's no surprise that much of the QRM we deal with in a mobile installation comes from within. The big question is whether RFI can hinder your ability to operate? The only way to know for sure is to survey the candidate vehicle.

This is especially true if you plan to operate an HT on the VHF or UHF bands with only the attached flexible antenna. In some cases, RFI can be generated from somewhere inside the vehicle that will cause interference to an antenna inside the car. An outside antenna is often the cure.

If you plan to operate the VHF or UHF bands, and especially if you are going the HT/rubber duck route, try using that setup to scan the entire band for interfering signals originating from inside the vehicle. What you find may affect your mode of operation, choice of vehicle or both.

What about cases where our radio equipment adversely affects the electronic systems in our vehicles? With the potential

for expensive damage, coupled with the fact that not a lot is known on the subject, this is a pretty hot topic. For now, let's just say that if you are not sure what effect your radio installation will have on a particular vehicle, do some research. Start with the manufacturer.

Then carefully read the sections later in this chapter on installing equipment and dealing with RFI problems. The ARRL Web site contains contact information for many manufacturers, plus a description of their stated policy statement given in response to an ARRL survey. See *Automotive Interference Problems: What the Manufacturers Say* at **www.arrl.org/tis/info/carproblems.html** for details.

NEW OR USED EQUIPMENT?

Although our primary focus in the chapter has been on the selection of new mobile equipment, many cost-conscious hams will consider the purchase of a pre-owned radio. If you choose to go that route, all the previous guidelines still apply. However, if the radio you are considering is no longer being produced you might have difficulty getting information about its particular features. Usually, a call to the manufacturer will provide you with a general overview of the radio's capabilities. (Many factory technicians will also make you aware if the model you are considering has exhibited a record of poor reliability.)

Of course, if you are doing your shopping at a hamfest flea market (which presents its own hazards and rewards!), you'll want to do some prepurchase research for rigs that interest you. *QST* Product Reviews and related information are on the ARRL Web Technical Information Services (TIS) page at **www.arrl.org/tis/**. The *QST* Product Review List contains a list of product reviews going back to 1970. There are also links for *QST* Product Review Downloads to read or download product review columns going back to 1980. These reviews are an invaluable aid to the buyer of used equipment, detailing the specifications of each radio as well as the reviewer's impressions of its operation.

There is also a link for the *QST* Product Review Transceiver Summary, which provides a list of transceivers and compares them head to head. Both the *QST* Product Review Downloads and Product Review Transceiver Summary are Members Only web pages.

Although it is no longer in print, the *ARRL Radio Buyers Sourcebook* (or other buyer's guide for used equipment) can provide an excellent source for product reviews and related information.

Installation

Proper installation is the key to a safe and successful mobile setup, guaranteeing many years of enjoyable operation. In this chapter, we'll look at the various steps involved in performing mobile installations—both temporary and permanent. But before we drill that first hole or make that first connection, let's consider a subject important to *all* mobile installations.

Location Is Everything

No matter how you plan to operate mobile, your rig must be located so that it does not compromise your ability to drive safely! Remember, operating your radio is a *secondary* operation. *Driving* is your primary responsibility. To this end, both where and how you mount your radio equipment will be the most important aspects of your mobile installation. You should precede any radio installation with a thorough survey of the passenger compartment.

Does the prospective mounting location offer you a clear, unobstructed view of the display and controls? Not only should you be able to see the controls, it's imperative you also be able to easily reach them. Leaning over in the seat as you stretch across the car to reach your radio is a quick way to make unintended lane changes. If you cannot find a place to mount your rig where you can easily operate it as you drive, find another vehicle, another rig or limit your operating to times when you are a passenger.

As you scout for a suitable location for your rig, be especially observant for any potential interference with your vehicle's safety equipment. Air bags, for example, are found in all US built passenger cars since 1993. You must be absolutely certain you don't mount equipment where it can hinder or prevent an airbag's deployment. If you are unsure of the location of your vehicle's air bag—or not sure if your vehicle even has one—contact your car dealer for assistance. You might also wish to refer to the April 1993 *QST* article "Don't Get Blown Away by Your Mobile Rig," by Brian Battles, WS1O, for some helpful information on the topic of airbags and mobile-radio installations.

Air bags aren't the only safety feature found in the interior of an automobile. Padded dashes used to be offered only as optional equipment, but they've been around so long now we pretty much take their face-saving qualities for granted. Over the years auto safety engineers have taken great strides designing dashboards and other interior components with impact-absorbing materials. Be careful you don't negate the effectiveness of those materials by mounting a rig in a location where you or a passenger might strike it during an accident.

Physics First

Consider this. A radio falling off the roof of a twelve-story building hits the ground at about the same speed it has in your car while driving down the highway. If your vehicle should become involved in a collision, you might find your nice friendly radio coming at you (or a rear-seat passenger) with a vicious force equal to that of a sledgehammer pounding spikes into seasoned railroad ties! But *not* if you have taken the times to secure it properly. Sure it's tempting, especially in temporary installations, to forego the proper mounting of a rig. Resist that temptation as though your life depends on it. And be sure the mount really is secure. A 12-pound HF rig can exert *hundreds of pounds* of force upon its mounting hardware in the milliseconds of extreme deceleration caused by a collision. Don't let your rig become a deadly missile. Make it secure.

Do It Yourself?

It's been said the best way to get a job done right is to have someone else do it. Does this axiom apply to the job of installing a mobile rig? It depends on you. Are you confident in your ability to route cables, mount equipment and perhaps even drill holes in your automobile? Some people are understandably reluctant to attempt those sorts of tasks, especially with a new vehicle. But you shouldn't let the prospect of making a costly mistake while performing a radio installation deter you from operating mobile. Instead, you might want to consider having your radio equipment professionally installed.

Admittedly, you won't find a listing for "Amateur Radio Installers" in the yellow pages of your telephone directory. Few of the radio dealers selling amateur equipment even offer installation service. (The long-held reputation of hams as do-it-yourselfers must be a factor.) Who're you gonna call? The people who install and service the radios in police cars, fire trucks and other emergency vehicles, that's who.

Actually, any shop installing commercial or land-mobile two-way radio equipment should be capable of installing amateur gear. After all, most of them have done hundreds of mobile installations, some of them possibly in vehicles like yours. And that's valuable experience. Not only that, if interference problems crop up most shops are willing (for a fee) to seek out and correct them. If you check around, you might even find a shop in your area employing one or more of your ham friends. For the ham who doesn't want to perform his or her own installation, these shops are a good alternative. If you decide to have your equipment professionally installed, here are some suggested guidelines:

- Be sure the shop you select understands what type of equipment they are dealing with. (Some shops are willing to do Amateur-Radio installations but won't install CBs or consumer electronics.)
- Discuss beforehand where you wish to have everything mounted.
- Establish who will be liable for any possible damage to the radio or vehicle.
- If on-the-air testing of the completed installation is planned, make certain a licensed ham is present as control operator. It is illegal for an unlicensed person to transmit on the amateur bands for any reason.

You *Can* Do It!

Perhaps the easiest, and arguably the most common, way to operate mobile is with a hand-held transceiver, an HT. Powered by its own battery and equipped with a rubber ducky antenna, installation of an HT is as simple as climbing in your car and driving away. Besides easy installation, the HT route has other advantages. Security is one. You can take your rig with you when you leave your car. And if you frequently drive different vehicles it's not necessary for you to install a radio in each one. This is a real advantage for the ham who often rents cars.

Unfortunately, some of the features that make an HT so convenient can also pose a problem for the mobile operator. Using an HT with its rubber ducky antenna inside the car typically restricts your communication range to nearby repeaters. Fortunately, something as easy and simple as a mag-mount outside antenna can dramatically improve the situation. We'll talk more about this in the next section dealing with antennas.

If you plan to use an HT while you are mobile, carrying a spare battery can prevent you from having to sing the "There Goes the Battery Blues." A spare battery can rescue you in a pinch, but you might want to go one better and connect your HT directly to your car's electrical system. Before you do, determine if your radio allows a direct connection to 13.8 V. A nominal voltage of 13.8 V is the rating of an automotive electrical system, but a maximum system voltage of 14.5 V is common. The owner's manual of the radio should provide you with the needed information. By the way, don't assume

that just because the wall charger for your HT is 13.8 V, you will be able to connect the HT directly to 13.8 V in your car. Not only can excess voltage damage your radio, an overcharged NiCd battery can explode! Check that owner's manual.

Choose carefully your method to tap into your car's electrical system. The fuse panel is a popular place to extract the needed power, but making good connections can sometimes present quite a challenge. (It also requires you to locate a good grounding point nearby, which can be a real exercise in frustration.) Perhaps you'll find it more convenient to make your connection at the accessory or cigarette lighter jack, using a plug made for this purpose. That will not only spare you from doing an impromptu "Pretzel Man" impression as you grope under the dash for the fuse panel, it also makes moving your HT connection from one vehicle to another as easy as pulling a plug. Having your HT wired to the vehicle electrical system can ensure that your HT battery doesn't go flat at the worst possible moment. It will often enable your HT to produce full output power, something many HTs don't do when operated from a standard portable battery.

Although convenient, several important caveats should be observed whenever using an accessory jack. As a general rule these jacks are not recommended for anything more than an HT or low-power transmitter. Higher power means higher current, and typical plugs for these jacks are limited to 8 or 10 A. Voltage drops can also develop, causing serious and unexpected deficiencies in transmitter performance.

No matter where you make your connection, however, be sure the power leads coming to your radio are properly fused. Don't depend on the existing automotive fuse for protection. It's common for a cigarette lighter circuit to be equipped with a 20 to 30-A fuse, and that's sufficient to current to cook even the toughest HT. Also be absolutely certain that you check your connections for the correct polarity *before* plugging in your HT. HTs have been known to vary in polarity *even for the same type and size polarity jack*. Never depend on a fuse to verify your HT's polarity.

Reaching Out

Even having skirted the dead-battery syndrome, it's possible you'll still find the range of your HT to be inadequate at times. This is when the addition of an outside antenna can help it behave like one of the big boys. While some hams use a permanently mounted antenna to boost the

range of their HTs, a magnetic mount (or *mag-mount*) antenna is more in keeping with the temporary flavor of HT mobile operation. We'll be taking an in-depth look at the selection and installation of antennas in later parts of this chapter. If you opt to use an external antenna with your HT, choose one using coax cable with a *stranded* center conductor. The constant flexing of the coax where it connects to your HT will inevitably break a solid center conductor, putting you off the air without warning.

HT stands for "Handy-Talkie." Having power and antenna cables dangling from your HT as you attempt to operate it and drive at the same time can seem about as handy as wrestling a rattlesnake. How do you tame such a beast? The best way is with the addition of a *remote speaker/mic*. With one of these neat devices, you'll be able to leave your HT stowed safely away in its mounting bracket while you are engaged in a QSO.

You are planning to provide a mounting bracket for your HT aren't you? Unsecured HTs are notorious for slinking around the inside of a car like hungry rats, hiding beneath the seat, munching on old French fries and sometimes dancing between the driver's feet. Seriously, you should always provide some means of securing your HT, no matter how you plan to use it. Several companies manufacture suitable mobile brackets for HTs, or perhaps you'll opt to fabricate one yourself. In his March 1994 *QST* article, "An Over-the-Dash H-T Mount," Herbert Leyson, AA7XP, describes a simple mount he devised to secure his HT while mobile.

And by the way, a brief mention should also be made about using your car's existing sound system for better mobile receive audio. There have been several schemes used successfully over the years, one of which uses a CD-to-cassette adapter. This device consists of an audio coupler in a cassette shell that plugs into a portable CD player via a cable with a $1/8$-inch stereo plug. This scheme, of course, requires that your car have a cassette player. Another scheme uses a device that connects auxiliary inputs to a stereo through its CD-changer connector. See the Hints & Kinks column in February 2005 *QST* for details, starting on page 72.

Going for the Long Haul

Although using an HT is fine for the car-hopping ham, you may find the day-to-day rigors of your style of mobile operation are better suited to the use of a conventional mobile rig, mounted in a more permanent location. If you are one of those fortunate hams who own a vehicle with generous interior proportions, an *under-dash* installation will most likely be the easiest and most sensible way to go. Tucked safely away under the dash, your rig will not only be less likely to be in harm's way, it also will be much easier to see and operate. As an added bonus, if you've equipped your rig with the proper mounting hardware and cable connectors, it will be a simple chore to remove it from the vehicle when you leave, providing increased security or making dual-usage of the rig possible.

If you are planning an under-dash installation, be aware some automobiles have dashboards that are very unsupportive. This can be especially true if you are installing a heavier HF rig. Not only does the dash have to support the weight of the rig, it also has to contend with the leverage produced by the rearward mounting location of most rigs in their brackets. To circumvent possible damage to the dash, choose a metal point of attachment. If you cannot find any metal along the bottom of the dash (this is not uncommon), try to locate an area having some type of reinforcement above it—frequently, this will be where a brace is attached. Should that fail, it may be possible to place a narrow strip of metal on the upper side of the lip of the dashboard, providing the needed reinforcement.

On larger rigs it's advisable to provide a rear hanger or bracket for additional support. **Fig 1-11** shows how K4JQA made a wooden base for his HF rig, securing it with heavy-duty rubber bungee cords.

To Drill or Not to Drill...

If the automotive engineers have been kind, there's a chance you'll be able to use existing holes—and if they've been really kind, existing screws or bolts—to attach a bracket for your rig. Lacking such good fortune, some soul-searching may be in order as you struggle to decide if going mobile is really worth giving your car the Swiss Cheese treatment.

Why should you drill holes? Simply put, in many cases it's the *only* way to adequately secure a rig. But won't drilling holes in your car destroy its resale value? Maybe not. Sometimes, especially with under-dash installations, you can drill *invisible* holes. An invisible hole is simply a hole that won't be visible (at least not to

Fig 1-11—Richard Scott, K4JQA, built this wood base for his rig. The rig is held in place by a bungee cord. The raised front gives him a good viewing angle of the rig's front panel. *(Photo courtesy of K4JQA.)*

a prospective buyer of your vehicle) when you have removed your radio equipment. If the invisible hole is a little more visible than you'd like it to be, a bolt, screw or push-in plug of the right type and finish can often leave things looking factory original.

Isn't it risky to drill holes in today's cars? It certainly can be. Even professional two-way radio shops have made some very costly mistakes. In one such case, a technician drilled through the floorboard and into the transmission case of a new car. If you find it necessary to drill holes anywhere in your vehicle, follow some simple precautions.

- Check thoroughly *behind* where you plan to be drill to be sure there is nothing that can be damaged by the drill.
- Use a center punch to mark the hole—this prevents the bit from "walking" and doing damage.
- Use a stop collar to prevent the bit from penetrating any deeper than necessary.
- Before drilling through carpet, make a short slit with a sharp knife. This will prevent the bit from grabbing the yarn and unraveling it.
- Be especially careful when drilling holes in the trunk—gas tanks aren't always located where you expect them to be.
- Pick your drill carefully. Use either a 3-wire drill with a 3-wire extension

cord or a 2-wire double insulated drill.
- Whenever practical, secure a block of wood behind sheet metal when drilling through it. This can help provide a smoother transition whenever a bit breaks through the metal. It can also shield precious parts of your vehicle from unintentional and damaging contact with the bit.

Consoles and... Cubbyholes?

While floor shifts and consoles stretching from the dash to the back seat give a car that nice, sporty look, they can sure give fits to the ham who's bent on installing a mobile rig. With an under-dash installation pretty much out of the question in this type of auto, what's a ham to do?

Obviously, if you already have the car and you are shopping for a rig, then one of the best options is to choose a rig with a removable control panel. Not only will you be able to place the control panel just about anywhere that suits you, the companion mounting bracket can usually be affixed with Velcro, double-sided tape or something similar, eliminating the necessity for drilling holes. Of course, it will still be necessary to find a good, out-of-the-way place location to mount the main chassis of the radio—the trunk is usually the best choice. Avoid mounting any radio equipment—transceivers, amplifiers, et al—under the seat. Poor ventilation there can cause your equipment to experience potentially damaging high temperatures, not to mention that some power-operated seats can crush an improperly located under-the-seat rig. In addition, some automakers place electronic control modules under the seat—all good reasons not to mount equipment there.

The newer, remotable rigs are nice, but if you don't own or plan to purchase one, you may choose to do an *in-dash* installation of a conventional rig. There's no question mounting a rig in the dash provides for a very unobtrusive and aesthetically pleasing installation. Properly installed, the in-dash mobile rig can blend in with its surroundings so well that it is often mistaken for a factory-installed item. Nevertheless, some hams are reluctant to chop up the interior of an automobile to install a radio. There may be a solution.

Many vehicles have small storage areas or pockets built into the dash or console. Designed to hold tissue boxes, tapes, CDs or other items, these compartments (one automaker calls them *cubbyholes*) are often ideal places to mount a VHF/UHF rig. They are usually both tall and wide enough to allow installation of a rig, but may be a bit shallow, not providing enough depth for the radio to fit flush with the front. Even if there is enough depth, you still must provide an exit point for power and antenna cables (be sure your rig's heat sink or cooling fan isn't obstructed). You can often overcome these obstacles by carefully cutting off the vertical portion of the back wall of the compartment, allowing the rig to extend out the rear. Don't discard the piece you cut out. It can be glued back in place when you remove the rig, maintaining the resale value of your vehicle.

Or better yet, cut the piece out of someone else's vehicle. A quick trip to your local auto salvage yard can often yield parts from a similar vehicle that you can modify for your rig's installation without care or concern—and for a reasonable price too. (All mobile hams should be on a first-name basis with the propriety of the local auto-recycling emporium.) Simply swap the modified part for the original when you install rig. And your vehicle's original parts can be stored safely away for later installation at trading time. You'll be happier and so will the car.

Even if your car doesn't offer you the convenience of a cubbyhole for mounting a rig, there are other possibilities. Sometimes you can remove a map light, or perhaps an ashtray, to make space for a rig. And don't forget the accessories, like a microphone clip or external speaker. Removal of something like a conveniently located ashtray, or a modified junkyard substitute, can often help in these cases too. See **Fig 1-12**.

Regardless of how you approach an in-dash installation, be sure to take time and plan *before* you begin. Accurate measurements are a must. The rig obviously has to have enough room to fit where you want it and you want to be absolutely certain you don't remove more material than required, leaving your rig framed by nothing but open space. As you measure, be sure there is enough clearance behind the dash for your rig and any required connections. Remember too that while it may have taken quite a while for you to install your rig, it will take a thief much less time to remove it. You may want to protect your rig by using an auto-security system or by fabricating a cover

Fig 1-12—N1II removed the unused ashtray from his minivan and ,with a little judicious filing, replaced it with his Kenwood rig. A block of wood, cut to fit the cup holder, has a microphone clip screwed to the top to hold the mike. *The ARRL Repeater Directory* rests on the utility shelf above the broadcast radio. *(Photo courtesy of N1II.)*

plate to hide it when you leave the vehicle.

If neither an in-dash or under-dash installation is feasible, you may have to resort to mounting your rig on the floor (common for larger HF rigs). If this is the case, you'll need to either purchase or fabricate a suitable mount for your rig. See Fig 1-11 again. As with any installation, make sure the rig and its mount are suitably secured. Screws or bolts through the floorboard are the best plan here... but watch out for those transmissions! (Automotive, not radio.) Remember to slit the carpet before drilling as previously mentioned, and the holes won't be visible when the equipment and screws are removed later. Also, it's a good idea to place an anti-seize compound on any bolt or screw that extends outside the passenger compartment, just to ensure you can remove it when you are ready.

Getting Wired

A neatly installed rig is a beauty to behold, but it's only so much window dressing until the necessary power and antenna cables are routed and connected. Although this phase of the installation will leave you more intimately acquainted with

An Ashtray Speaker Mount

Installing mobile equipment is today's modern vehicles is often challenging. Safe and secure mounting must not inhibit accessibility and visibility. The mounting location must not interfere with any of the vehicle's controls or operation either. Now add a "no holes" requirement and the situation may seem impossible.

I was fortunate to have convenient panel screws for the radio in my Saturn. An external speaker, however, was another matter. I solved this problem by obtaining a second ashtray at my local junkyard. Its console location is ideal (**Fig A**), and excess wire can be coiled inside the tray. I added a hole and grommet just behind the speaker bracket to conveniently pass the wires (**Fig B**). Since the new ashtray was no longer the same color as my car's upholstery, I used black shoe polish to make it match the speaker.

A second ashtray is relatively inexpensive, and the original intact ashtray can be returned to service in a minute. Depending on your vehicle and specific requirements, this concept can be used for other accessories as well.—*Mike Gruber, W1MG.*

Fig A—A speaker mounted to an ashtray lid saves drilling holes in the car interior.

Fig B—A hole behind the bracket permits storage of excess speaker cable inside the ashtray.

the inner being of your automobile than you may have expected or desired, don't worry. By following a few simple ground rules, your success is all but guaranteed.

You *Can* Get There From Here

The ideal installation is one leaving no wires or cables exposed where they can be damaged or present a trip hazard to passengers. If you'll be running any cables front-to-rear through the passenger compartment, it's preferable to place them *under* the carpet, along the bottom of the transmission or other convenient tunnel. Such a location has some advantages. It's typically an area of low wear and impact, so there's less chance of damage to a cable.

It helps to provide greater separation from vehicle wiring, which is usually routed along the doors under the *sill plates* (those are the plastic or metal strips running along the top of the rocker panels to cover the edge of the carpet).

Above all, be sure to avoid running power or antenna cables near any electronic control modules. You'll most often find them located under the passenger side of the dash, usually behind the kick panel. Unfortunately, they can also be lots of other places. If you're not sure of their location in your particular vehicle, contact your dealer's service department before you make your cable runs. Your dealer may also be able to provide you with valuable technical information from the manufacturer pertaining to mobile radio operations. Or you may wish to obtain a copy of the General Motors *Mobile Radio Installations Guidelines* by sending your request, along with an SASE to the EMC Department, Mail Code: 483-340-111, General Motors Proving Ground, Milford, MI 48380. This document is also available from the GM Web site at **service.gm.com/techlineinfo/ radio.html**.

Getting Connected

Not too long ago, power hookup for an add-on electronic device (such as a tape player) may have seemed a relatively straightforward process. The typical procedure was so simple and easy that it may have seemed obvious at the time. It went something like this.

First, strip off about an inch of insulation from the negative power lead. Slip the lead between the dash and your newly installed mounting bracket just before you tightened the last bolt, clamping the wire firmly in place. Next, snake the positive power lead along the back side of the dash, across the steering column and over to the fuse block. Once there, pry one end of a glass fuse out of its socket (the one labeled "radio" was a

favorite), fan out the strands of the lead, and insert them into the vacant fuse clip. Finally, to complete the installation, press the fuse gently back into place. Presto! Like magic, in only a matter of minutes you could groove to the stereophonic sounds of your favorite group as you cruised the local drive-in.

If this was you, rethink your approach. While wiring equipment that way may have worked for a tape player, in yesterday's automobile, today's mobile rigs require a much different approach. For starters, each piece of equipment should be individually and directly connected to the vehicle battery. Connecting to the battery instead of to existing vehicle wiring has several advantages. First, it can help to eliminate reception of electrical system noise. Also, the battery tends to dampen voltage spikes and surges (after all, it's really just a giant capacitor). You've probably heard that your car's starter induces a very unhealthy voltage spike into the electrical system when it is disengaged. What you may not have known is that when a malfunctioning alternator that's stuck in full-charge mode loses its connection to the battery (and thus the load), it can pump more than 75 V across the electrical system of the vehicle. So go straight to the battery, with both the negative and positive leads. This means you'll be running *two* leads to the battery for each unit to be powered. (A transceiver and outboard amplifier will require a total of four power leads, for example.)

It might appear to be easier to simply connect the ground lead to the chassis of the vehicle instead of to the battery, but it's not advisable. *"Okay, but can't I just run one set of power leads from the battery, then feed all my equipment from them?"*

You could. Here is why you shouldn't: *voltage drop.* The power leads for your equipment are sized according to the amount of current the need to carry. Try to power several devices using conductors that aren't large enough to carry the combined current, and they all end up starved for voltage. Modern microprocessor controlled rigs can do very strange things when the supply voltage drops. If you aren't using the factory-supplied power cable for your equipment—or if you have to increase the length of the leads, be sure to use wire of sufficient size.

If uncertain about a particular cable or hook-up, you can always measure the voltage drop. Connect a voltmeter across the cable's conductors and measure the drop across each at full load. Alternately, you can measure the voltage at the radio end of the cable. Drops will be evident by comparing the full load voltage at the radio end versus the battery end. You can also compare the full load versus no load voltage at the radio end, but verify there is no significant voltage sag at the battery under full load.

Breaching the Wall

Does the firewall appear to be an impenetrable barrier as we journey toward the battery? Take a second look. On most vehicles, you'll find lots of holes in the firewall. It's simply a matter of picking one to use. If you can, choose an unused hole that has been fitted with a rubber or plastic plug. Those are the holes for equipment not installed on your particular vehicle (for example, many automatic transmission-equipped vehicles have an unused hole in the firewall where the clutch cable or linkage would have passed through.) If you put one of these idle holes to work, don't throw away the plug. By drilling a small hole in it, you can use it as a *grommet* to protect your wiring and to reseal the hole, as shown in **Fig 1-13**.

If you cannot locate an unused hole, it's usually possible to persuade existing hoses or cables to share their grommets with your cables. Perhaps the most obvious (and visible) choice will be where the factory wiring harness, a large bundle of perhaps as many as 50 wires, passes through the firewall. Unfortunately, it's also the least desirable choice for a couple of reasons. Many of the wires in those bundles are connected to electronic devices, posing the possibility of harmful interference. (If

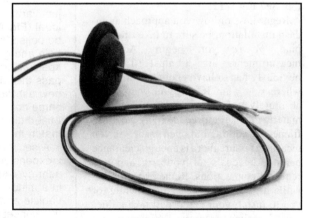

Fig 1-13—You can use a rubber plug, with a hole drilled in its center, as a firewall feedthrough grommet for your cables.

there are two bundles passing through the firewall on your vehicle, the one on the passenger side is especially a must to avoid. It usually connects to the engine control module.) In addition, the grommet is often molded into the harness, leaving very little room for extra wires to pass.

Holes for hood-release cables, clutch cables or vacuum hoses all offer a better point of entry for your power cables. Of course, if all else fails, you can make your own hole in the firewall. If you must do that, avoid possible damage by accurately locating the place you plan to drill. Make the hole no larger than necessary and be sure to seal the hole with the correct grommet, to prevent the entry of fumes and noise into the interior of the vehicle.

Once they are through the firewall, route your cables carefully along the driver's side of the engine compartment, avoiding exhaust system components, belts, pulleys and anything else that moves. Use wire ties to keep your cables in place while maintaining as much distance from the vehicle wiring as possible.

Getting Disconnected?

Radio equipment manufacturers stipulate that to prevent possible damage to your rig, you should always turn it off before cranking the car engine. Since you are connecting directly to the battery, you can't rely on the ignition switch to cut the power to your rig when the starter is engaged. To eliminate having to always remember to turn the rig off, you may want to install a cut-out relay in the power lead to disconnect the rig automatically when the starter is engaged. Check the August 1994 *QST* "Hints and Kinks," page 67, to see how John Conklin, WDØO, built a cut-out circuit for his mobile rig. Be sure to choose a relay having contacts rated to handle your rig's input current.

The best way to make the connections to the battery is with a suitable terminal, attached to the existing cable clamp hardware (see **Fig 1-14** and **Fig 1-15**). Although you've probably seen it done, it is *not* advisable to remove the cable clamps, place the ends of the power leads into the clamps and then replace the clamps on the battery posts. Here's why. The posts on a top-post battery are tapered, and so are the holes in the cable clamps. When properly mated, the post and clamp form a tight fit with a large contact area—necessary for the large amount of current they must carry. When you place a wire between the post and clamp, the misalignment you cause not only reduces the mating surface area and makes the clamp difficult to keep tight, it also allows acidic fumes to corrode the post-to-clamp

Fig 1-14—Here's a good example of how to make your connection at the battery. Sandwich the ring terminals between two flat washers and use an additional nut to hold them in place. Coat the surfaces with plenty of conductive anti-seize compound.

Fig 1-15—These convenient and inexpensive battery terminal adapters were purchased at a local automotive supply retailer. They are ideal for connecting a mobile station directly to automobile's battery terminals without compromising your battery's connections.

junction. Eventually, the rapidly deteriorating connection will leave you with a car that won't start.

Be sure to observe polarity when making your connections to the battery. It's also a good idea to leave all equipment disconnected from the power leads until those connections are made (pull the fuses if necessary). This will not only save you much grief if you inadvertently hook things up wrong (double check your connections), it will also help to prevent unwanted sparks near the battery.

There's a Bomb Under the Hood?

Your car's battery looks harmless sitting there in its black plastic wrapper, doesn't it? Don't let it fool you. Just ask my friend Joe who experienced the effects

Outgassing and Maintenance-Free Batteries

Outgassing in lead acid batteries is primarily a function of what metal is added to the lead plates to strengthen them. In the "good old days," automobile batteries had antimony added to the plates and grids. This acted as a catalyst and caused a lot of outgassing. The average automobile owner had to add water to the cells three or four times per year to replace the water lost through electrolysis and outgassing.

About thirty years ago, calcium was substituted for antimony, and the first maintenance-free batteries appeared. The calcium had much less of a problem with outgassing and so the need for refilling was greatly reduced. However, calcium had its own problems in that the batteries were less able to handle a deep discharge, and also the battery needed a bit more voltage during charge to bring it to the full charge state. That's why there are two switch settings on the better battery chargers—regular and deep discharge (or deep cycle).

Combining the two technologies yielded a third approach with antimony added to the positive plate and calcium added to the negative plate. This compromise is better able to handle a deep cycle but still has some gassing, so the battery usually has fill caps for addition of water from time to time.

As with any lead acid battery, including maintenance-free types, safety should always come first. Most hydrogen evolution from the lead in the positive plate is recombined at the negative plate during the charge cycle. It is not a problem during discharging. A maintenance-free battery is usually a calcium-added type and does not produce high levels of hydrogen. What hydrogen is produced passes through pressure-relief valves built in the tops of the cells. Remember also that hydrogen is very light and will quickly rise into the atmosphere. The problem with the old antimony battery charging was that there was often a car hood which could have been pulled down over the engine and battery, thereby forming a trapped hydrogen bubble. A person charging the battery in a car might pull down the hood over the battery and charger to protect against rain or snow while charging, for example. Same with charging in a small, unvented enclosure.

A sealed, maintenance-free battery can still emit some hydrogen, but the cells are not under high pressure when this happens. The relief valve will allow venting at a low pressure. But you cannot add water to a sealed battery.—*Ken Stuart, W3VVN.*

of an exploding car battery first hand several years ago. Joe was adding oil to his car's engine when he accidentally knocked a container across the battery terminals. The battery suddenly exploded, spewing acid and pieces of the battery in his face. Joe and I, along with our wives and several friends, were scheduled to go out that night. Instead, we rushed Joe to the nearest hospital. Joe had lost his vision and concern ran high.

Fortunately, this story has a happy ending. Thanks to expert treatment at the emergency room (which included an eye flush), Joe completely regained his vision and suffered no permanent effects. He was lucky!

Still not convinced? Consider another case involving my neighbor Kim. She was a high school student at the time the incident occurred. Kim was startled by a loud explosion under the hood of her parent's car while driving it in the school parking lot. Fortunately for her, she was able to drive the vehicle home. Later that evening, Kim's Dad, Jim, investigated under the hood and found the entire top of the battery was missing. The battery had suddenly and unexpectedly exploded for no apparent reason! In this case, a simple trip the local garage cured the problem. Fortunately no one was injured in this case, *but don't fail to learn the lesson*. Treat your car's battery with the same respect as you would a live hand grenade.

Lead-acid battery explosions such as these are typically a result of hydrogen gas. The gas is generated as a result of a chemical reaction when they are being charged. Normally, the hydrogen is harmlessly dissipated into the atmosphere. But, given the right conditions, a spark can ignite the gas and cause the battery to explode—hurling battery fragments hundreds of feet and showering the area with sulfuric acid. Lighter-than-air hydrogen can also be trapped under the hood of your car like a bubble, another particularly dangerous situation. There have been many instances where lost eyesight and bodily disfigurement resulted from battery explosions. Not everyone has been as lucky as Joe or Kim.

Keep in mind too that batteries don't have to explode to do lots of damage. Even though they supply relatively low voltages, they can supply tremendous currents that rival most arc-welding machines. A misplaced tool or bare wire bridging the terminals of a battery can be heated to a searing white hot in milliseconds. Take these basic precautions when working around batteries.

- Wear eye protection, or better yet, a face shield.
- Don't place tools or other conductive objects on top of a battery.
- Avoid creating sparks when making connections to a battery.
- Keep spectators at a safe distance.
- Do not remove insulating caps from battery terminals.

Keeping Your Car Smoke-Free

Fuses are cheap life insurance for your rig and your car. Make sure you properly fuse *both* power leads. Think a fuse in the negative power lead is a waste? It really isn't. If you examine your vehicle's battery cables, you'll find the large negative cable is connected directly to the engine block. That's because it must handle the return current from the starter, which can easily exceed 100 A. If the cable should lose contact while the starter is engaged, your rig's case and negative power lead may attempt to complete the circuit. Just one additional 10-cent fuse could save your rig.

Make sure the fuses are located as close as possible to the battery connections. This prevents the possibility of fire in the event a power cable shorts to the chassis. The power cables for most mobile rigs now come with the fuses at the battery end; if yours didn't, you'll need to relocate the holders. But don't let fuse holders rest on top of the battery. Fumes from the battery will corrode the terminals, making disassembly of the fuse and its holder difficult or impossible.

Stay Connected

Vibration, moisture and temperature extremes can team up to deliver a knockout punch to your mobile rig's power and antenna connections. Over time, connections that were once sound can deteriorate, causing electrical interference, equipment malfunctions and possibly even a fire. A few simple steps can help keep your connections secure.

Make sure all mechanical connections are tight. Use lockwashers where needed and periodically check nuts, bolts and cable connectors for loosening. If you find it necessary to splice wires, it's best to solder them and cover the splice with heat-shrink tubing. If you choose to use crimp-on terminals, be sure to use the proper crimping tool to install them. Where possible, take time to also solder the wire to the terminal to ensure a good electrical connection. Crimp terminals that cannot be soldered, as well as any nut, bolt, screw or washer used for power or antenna connections should have a conductive anti-seize compound such as *Kopr-Shield* applied to prevent corrosion. You may also want to add some further insurance against corrosion by using stainless steel or brass hardware.

Sound Solutions

You've seen it on television. The private detective places a drinking glass against a wall, puts his ear to the glass and eavesdrops on the conversation in the next room. If you've ever tried it yourself, you would probably agree the sound could best be described as being somewhat hollow, having a far-away quality.

Unfortunately, the audio from many

The Traveling Ham's Survival Kit

Odds are you can motor around your local neighborhood for the next 50 years and never experience a problem with your mobile rig or antenna. But hit the road for places far away and more than likely some part of your mobile installation will hold your QSOs hostage with an intermittent connection, loose mounting hardware or perhaps just a blown fuse. If you have packed your Traveling Ham's Survival Kit, you can quickly pay the ransom and be back on the air. If you haven't, you may find some towns don't have a radio store.

The list here is far from all-inclusive; you'll want to modify it to suit your own installation, travel agenda and determination to get back on the air. The idea is to take along the most basic items you might need, and hope Murphy's law does not apply—after all, a screwdriver is a screwdriver, right?

- A VOM and miscellaneous tool to fit battery connection, rig mounts and antenna hardware. Include Allen wrenches and metric sizes.
- Electrical tape—don't leave home without it! The same for the manual for your rig, unless you have memorized all 2461 programming commands for your dual bander.
- Spare fuses—both for your rig and for the auto. If the air conditioner fuse blows while you are operating, your ham gear will be blamed.
- The *ARRL Repeater Directory*. This pocket-sized book comes in very handy in a strange area if you need help finding a motel or restaurant by asking someone on a repeater. *The ARRL Repeater Directory* also comes in electronic form (*TravelPlus for Repeaters*), perfect for trip planning and laptop computers.

mobile installations has very similar characteristics. Granted, tiny speakers in tiny rigs are obviously a *big* part of the problem. Further compounding the problem is that most rigs have only two possible locations for the built-in speaker—on the top or bottom cover of the rig. Consequently, the sound coming from your under-dash rig either gets directed toward the floor, where the carpet absorbs it very nicely, or it gets bounced off the bottom of the dash and subsequently ricochets haphazardly about the car, creating a not-so-pleasant reverb effect. The in-dash rig fares even more poorly, it's audio sounding like it's coming from the car in front of you.

Unless your mobile operation is casual and very infrequent, you'll probably want to provide an outboard speaker for your rig. Many companies manufacture remote speakers, some with built-in amplifiers, suitable for mobile use. You can also find used or surplus commercial mobile commercial mobile two-way radio speakers at many hamfests. These speakers are ruggedly designed and provide good quality sound, and are usually very reasonably priced. Be sure the speaker you choose is the correct impedance for your rig.

When installing an outboard speaker, keep in mind the best results will be obtained by mounting the speaker as close as possible to your ears, aimed so the sound is directed toward you—especially important if you frequently travel with non-ham passengers who haven't developed a love for the musical quality of ham radio.

In fact, you may want to mount the speaker right over your head. In some vehicles, there is enough space between the interior headliner and the roof to install a speaker. Normally, all that's necessary is to remove the trim along the driver's side edge of the headliner and gently pull the headliner down far enough to insert your speaker. Route the speaker leads behind the windshield pillar trim and then to your radio.

Unless you've got lots of room, you'll probably have to use a small speaker with no enclosure. Be sure to secure the speaker so it doesn't wander about. (The overhead speaker approach really works great for mobile CW operators. Family members and other passengers aren't subjected to the sound of the CW, especially important on long trips.)

Stealth Speakers

Before you dash out to purchase an outboard speaker for your mobile installation, you may want to investigate the possibility of using components of your existing vehicle sound system.

Several approaches are practical. Some hams provide an interface between their rig and the car radio, allowing them to use the entire audio system, as Tim Brown, NM3E, described in a *QST* "Hints and Kinks" column. Alternately, you may wish to feed your rig's audio to only one of the speakers, but if you do that, provide some means of isolating your rig from your car radio, to prevent possible damage.

If you are hesitant to risk making connections to your car's sound system, you may be able to install your own speaker in an existing but unused factory speaker location. Since most automobiles

---

Overhead Consoles

Many new minivans and other autos now come with an overhead console. Some, such as the vans from Chrysler Corporation, include a sunglass holder and a bin for a garage-door opener transmitter. If you are not using one of these, the holders can be used to contain a small speaker.

Even without using one of these bins, the overhead console makes a good mounting place for an external speaker, but be careful to select a position away from your head. If the speaker is mounted on the outside surface of the console, you should be able to swing your head toward the passenger side without encountering the protruding speaker.

---

have many sound system options available, there is almost always a place where you can install an additional speaker. For example, many autos with factory-installed sound systems will have an unused speaker location, complete with grill, in the center of the dash. (This is an excellent location. The windshield focuses the sound back toward the passenger compartment.) A trip to your dealer or the local auto salvage yard may net a factory correct-fitting speaker.

Don't Forget Your Key

Anyone who has operated mobile CW can tell you it opens up a whole new world of ham-radio excitement. For long-lasting enjoyment of this unique form of mobile operation, it's essential you mount your key or paddles properly. Unlike a microphone, the key must remain stationary during operation. If the seat of your vehicle has a fold-down arm rest, it may be possible for you to mount your keying device there. This can provide the added convenience of being able to fold the key up and out of the way when it is not in use.

Fig 1-16 shows a mount that has been used successfully by Roger Burch, WF4N, in three different vehicles equipped with bucket seats. The platform keeps the paddles firmly in place and its added length in front of the paddles provides a convenient forearm rest—a must to prevent operator fatigue during extended periods of high-speed CW operation.

As you can see, setting up a ham station

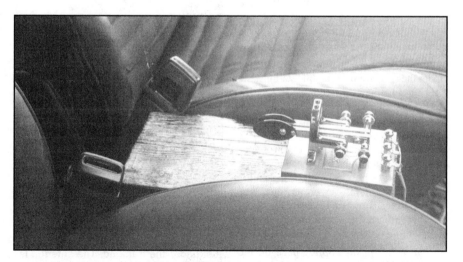

Fig 1-16—Roger Burch, WF4N, has used this arrangement to hold his keyer paddle in several cars equipped with bucket seats. The platform is made from 2×6-inch wood stock. Not visible are two feet, also made from 2×6-inch stock, extending from the platform to the floor. A short length of metal strapping is attached between the legs and screwed to the floor with a sheet-metal screw. One of the rubber feet on the paddle base has had its retaining screw replaced with a longer screw. This passes through the platform and holds the paddle in place.

on wheels isn't a plug-and-play proposition. So take your time, plan carefully, proceed even more carefully, and your mobile installation will be a source of pride an operating enjoyment for a long time to come.

Stay tuned as we go on to address one of the most talked-about topics in ham radio: Antennas.

VHF/UHF ANTENNAS

Buy the best, most expensive, gizmo-of-the week radio equipment you can find, install it in your car, hook it to a poor antenna and most likely you'll be disappointed in the results! The antenna provides the launching pad for your signal, so the better the antenna and its installation, the greater the reward—more effective and enjoyable communications. But first, let's take a closer look at an important topic before we consider any antenna or mobile installation.

Radiation Hazards and Antenna Placement

On August 1, 1996, the FCC announced a significant new rules change. Most radio services, including the Amateur Service, are required to comply with new requirements regulating human exposure to RF radiated fields. Although *all* Amateur Radio stations must comply with the Maximum Permissible Exposure (**MPE**) limits, regardless of power, operating mode or station configuration, the FCC presumes that certain stations are safe without an evaluation. Those are:

- Amateur stations using a transmitter power of less than 50 W PEP at the transmitter output terminal.
- Mobile or portable stations using a transmitter with push-to-talk control.

Although an evaluation is not required under the rules for most mobile stations, the MPE limits must still not be exceeded. Before considering any mobile antenna and mounting scheme, you should be certain it your station will be in compliance with these limits and not present a hazard to you or your passengers. Although it is beyond the scope of this book to provide a full and detailed treatment of this all-important topic, the ARRL provides several excellent resources for this information, including the limits. Read before you install a new antenna:

- *The RF Exposure Safety* page on the ARRL Web site at **www.arrl.org/tis/info/rfexpose.html**.
- *RF Exposure and You* is an ARRL publication is dedicated entirely to the subject of RF safety and FCC RF Exposure requirements. This is

mandatory reading for exposure guidelines.

- *The ARRL Handbook* safety chapter contains information on RF exposure.
- *The ARRL Antenna Book* safety chapter contains information on RF exposure.

Although an evaluation is not required under the rules, common sense dictates that RF safety be heavily considered when you install your mobile equipment. Most mobile antennas, for example, can be placed in several locations on your vehicle. Placing it in the center of the roof usually gives you optimum performance and the best radiation pattern. Aesthetic considerations, as well as the style of mount you are using, however, might prompt you to place the antenna elsewhere. If so, make sure you aren't exposing yourself or your passengers to dangerous or excessive levels of RF radiation. A cowl or trunk lip mounted

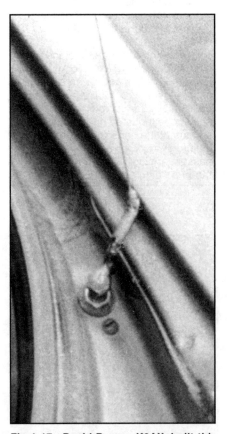

Fig 1-17—David Brown, K8AX, built this almost invisible 2-meter antenna. A banana jack is mounted to the car body, just under the trunk lid. The antenna is made from piano wire and a banana plug. Worried about losing the antenna? Just open the trunk, unplug the antenna, close the trunk and walk away. See the text for a discussion of antenna location and radiation safety. *(Photo courtesy of K8AX.)*

antenna might be as close as 1 to 3 feet from the heads of your rear seat passengers. See **Fig 1-17**. Mobile installations with a high power amplifier pose an even greater risk.

As you peruse the manufacturers' catalogs or the ads in Amateur Radio magazines, you'll see a mesmerizing variety of VHF/UHF mobile antennas. Which one should you choose? Your choice will be affected by many factors, such as cost, method of mounting, performance, appearance, and even the height of your garage door. While all antennas are a compromise, this is especially true with mobile antennas. Let's now take a closer look at some mobile antenna considerations.

Dual Antennas vs Dual-Band Antennas

If you will be using a dual band rig, you'll need antenna for both bands. Should you use separate antennas for each band or one for both? If your rig has a *diplexer* built in, this decision is made for you. A diplexer combines the signals for both bands into a common feed line. If your rig doesn't have a diplexer cost may be a factor. Using separate antennas not only saves you the expense of buying a diplexer, the combined cost of two antennas may be less than the cost of some dual-band antennas. But you will have to mount two separate antennas.

On the other hand, if you prefer to mount only one antenna, you'll need a dual-band antenna. Of course, everyone knows dual-band antennas are a compromise, offering lackluster performance. Well, if you compare specifications, you may be impressed with the gain figures posted by some dual-band antennas. See the sidebar "The Gain Game" for a discussion of advertised antenna gain. By using a long radiating element (some 2-meter/440-MHz antennas approach 5 feet in length) and appropriate loading coils, some dual-band antennas boast greater gain than that of many single-band antennas.

Be aware, however, that in addition to being rather tall, many of these high-performance dual-band antennas are very rigidly constructed, making them extremely unforgiving of collisions with low-hanging fixed objects. Rigid construction and lack of a spring base helps to keep the antenna vertical at highway speeds, thus enhancing antenna performance.

If you choose one of these large antennas, you may either want to select one with a foldover mast, or use a foldover mount to allow the antenna to be lowered out of harm's way.

The Gain Game

As you shop for a VHF/UHF antenna, you'll often see references made to the term *gain*. What is antenna gain and why is it important?

Gain is a relative measurement of an antenna's performance, stated in decibels or dB, as compared to a standard, or reference antenna. If the reference antenna is a half-wave dipole, the antenna gain will be listed as *dBd*, or gain over a dipole (a half-wave dipole and a quarter-wave vertical have equal gain). On the other hand, if the reference antenna is an isotropic radiator (a theoretical antenna that radiates equally in all directions), the gain will be expressed as *dBi*—standing for gain over an isotropic radiator. If the antenna you are considering has an advertised gain of 3 dBd, your signal should be twice as strong as it would be if you were using a quarter-wave vertical. No antenna design can increase your power output; they can simply concentrate the signal in the elevation plane, radiating less power skyward to wastefully warm the ionosphere. The law of reciprocity indicates you'll realize the same improvement on received signals as well. Not only will an antenna with higher gain increase the range of your signal, it might help reduce the *picket fence* effect—a rapid chopping or fluctuation of signal strength often experienced by mobile stations.

As you compare advertised gain figures for various antennas, be sure to know the standard of comparison (dBd or dBi) being used. This is important when you remember a dipole has a theoretical gain of 2.14 dB over isotropic. Antennas referenced to an isotropic radiator have a gain figure of about 2 dB more than antennas referenced to a dipole. Although reputable antenna manufacturers strive to provide accurate gain measurements for their products, beware of imaginative claims! If the 19-inch long 2-meter antenna that has caught your eye is rated for 10 dB of gain, don't be surprised if it's 10 dBwdr—that is, gain over a wet dish rag! Because of the difficulty in verifying advertised antenna gain, the ARRL does not publish gain figures without a whole list of reporting requirements.

Wavelength vs Gain

New hams are often confused by the various sizes available to them when selecting a new antenna. First and foremost, don't buy an antenna that's longer than the practical limits imposed by your vehicle. A 3/4-wavelength antenna, for example, might not be a good idea if you have a high van and like to frequent the local car wash. You may also have some exciting times with low garage doors and overpasses!

If your operating habits confine you to an area of strong repeater coverage, you probably won't notice any difference in the performance of these antennas. The shorter and typically less expensive 1/4-wavelength is as good a choice as any. If, however, you operate in a fringe coverage area, a 5/8-wavelength model may provide some badly needed gain. The 3/4-wavelength antenna is also a good choice, although it tends to waste signal energy at high angles.

Fig C compares the patterns of each of these mobile antennas. The 1/4-wavelength antenna and its 3/4-wavelength big brother provide a close match for coax. A 5/8-wavelength antenna, however, requires a matching network that's typically located in its base.

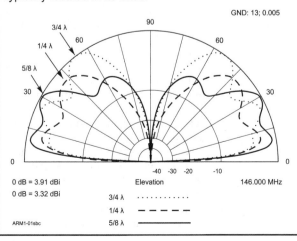

GND: 13; 0.005

0 dB = 3.91 dBi
0 dB = 3.32 dBi

146.000 MHz

3/4 λ ....
1/4 λ – – – –
5/8 λ ——————

ARM1-01sbc

Fig C—Typical antenna radiation patterns for 3/4-, 5/8-, and 1/4-wavelength mobile antennas. Notice that the patterns for the 3/4 and 5/8-wavelength models bulge outward, indicating somewhat improved gain toward the horizon.

The Ubiquitous Mag-Mount

Magnetic mount or *mag-mount* antennas are a popular choice for many hams. With a powerful magnet contained in the base, these antennas maintain a firm grip on the body of your car, eliminating the need to drill holes. In addition, their ability to be mounted and removed quickly makes them attractive to hams who frequently switch vehicles or who don't want to advertise the presence of their radio when they leave the car. Some new cars, however, cannot use mag-mounts because the car top is plastic or fiberglass.

If you choose a mag-mount antenna, a few simple precautions will minimize the possibility of damage to your car's finish. Before mounting the antenna, make sure the vehicle surface is clean and protected with a coat of good wax. If the antenna doesn't come with one, provide a protective pad between the antenna and the car—hams have reported success with various materials, including plastic sandwich bags, wax paper, even balloons. But be careful! Some materials can damage paint if left in contact for a long period of time.

If the material is too thick, it can seriously reduce the holding power of the magnet. Before mounting, check the base of the antenna to make sure it is clean and there are no burrs or protrusions—especially important if the antenna is pre-owned.

When placing the antenna on the vehicle, tilt the base and set one edge down first, the gently allow the remainder of the magnet to make contact. Once the antenna is in place, never slide it across the vehicle surface. When you remove the antenna, tilt it over onto the edge of the base and lift it free. Whenever possible, route the feed

Coax Routing Caveat

I typically ran the coax from my mag-mount antenna on the roof through the window in the rear door of my car. With the door locked, I was confident no one would try to open the door with the coax tightly wedged between the window and the top of the door. I was wrong! Within a few months, another family member unlocked and opened the door to exit the vehicle. The mag-mount base was dragged across the roof and scratched deep into the paint. Now, whenever I use a mag-mount long term, I put tape over the lock (as a minimum) to ensure this never happens again.— *Mike Gruber, W1MG.*

line lengthwise along the roof of the vehicle to the point of entry to minimize wind-induced movement that can scratch paint. If your mag-mount installation is long term, it's a good idea to remove the antenna periodically and clean accumulated dirt and moisture from under the base. You may also want to reposition the antenna occasionally to lessen the possibility of paint discoloration. What appears to be discoloration may actually be where your antenna has shaded the vehicle surface from the sun's powerful paint fading rays. For a seriously strong mag-mount capable of holding a small HF whip, see **Fig 1-18**.

Glass-Mount Antennas

If you consider a mag-mount antenna to be lacking in visual appeal, yet your conscience won't allow you to drill holes for a permanent antenna, a glass-mount antenna may be the solution. Common among cellular phone users, they can work well on the amateur bands too. If you own a sports car, the glass-mount antenna can provide a reasonable compromise between aesthetics and antenna performance.

Although the concept of transmitting power to your antenna without a hard-wired connection might seem mysterious, the operation of a glass-mount antenna is really quite simple. Two plates, one at the base of the antenna and the other on the end of the feed line, form a capacitor to couple the RF between the antenna and the feed line. Because this capacitor adds reactance that must be tuned out, a longer radiating element, or one with inductive loading—and sometimes a matching network at the end of the feed line—are employed to achieve a low SWR.

Build or Buy?

Many manufacturers make glass-mount antennas, but given the simplicity of this antenna, why not build your own? If this idea appeals to you, you want to refer to two *QST* articles: "A Glass-Mounted 2-Meter Mobile Antenna," by Bill English, N6TIW, in the April 1991 issue, p 31, and "An Easy, On-Glass Antenna with Multiband Capability," by Robin Rumbolt, WA4TEM, in March 1993 *QST*, p 35. Both articles have construction details and other useful information about glass-mount antennas. Members-only links for both articles appear on the ARRL Website at **www.arrl.org/tis/info/Mobile-V.html**.

You will generally get best performance by mounting the antenna at the top center of the rear window. Yes, you can mount your antenna at the top of the windshield, but be sure to check for windshield wiper clearance. Don't let your handiwork fall

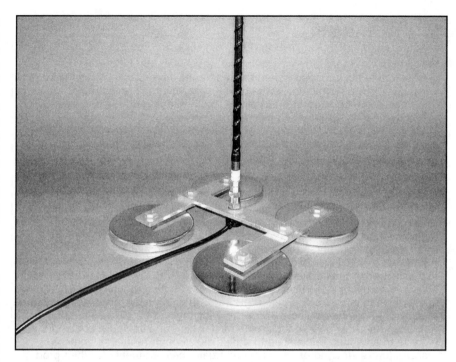

Fig 1-18—Mag-mounts are popular no-holes option, especially when considering multiple vehicles. Here is a very strong four-magnet mag-mount suitable for an HF whip. This mount uses the $^3/_8 \times$ 24-thread connector typical of HF and some VHF antennas. The NMO and $^3/_8 \times$ 24 connectors are the two most common connectors you'll likely encounter.

victim to a windshield-wiper assault the first time it rains! Besides allowing the greatest possible height above ground, placing the antenna at the top of the window provides the ground plane specified by some antenna manufactures as necessary to achieve advertised gain. Remember, the nonmetallic roof found on some cars and motor homes will not act as a ground plane.

When mounting the antenna, avoid placing it over or in close proximity to widow defogger/defroster grids or elements—these will degrade the antenna's performance and make it difficult or impossible to tune. You cannot use this type of antenna with tinted windows (or some other coated window types) made with a metal layer.

If your antenna requires grounding of the coax shield at the antenna end, make the connection directly to the vehicle body, as near as possible to the antenna mount. Use a volt-ohmmeter to ensure the metal where you make your connection is indeed grounded to the vehicle body. Although glass-mount antennas are removable, be sure you are satisfied with the location you have chosen before proceeding with the installation. If you find it necessary to remove the antenna, follow the manufacturer's instructions. Never use a metallic tool to pry the antenna base from the window.

Permanent-Mount Antennas

Glass-mount antennas have a trendy appearance that is truly chic, but their method of attachment to the vehicle limits them to a relatively short radiating element, typically a quarter wavelength. If you need better performance from your mobile antenna than a glass-mount can offer, you will probably want to consider a permanent-mount antenna. In addition to providing superior performance, when it comes to appearance permanent-mount antennas are hard to beat.

Although the term *permanent mount* is a bit of a misnomer (obviously, you can remove a permanent-mount antenna), it's generally used to describe any antenna requiring a mounting hole. The decision to take a drill to the body of your auto can be a difficult one. Consider it carefully before drilling. Aside from affecting resale value, drilling a hole in your auto can set the stage for future problems with rust—and provide water with a point of entry into the vehicle. Proper installation and maintenance are essential.

As mentioned previously, the antenna should be mounted in the center of the roof, which provides the largest, most-uniform ground plane. On some vehicles this location will be over the dome light. When you remove this light, you will have access to the lower side of the roof. Truck owners often locate the antenna at rear center of the roof to take advantage of dome light placement in those vehicles. Many minivans will give you convenient access by removing the overhead console. On vehicles with a reinforcement channel located in the center of the roof, you will probably want to shift the location of the antenna slightly to avoid having to drill the channel.

Take careful measurements and double check them before you drill. If your vehicle has a retractable sunroof, determine its area of travel before choosing a location for your antenna. If the antenna won't be installed over the dome light, it is a good idea to detach and lower (or remove) the headliner before you begin, providing you more room to work and possibly preventing damage to the headliner. Drill bits and hole saws have a voracious appetite for headliner material. You don't want them to indulge in a feeding frenzy on your vehicle.

Exert light pressure when drilling the hole, especially if you are using a hole saw. The thin metal roof of most modern automobiles is easily deformed, especially if the drill breaks through unexpectedly while you are balancing your full body weight on the handle. If you have access to a *chassis punch*, use one; it makes a neat and precisely dimensioned hole. De-burr the finished hole and lightly feather the paint along the edge of the opening with fine sandpaper to prevent chipping and subsequent rust. As you complete the installation, be sure to install the weather seal or O-ring provided with the antenna.

If you aren't comfortable with the idea of doing your own through-hole antenna installation, you may want to enlist the services of your local two-way radio shop. Most of them will be glad to mount the antenna for a nominal fee, relieving you of the work and worry.

Removal of a Permanent-Mount Antenna

As soon as you install a permanent-mount antenna, your car's resale value will plummet—or will it? Depending on the vehicle, the antenna and your savvy, perhaps not. If you are concerned about the effect of a permanent-mount antenna on the value of your automobile, consider the following tips:

Plug the Hole

If your antenna uses a standard size mounting hole, your local two-way radio shop can provide a snap-in rubber plug made just for filling antenna holes. If your vehicle is tall (a van or tuck, for example),

it's possible the plugged hole won't be visible to anyone who's less than 7 feet tall. This isn't a suggestion to deceive— it's just that some prospective owners of your vehicle won't object to the plugged hole if it isn't readily visible.

Patch the Hole

Before you drill the hole for an antenna, check with your dealer or a body shop to see what the charge will be to fill the hole and repaint the surrounding area at trading time. Often the charge is less than the deduction in vehicle value that the dealer or prospective buyer will assess.

Remove the Hole?

Drilling a hole in a new vehicle is almost always a soul-searching decision. If you find yourself resisting this option but still like the idea of a permanent mount antenna, take a closer look at your vehicle before resorting to the drill. Any parts that are easily removed and could be used to support an antenna might offer a more palatable solution to this dilemma. Simply replace the body part with one obtained at your local salvage yard—*after* drilling the hole of course. When removing the antenna, you can reinstall the original unmodified part.

Fig 1-19 shows a permanent-mount antenna that can be removed when you want to sell the car. This setup doesn't require you to even plug or patch a hole.

How About the Cell-Phone Antenna?

Having a cellular phone in your car is no longer an elite status symbol it once was.

Fig 1-19—While his antenna is not placed optimally in the center of the roof, Mark Goff, WA4JSN, did not have to drill a hole in his car's metal roof. He used the roof-mounted stoplight (see text). At trade-in time he will simply replace the light housing. (Photo courtesy of WA4JSN.)

Most cell phones no longer require an external antenna. Some phones, however, have provisions for such an option. Don't discount the possibility a future owner of your vehicle may also want a hole for an external antenna—and possibly even the mount. CB and other radio services also typically require an external antenna. GPS and satellite radio are also on the rise. There are many possibilities!

If considering this approach, check your antenna catalogs and you'll find many manufacturers make non-amateur antennas using the same type of mount as their amateur cousins. When it comes time to trade or sell, it's a simple job to swap antennas.

Trunk-Lip Mounts

You say you want the good looks and the performance of a permanent-mount antenna, but without drilling holes? The trunk lip mount may be your answer. Available in a variety of styles and configurations, the trunk lip mount offers a sturdy and attractive support for the largest VHF/UHF antennas-many models also allow the antenna to be folded over for clearance when necessary.

Placement of a trunk lip-mounted antenna is often governed largely aesthetics, but keep in mind mounting it is in the center of the forward edge of the trunk lid will skew the radiation pattern and possibly make tuning difficult—due to the close proximity of the roof. Cars with a hatchback fare better in this regard since the antenna can be mounted near the top of the hatch, clear of surrounding metal. When contemplating the use of a trunk-lip mount, take into consideration the amount of radiation rear seat passengers may be exposed to—as previously mentioned.

Installation of a trunk-lip mount is straightforward, but follow the manufacturer's instructions carefully. Remember, the trunk-lip antenna depends on a good ground connection to the body of the vehicle for proper operation. This means you will have to remove the paint where the mount's setscrews contact the trunk lip. Mark contact points carefully and remove the minimum area of paint necessary to allow setscrew-to-trunk lid contact This is a good time to use your VOM to determine if the trunk lid is electrically connected to the rest of the vehicle.

Before installing the mount, coat the bare spots of metal with a good anti-seize compound, and be sure to periodically check the mounting point for any signs of rust. With the antenna mounted, carefully open the trunk or hatch while observing for clearance problems between the antenna and the

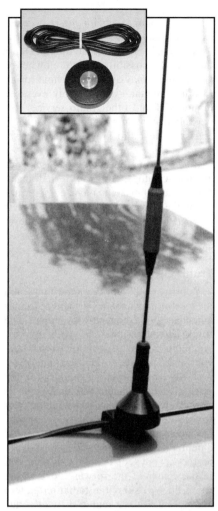

Fig 1-20—A trunk gutter VHF-antenna mount is easy to install. As with any deck-mounted antenna, the radiation pattern will not be symmetrical. It will favor some directions. This mount uses the popular NMO connector typically used with smaller VHF and UHF antennas. See insert for NMO close-up on a mag-mount.

vehicle body. On some vehicles it may be necessary to use an antenna or mount with fold-over capability.

Trunk-Gutter Mounts

No relation to the trunk-lip mount, the trunk-gutter mount requires two or three small holes to be drilled to facilitate installation. Fortunately, because the holes are made in the gutter under the trunk lid, they aren't visible when the trunk is closed. When the time comes to remove the antenna, the small holes are easily filled, requiring no expensive body and paint refinishing. See **Fig 1-20**.

Contrary to their name, they aren't limited to trunk-gutter mounting, but can be used about anywhere there is a gap in body panels. Due to the simplicity of the

trunk-gutter mount, if you are handy with metal-working tools, you might want to fabricate your own. If the mount requires you to drill holes, the finished installation should be checked for the appearance of rust or water leakage into the vehicle.

HF ANTENNAS

Everyone knows the most important rule of HF antennas: Make them big and put them high in the sky. This is quite a challenge for the prospective mobile HF operator!

After all, with the possible exception of 10 meters, your HF mobile antenna will always be shorter than a quarter wavelength, and it certainly isn't going to be very far off the ground. Since mobile HF antennas are such a compromise, their performance will always rank somewhere between poor and terrible. Right? Well…you may be surprised. Mobile stations have been known to work plenty of DX and maintain reliable skeds with other hams across the country.

Choosing the Right HF Antenna

Your choice of antenna will be influenced by many considerations, such as appearance, cost, performance and the ability to QSY within a band and from band to band. Several manufactures offer mobile HF antennas, or perhaps you'll want to build your own.

Fig 1-21 shows a bumper-mounted HF mobile antenna used by Jack Schuster, W1WEF.

Monoband Antennas

If your plans for mobile operation don't include a lot of band hopping, you may

Fig 1-21—Jack Schuster, W1WEF, uses a tennis ball to keep the Hustler mast from banging into the car body, and the bungee cord restrains the mast from tilting back at highway speeds. *(Photo courtesy of W1WEF.)*

want to consider using a monoband antenna. Generally, they are cheaper, simpler and somewhat more pleasing to the eye than their multiband cousins. They come in two basic configurations.

In a design made popular by the Hustler antenna, a fixed-length mast is topped by removable loading coil-whip section, known as a *resonator*, which allows the antenna to resonate in a particular band. In addition to the standard resonator, the Hustler antennas are available with "Super" resonators, which provide greater power handling capability and more bandwidth. Because you only purchase one resonator for each band you operate, you won't be spending money for extra band capability you don't require. If you frequent areas with low clearance, you'll appreciate the Hustler's fold-over mast. Hustlers are now manufactured by the New-Tronics Antenna Corporation.

Another approach is to make the entire antenna function on one band only. Some designs utilize a fiberglass mast helically wound with a copper radiator and topped off with an adjustable stainless steel whip. With performance comparable to the Hustler-type antenna on some bands, many hams choose this style of antenna for its sleeker appearance.

If aesthetics are a top priority, check out the line of diminutive monoband antennas made by some manufacturers. One design, offered in versions for 10, 20 and 40 meters, consists of a short, stainless steel whip with a center-loading coil. How short? The 40-meter version is only 40 inches in length. No matter what brand you choose, you'll need separate antennas for each band you plan to operate.

Multiband Antennas

Since you'll probably be using a multiband rig, why not complement it with a multiband antenna? As the name implies, multiband antennas can be operated on several of the HF bands, with most models covering 10-80 meters. In addition to appearance and size, the method of bandswitching will be factor in your choice of antenna.

Manual Bandswitching

If you don't mind having to get out of your car to change bands, you have a choice of several multiband antennas. One very popular one is the *Bug Catcher* antenna, marketed under several trade names, including the Texas Bug Catcher. Believed to have been developed and named by the Texas Army Corps of engineers in the 1920s, the Bug Catchers use a tall mast with a large, air-wound center loading coil, topped off with a

Double-Duty HF Mobile Antennas

We had just acquired a new Dodge minivan about a month before our vacation. The destination was our home away from home, a small beach condo overlooking Cape Cod Bay. The condo's top floor balcony also provided a clear shot across the Bay back to the mainland. A 2-meter hand-held worked wonders, but this location begged for more—and the urge for HF was getting stronger with every trip I made!

In terms of distance, I knew 75 meters would provide excellent propagation to my ham friends back home in Connecticut. While the location couldn't be better, the balcony was tiny, way too small for anything practical... or was it? A local Hamfest soon provided the answer. But meanwhile, back at the minivan...

For mobile use, I acquired a heavy-duty mag-mount for HF. I didn't have time to get involved in a complicated installation project—and I knew my wife would never approve of my drilling holes in *her* new vehicle. The mag-mount was an obvious choice in this situation. It was a fast, easy-to-install and no-holes option that kept everyone happy.

Next, at a nearby hamfest I wanted to purchase a monoband 75-meter mobile antenna. They were under $20 each and antennas were available for most of the HF amateur bands. While at the counter, I saw an adaptor that enables you to fabricate a dipole from two monoband mobile antennas. I quickly surmised this was the answer for my problem. I could use a vertical whip antenna for mobile use, and then at the condo use the same antenna as part of a dipole on the condo's balcony. The adapter and second half of the

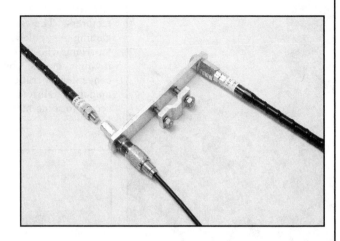

dipole conveniently fit inside the minivan for the trip out to the Cape. .

So how well did it work? I kept my expectations low. And as expected, the antennas required retuning when switched from mobile to dipole use. (I used color-coding to set the lengths, which is a pretty simple process.) By careful selections of times and frequencies, however, I was able to keep daily skeds with friends back home. A couple days I even received S9 or better signal reports. Antennas for other bands are also easily stored in both the van and condo. Sure, this solution is a compromise in performance, but it was economical, convenient, and it worked! —*Mike Gruber, W1MG.*

---

capacitance hat and stainless-steel whip. Band changes are made by adjusting the location of a tap on the loading coil. While the Bug Catchers are considered to be top performers, their imposing presence, especially when used on smaller vehicles, might be aesthetically stressful. These antennas are heavy and their size subjects them to quite a bit of wind loading, so they'll need a high-strength mount and possibly a guy rope or twine to prevent excess movement. See **Fig 1-22**.

Since the construction of the Bug Catcher is readily apparent to anyone who looks at one, many hams choose to build their own version, incorporating changes or modifications as they see fit. See *Another Look At An Old Subject: The Bug Catcher*, by Charles Frazell, WD5FRN in the December 1980 issue of *QST* for Bug Catcher construction details. *The HF Mobile Antenna* page on the ARRL Website contains a link for this article at

Fig 1-22—W5KFT's beefy "Texas Bug Catcher" antenna uses a large-diameter, air wound coil with a moveable tap to change bands. A rugged mount attaches the antenna's base solidly to the bumper of Bryan's vehicle. (Photo bt W5JUV.)

www.arrl.org/tis/info/Mobile-H.html. Commercial versions and parts are available from GLA Systems. See their web site **www.texasbugcatcher.com/** for more information.

Generally acknowledged as one of the most rugged multiband antennas on the market, the *Outbacker* antenna design is appearing on vehicles all over the world. Offered in several versions in lengths up to 7.5 feet, the Outbackers cover 10 to 75 meters with power ratings up to 300 W. The antenna's two piece construction consists of a fiberglass mast—wound with a helical radiator covered with an epoxy resin coating—and an adjustable stainless steel whip. Band changes are accomplished by connecting a jumper (the *Wander Lead*) to the appropriate tap on the mast. Hams report good performance from the Outbacker.

Automatic Bandswitching

Some hams want instant band changing capability—no delays—and no stopping to get out of the car to shuffle antenna parts around. For them, a multiband antenna with automatic bandswitching is the *only* way to go. Such requirements impose some interesting challenges—and good

Fig 1-23—John Gerlach, K6BRD/7, built this HF antenna array. It consists of a standard 52-inch Hustler fold-over mast and a home-brewed spider. John built coils for 17, 15, 12 and 10 meters, each with a capacitance hat. The tennis balls keep the mast from banging into the car body, and the bungee cord restrains the mast from tilting back at highway speeds. *(Photo courtesy of K6BRD/7.)*

old ham ingenuity has met those challenges with some equally interesting antennas.

Although hams have tried many forms of automatic antenna bandswitching, one of the most popular methods uses multiple resonators in what is often referred to as a *spider* design. See **Fig 1-23**. The spider uses a resonator for each band, spaced radially around an adapter attached to the top of the mast. Because the resonators for the bands you aren't operating don't present a match for your rig, they are electrically divorced from the system and don't accept any power from the transmitter.

Several companies make spider-style antennas. Multi-Band Antennas makes the original Spider, or if you own a Hustler antenna they can supply an adapter to mount several resonators at once on the Hustler mast—à la the spider.

Another very popular antenna you may have seen or heard about is the *screwdriver* antenna. Although in its basic form it doesn't automatically chance bands, it can be adjusted from the driver's seat to operate *anywhere* within its designed frequency range, usually the 10- to 80-meter bands. Controllers to automatically resonate the antenna are available.

The brainchild of Don Johnson, W6AAQ, commercial versions are advertised in *QST* by several manufacturers. The screwdriver is rapidly gaining in popularity and for good reason too. Hams who have tried this antenna give it high marks for performance and convenient operation. Although it requires moderate fabrication skills as well as access to some machine tools, it is also a

Fig 1-24—The "screwdriver" all-band HF mobile antenna requires a rugged mount. *(Photo courtesy of W3IZ.)*

popular homebrewed HF mobile antenna design. See **Fig 1-24**.

Spanning the Band

At 14 MHz and above, most mobile antennas have sufficient bandwidth to allow reasonable frequency excursions without the necessity of retuning. At lower frequencies, antenna bandwidth diminishes, and even moderate changes in frequency can cause SWR to rise to a level your rig won't accept. If you desire the ability to cover much or all the bands you plan to operate, what are your options?

For an antenna tuned to resonance with an adjustable whip, you can use a permanent marker to identify the points on the whip corresponding to resonance, on the frequencies you plan to operate. This will allow you to quickly change the resonant frequency of the antenna with the need for an SWR meter. Likewise, antenna using tapped coils for tuning can be marked for the appropriate tap locations.

But what if you'd rather not have to stop and make adjustments just to move from one portion of the band to another? In this case, you can use an antenna tuner to achieve a low SWR when you are operating you antenna at a nonresonant frequency. If you haven't yet selected a mobile rig, you may want to consider one of the many models that contain a built-in antenna tuner. Not only will this save you valuable space inside your vehicle,

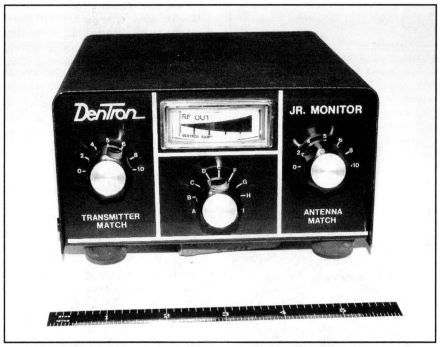

Fig 1-25—This Dentron Jr Monitor Tuner was a prize acquisition for Paul Galica, W1EPG at a hamfest flea market. Although the manufacturer has out of business for many years, Paul purchased this tuner for $25. Its small size makes it ideal for his mobile HF installation. Never overlook flea markets and similar to supplement your mobile station equipment and accessories. The 6-inch rule is for scale.

internal tuners adjust themselves automatically—all you do is key the rig and your tuner quickly achieves the best possible match. If your rig doesn't have an internal tuner, small outboard tuners, suitable for mobile use are available from several manufactures. See **Fig 1-25**.

Make all adjustments to an outboard tuner when the automobile is stationary—watching the SWR meter while you tweak the tuner and drive is dangerous. Don't try to use a tuner to make your 10-meter antenna work on 75 meters. Not only will increased loss make the antenna very inefficient, both resonators and the tuner can get very hot when force fed out-of-band RF, possibly sustaining permanent damage.

MOUNTS

Perhaps one of the biggest challenges faced by the would-be mobile HF operator is how to attach the antenna. Back in the days when cars had real bumpers, it was a simple matter to bolt on a spring base, or clamp on a bumper strap mount. No more. So what's a ham to do? Let's have a look.

If you are totally dedicated to HF mobile, the optimum installation can be had by drilling a hole and mounting the antenna in the middle of your car's roof. And if you are using one of the really tall antennas, you might be able to get the job of trimming low tree branches along the streets of your city. Seriously, in addition to the possible effects on the property of others (one ham reported his roof-mounted Outbacker antenna shattered fluorescent light bulbs under the canopy of his local gas station), consider the antenna's effect on the area where it is mounted. You may need to provide some type of reinforcement to prevent fatigue and fracture of the metal surrounding the mounting hole.

If you're squeamish about drilling holes, there are other ways to get your antenna on the roof—and keep it there. If your vehicle has a luggage rack, there are several companies that make mounts that clamp to the rack, or you can build your own luggage rack mount from some flat stock.

Be sure the rack is grounded, or add a good, low-impedance ground connection from the antenna ground connection to the body of the vehicle. What's low impedance? A good RF ground is almost always hard to achieve but there are some basics. Always make the connection as short as possible. Use heavy braid and avoid bends and curves. In an actual case involving a 20-meter mobile set-up, the ground was made with a strap routed down from the bed to the chassis of the truck. The inductive reactance of the so-called ground strap was more than $100 \, \Omega$ at 14 MHz. Not surprisingly, this set-up had

high SWR and the microphone was hot with RF.

Mag Mounts

Although mag-mount antennas were once considered to be practical only for VHF/UHF antennas, many hams now use them for HF antennas as well. Utilizing three or four powerful magnets attached to an "H" or "T" shaped metal framework, mag-mounts provide a no-holes-required, stable, wide-stance support for even the largest antennas.

Magnetic HF antenna mounts are available from several *QST* advertisers. If you prefer to build your own, refer to an article by Ed Karsin, W3BMW, titled "The Impossible Dream Whip HF Mag-Mount," in December 1992 *QST*. Remember, while the large area of the mount provides a capacitive electrical ground for the antenna, many installations work better with a direct ground connection to the vehicle.

Rear-Bumper Mounts

Mounting the antenna to the rear of the vehicle is a more attractive option for many hams, but rubber-covered bumpers offer no support for this approach. Well actually, there is an honest-to-goodness metal bumper lurking beneath the rubber cover. But since the bumper cover is usually not repairable, and replacing it is quite expensive, punching it full of holes just to mount an antenna isn't very practical. Fortunately, you may not have to. If you examine the lower edge of the bumper, you'll probably find one or more holes—sometimes where push-in plastic retainers are used to hold the rubber cover against the bumper. Of course, bumpers vary, but you can almost always fabricate a bracket—flat aluminum stock works well—that will extend beyond the edge of the bumper to provide a mount for your

antenna. Even if you have to drill holes, they won't be visible to anyone but ground-based insects and animals. Since the bumper is mounted on shock-absorbing brackets, it's a good idea to install a short ground strap from the chassis to the bumper to ensure a good electrical connection.

If you are unable to attach a mount directly to the bumper, it may be possible to use a longer bracket, attached to the frame and bent where necessary to clear the bumper. Either way, be sure your installation doesn't interfere with the operation of the bumper shock absorbers. See **Fig 1-26**.

Trailer-Hitch Mounts

You can also mount an antenna to some part of a trailer-hitch assembly. Although some hams mount their antenna in the existing hole provided for the hitch ball, this selection obviously limits your HF mobile activities to times when you aren't towing a trailer. If you mount the antenna to the slide-out section of a receiver hitch, you'll need to provide a ground strap from the hitch to the vehicle chassis to guarantee proper antenna performance. See **Fig 1-27**.

Fig 1-26—Pick up an antenna mounting point under the bumper. The plastic-covered bumper is bypassed with a piece of sheet aluminum. Notice the ground lug just to the right of the whip mount.

Fig 1-27—At A, underbody photo of elegant and rugged trailer-hitch assembly used for mounting a heavy screwdriver-type of mobile antenna. At B, topside view of trailer-hitch mount. *(Photo courtesy of W5ZAF.)*

License Plate Mounts

Concerned about the aesthetics of a heavy-duty magnetic antenna mount? Or worse yet, scratching the finish of your precious vehicle no matter how carefully you remove it? Consider a license plate mount as an elegant solution to your HF mobile antenna woes.

These antenna mounts attach to your car in one of the places least likely to suffer visible damage—behind the license plate.

Fig 1-28—At A, this license plate mount is ready for installation. Note the ground strap in the upper corner.

With strategically placed holes top, bottom and center, the heavy duty stainless-steel main plate attaches snugly between the license tag and your bumper or trunk lid. A horizontal tongue protrudes from the bottom or top of the mount. Typically, this extension has a large hole to accommodate a standard $^3/_8$-inch × 24-threaded antenna connector. See **Fig 1-28**.

With hundreds of different vehicles on the road, it isn't practical for a single license plate mount to be a one size fits all solution. You may find that you may need to do a little tweaking to achieve the best installation. And be sure to install your license plate mount and antenna in such a way that it complies with all applicable laws in your state. Obscuring a license plate with antenna hardware has been an issue in some states.

Towing-Loop Mount

Many cars have a heavy duty loop bolted or welded to the frame at the rear under the bumper. This allows the car to be towed and can be a convenient place to clamp a mount suitable for a mobile antenna. **Fig 1-29** shows how a pair of thick scrap aluminum plates can be used to clamp around a towing loop to make a mount for a mobile whip antenna.

Other Thoughts

A trunk lip mount may be adequate for one of the smaller HF antennas. Consult with the antenna manufacturer for recommendations, and refer to the VHF/UHF antenna section for installing guidelines.

So-called *quick-disconnect* adaptors make it quick and easy to install a screw-on type of mobile whip—a quarter turn and you're done! See **Fig 1-30**.

Recreational vehicles (RVs) can be a challenge to install permanent-mounted HF antennas. They are tall vehicles, so a long mobile whip mounted on the roof cannot usually be deployed while on the move — not unless you enjoy hearing the distinctive sound of a whip striking a bridge or underpass! Tilt-down whip mounts are sometimes employed so that a top-mounted whip can be lowered when preparing the RV for travel. Or the whip can be unscrewed from its mount before setting off. See **Fig 1-31**, which shows how the ladder on an RV can serve as a place on which to mount a simple HF antenna mount.

Jack Schuster, W1WEF, has taken the ladder to another dimension on his RV. He uses the ladder itself (which is insulated from the metal body of the RV) as an antenna! Jack feeds the ladder through an automatic antenna tuner mounted under the rear bumper of his RV. See **Fig 1-32**.

ANTENNAS AFTER INSTALLATION

The Care...

Life on the road is no joy for your mobile antenna. Wind loading, continual flexing, vibration and occasional blows from low obstructions can test the mettle of the best of antennas. Be sure to provide a secure mount for your antenna, and check it often—especially after antenna/low-flying object altercations. Some antennas use a whip-retention setscrew or collet. You should periodically check these screws for tightness. Having your antenna become a hood ornament for the car behind you on the freeway can have costly and potentially dangerous consequences.

Feeding

Feed your antenna with high quality coax, preferably one with at least 95% shield coverage. Even though most mobile feed line runs are relatively short, the higher quality coax will ensure that more of your rig's RF arrives at the antenna, and less of it ricochets around the inside of the vehicle—where it can cause automotive electronic systems to behave mysteriously.

As mentioned in previously, take care to

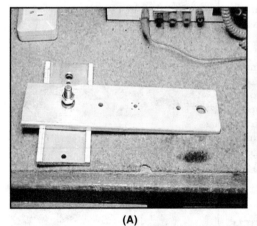

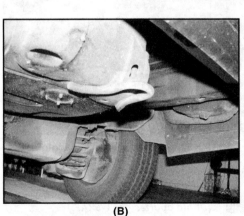

(A)

(B)

(C)

Fig 1-29—At A, a pair of scrap aluminum pieces clamps together around the heavy steel towing loop under a new Volkswagen Beetle to make a bracket for an HF mobile antenna. The towing loop is bolted to the car frame and is also present on many other vehicles. At B, photo of the towing loop under a 1996 Volvo. Lightly file both surfaces of the loop to clear away paint and dirt and use NoAlox or Penetrox to prevent corrosion between the steel and aluminum. At C, photo of Volkswagen with mobile whip antenna.

route the feed line away from vehicle wiring or electronics. If you must traverse a wiring harness with your feed line, make the crossing at right angles. Feed line for rear-mounted antennas is usually best routed through a hole in rear of the trunk or floor pan.

Check to see if you can use the existing plugged body holes—most vehicles have them. Feed line routed under the vehicle should be pulled snug with no excess slack left dangling. Secure it with cable ties or clips and use a rubber hose or a grommet where it passes through holes in the frame. Be sure to avoid exhaust and suspension components. Avoid making sharp turns or bends that deform the cable. These may shorten cable life and possibly increase the SWR by altering the characteristic impedance. With few exceptions (such as mag-mount and some glass-mount antennas) the coax shield should be securely grounded to the vehicle body or chassis as close as possible to the antenna. Remember to carefully seal the exposed end of the cable to prevent water migration between the jacket and center dielectric.

If your installation necessitates bringing the feed line into the vehicle through a door or the trunk, choose your point of entry very carefully. Although most doors and trunk lids can be safely closed on small (RG-58 style) coax without causing cable damage, the adjustment of latches and the thickness of weather stripping will be the determining factors. Besides causing high SWR, flattened cable can become shorted. Rule of thumb: With the door or trunk closed, you should be able to pull the cable through the gap with only a moderate tug.

Routing cable through the top of a rear hatch requires extra caution since the lip of the hatch usually recedes below the roofline as the hatch is opened, making a very effective feed line shear.

Tuning

The finished installation should be tuned for the lowest possible SWR to maximize performance. Although it may not be possible to obtain a perfect match, try to achieve a SWR of 2:1 or better, especially on VHF/UHF antennas. Adjust your antenna system with the vehicle in a clear location, away from power lines, trees and buildings, which can effect tuning. Make sure the trunk lid/hatch and doors are closed while you take SWR measurements. Use an accurate SWR meter designed for the band to be checked. Using an SWR meter outside its designed frequency range will give erroneous power and SWR readings.

Before you put the meter away and declare the installation finished, with an assistant driving, take your measurements again with the vehicle in motion. Although moderate fluctuations in SWR are normal as the antenna flexes and passes nearby objects, severe changes may reveal problems with the antenna or the need to reengineer your installation.

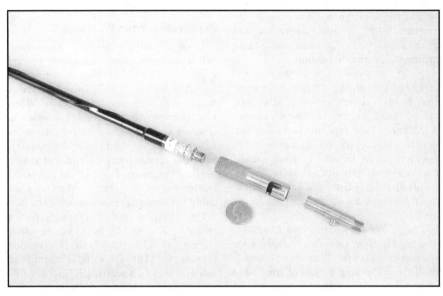

Fig 1-30—Quick-disconnect adaptors for HF-type mobile whips.

Fig 1-31—On many RVs, the ladder is about the only place strong enough to hold a conventional HF whip antenna. Simple mounts often prove to be the most effective. (Photo courtesy N6TST.)

Fig 1-32—W1WEF actually uses the insulated ladder on his RV as an HF antenna, feeding it with an automatic antenna tuner mounted under the rear bumper. (Photo courtesy W1WEF.)

Now Let's Hit The Road!

You're going to love going mobile. No longer will you have to cut short a good conversation because you have to be on your way somewhere. You can take the conversation with you. And should an emergency arise while you are on the road, your hobby might even save a life.

Speaking of emergencies, don't let your radio operation create one. If traffic or road conditions demand your undivided attention, stay off the radio. If you are involved in a conversation and you realize your ability to drive is compromised, either pull off the road or sign with the station you are working. No QSO is worth risking your safety or that of others.

HITTING THE ROAD ON VHF/UHF

Cars and VHF/UHF operation are made for each other. If you're sprinting around town or crisscrossing the country, the noise-free convenience of FM VHF/UHF radio is hard to beat—especially with all those repeaters out there.

Getting Maximum Mileage from Repeaters

When you are out on the road, perhaps one of the most discouraging things you can hear is the dreaded sound of silence right after you've keyed a repeater and announced "KB4AEY mobile, listening." Communicating is what ham radio is all about. So why isn't someone out there just jumping at the opportunity to communicate with you?

Well, first of all, you shouldn't take it personally. It has nothing to do with halitosis. Although several hams may have heard your transmission, they may be waiting to see if someone else is going to greet you. After an extended period of silence, they may even conclude you aren't really serious about talking to someone. (Or possibly they've been burned by a rude operator. See the sidebar, "Where's This Repeater Located?") How do you increase the odds of someone answering you? Ask any tournament-winning fisherman the secret to his success, and he will tell you that to catch fish, you got to use the right lure.

If the repeater you are monitoring has a conversation in progress, you might want to use a technique known as *tail-ending*. That simply means calling one of the participants when the conversation has ended. Make sure you don't call an operator who has indirectly indicated he won't be listening for further calls, for example, *"I've got to jump out and fill up with gas. 73 from KA6WAR."* (Calling the recorded voice ID of a repeater is also not considered good practice.) Of course you can join an ongoing QSO by interjecting your call sign between exchanges. Transmitting "break" is usually reserved for emergencies on most repeaters. Be considerate of the other operators, and make sure you can add something meaningful to the discussion.

But let's suppose you're on the road and have been monitoring a repeater for some time. Despite several attempts at a QSO, you've heard nothing but the occasional "kerchunker" and repeater identification. (And by the way, it's not at all unusual for repeaters to be located in rural areas.) You've already announced you're mobile, monitoring and listening. What now? How about adding a bit of bit of spice to the mix?

"This is WF4N mobile from Kentucky, entering Flagstaff on I-40 and looking for someone to chat with. Is anyone around?" Similar to extending a hand of friendship when meeting someone in person, a brief introduction of yourself to prospective contacts is an important first step in breaking the ice. Making it known that you are from outside the area tends to pique the interest of the traveler spirit that's intrinsic in most hams.

If you are planning to spend some time in the area, you might want to expand your invitation just a bit. *"This is WF4N mobile from Kentucky. We're in San Diego for a couple days and wondering if someone can recommend a good restaurant."*

Not only have you introduced yourself, you've given your fellow hams an opportunity to do what hams like to do most: Help others.

Preserving Your Welcome

Good repeater etiquette is important. When you use someone's repeater, you are a guest. So be polite. One of the easiest mistakes for the mobile ham to make is that of monopolizing the repeater. When you are engrossed in a good philosophical QSO with a ham you've just met, the miles can really fly by. And so can the minutes, or if the repeater has exceptional range, maybe even hours. Just like the party-line telephones of yesteryear, when you are using the repeater, everyone else waits.

Once you've established contact on a repeater, if possible, you and the other station should move to a simplex frequency. (The REVERSE button on modern rigs is specifically included to allow you to monitor the repeater input frequency while the other station transmits. If you can receive him on the input frequency, you should be operating simplex. Simple, no?)

But what if you can't work simplex? No problem. Repeater owners welcome the use of their machines by mobile operators. After all, this is the primary purpose of a repeater—to extend the range of mobile

---

"Where's This Repeater Located?"

Regardless of whether your journeys take you across the state or across the continent, you'll find *The ARRL Repeater Directory* an indispensable addition to your mobile station. At your fingertips you will have frequencies, locations, call signs, and other pertinent information (CTCSS tone frequencies, for example) for every repeater in the US (and Canada). When planning a trip of any distance, you can determine the repeater frequencies in advance of your trip for the path you'll be taking.

Of course you could save a few bucks by not having a repeater directory and just asking someone the location of each repeater you hear. No big deal, right? Well, it might be, especially if needlessly disrupt an ongoing QSO, or maybe some other unforeseen activity of the ham that answers you, possibly sensing a note of urgency on your call. It's eventually bound to happen if you do it often enough! Use common sense and good judgment whenever asking for information or help on a repeater.

With voice IDers becoming commonplace (and of course, with an *ARRL Repeater Directory* in hand), the traveling ham should never have a problem determining what repeater he is hearing. The rare commodity called patience helps too. Most repeaters ID every six or seven minutes when in use. In rare instances, where you cannot identify a particular repeater, by all means ask someone. The locals are always glad to make you aware of the features and coverage on their machines. But be courteous and invest a little time in some good conversation in return for the information you receive.

---

and portable stations. But be sensitive to the need of other hams to use the machine, especially during busy times of the day or evening, for example, rush hour and lunchtime. And don't forget to identify regularly.

If you are traveling alone, ham radio is a great way to combat the boredom of a long trip. But don't let your conversation grow boresome to everyone listening. How many times have you heard an extended monologue such as this one?

"Okay, yeah the traffic is pretty heavy here today, uh, there goes one of those uh, you know, big motor homes and he uh, he sure must have a lot of money in that rig, and uh, yikes, some @%! Just cut me off and uh..."

Avoid the tendency to ramble. If you can't think of anything more significant to say, leave your mike unkeyed. And keep the volume of your car stereo at a reasonable level while you are transmitting. If the folks on the frequency want to party, let them provide their own music.

Crossband Repeat and Remote Control

Enlisting the crossband repeat and remote control capabilities of your mobile rig can greatly add to your operating pleasure. By following some simple guidelines, you can keep your operation fun and legal.

The appropriate selection of operating frequencies for crossband repeat is important. First of all, you want to make sure you don't choose harmonically related frequencies. As an example, you program in 147.50 MHz for your mobile rig's transmit frequency and 442.50 MHz as your HT's transmit frequency. As soon as you key your HT on 442.50 MHz, your mobile rig receives it on this UHF frequency and starts to transmit your signal on 147.50 MHz. Problem is, when you unkey your HT, the mobile rig's VHF transmitter will continue transmitting because the third harmonic of its 147.50 MHz output is keeping the squelch open on its own UHF receiver (listening on 3 × 147.50 or 442.5 MHz). As if that isn't enough, it will be transmitting a tremendous howl due to audio feedback. Depending on the rig, you may find the only way to rescue it from its locked-up condition is to turn it off, connect a dummy antenna, turn the rig back on and take it out of crossband repeat mode.

It's also important to avoid the input, output, link and control frequencies of fixed-site repeaters in your area. A directory of repeaters and some listening will identify the input and output frequencies, but the link and control

frequencies are a little trickier since repeater owners choose wisely not to publicize them.

Considering that all repeaters (even mobile ones) should be coordinated anyway, your best bet is to contact your local repeater coordinator for assistance in choosing the proper frequencies for crossband repeater operation in your area.

Crossband repeat is useful and fun, but it isn't without its drawbacks. Fortunately, most are avoidable.

If you use your crossband repeater to relay your HT's signal to another repeater (one on 2 meters, for example), consider the effect of that repeater's *hang time* on your operation (hang time is the period that the repeater stays keyed after it has stopped receiving a signal—typically 2 to 6 seconds in duration). Because a crossband repeater will retransmit the first signal it receives for as long as it receives it—you might find it hard to get a word in edgewise on a busy repeater. If you are able to hear the repeater's output directly on your HT, you can reconfigure your crossband repeater for fixed direction crossband repeat. This simply means your crossband rig will only repeat the signals received on one band (the band where you are transmitting with your HT). This will make it possible for you to transmit without waiting for the 2-meter repeater to unkey.

When using your rig in crossband repeat mode, use the lowest power settings on both bands—those will provide reliable communications. This will not only help prevent you from returning to a vehicle that won't start because of a dead battery, it will also reduce potentially damaging heat buildup in your rig. (It also lessens interference on the bands.) Don't forget, anytime it's receiving, it's transmitting. If you have programmed a busy frequency, your rig may be transmitting nonstop for hours.

Of course, if you have the ability to remote control your crossband repeater, you can reprogram a different, less active frequency. Since the FCC considers the remote control of one radio by another to be *Auxiliary Operation*, you must follow the applicable rules. Primarily, this means your control signals must be transmitted within the amateur bands on frequencies above 222.15 MHz, with the exception of the segments 431-433 MHz and 435-438 MHz.

As the control operator, you are responsible for what is retransmitted by your crossband repeater. To prevent unauthorized use, most rigs allow you to utilize CTCSS tones or a DTMF tone sequence to activate a crossband repeater.

If your rig has a programmable time-out-timer, it's a good idea to enable it when you are using the rig to repeat. This can prevent a spurious signal from keeping your rig keyed until it reaches the meltdown point.

Identifying with your HT is adequate when your signal is being repeated, but it doesn't cover your mobile rig when it is repeating signals back to you. To be legal, you'll need to install an IDer in your crossband rig.

HF MOBILE

The difference between HF and VHF/UHF mobile operation can be a considerable. While repeaters and FM can provide excellent local area coverage, the operation tends to be channelized. Many repeaters also tend to have a high percentage of *regulars*, especially during commuting hours. HF, on the other hand, can provide direct worldwide coverage. Although there are some repeaters on 10 meters, which we'll discuss shortly, most HF operation is repeater-free and works by ionospheric skip. A response to a CQ can be someone in another part of the world you've never met, or an old friend that lives nearby in your community. You can get in on the fun too. HF and the road are calling.

Netting Contacts

Just getting on the air and calling or answering a CQ is a surefire way to make lots of HF contacts. But if you want to add a little variety to you operating, explore the various mobile nets in operation on the HF bands. The wide area nets, MIDCARS on 7.258 MHz, EASTCARS on 7.255 MHz and SOUTHCARS on 7.251 MHz perform a dual purpose. In addition to monitoring the designated frequency for calls from mobile operators in need of assistance, these nets serve as a meeting place of sorts for the traveling ham. Let's look at an example of how they work.

Suppose I'm traveling from New York to Texas and I want to periodically let Joel, N1BKE, in Connecticut know how my trip is going. At the prearranged time, I call the MIDCARS net control station with a request to make a contact. Permission granted, I then give Joel a call. If he answers, we move to another frequency to continue our QSO.

But what if Joel is working late and isn't able to meet me on time? In that case, I can leave a message for him on the net, letting him know where I am and when I'll call him again. Later, when Joel checks in looking for me, the net will deliver my message to him.

Or it could be Joel's on time, but poor propagation won't allow us to hear each other. If this happens, it's almost guaranteed at least one of the stations on frequency will be able to hear both of us and will perform a relay, enabling us to exchange necessary information, for a move to another band perhaps.

Of course, you don't have to be keeping a schedule with someone to get a lot of operating enjoyment from nets. You'll receive a hearty welcome to the County Hunter's nets on 3.865, 14.056 and 14.336 MHz. You've heard of WAS (Worked All States)? Well, the county hunters cut the pie into infinitely smaller pieces by attempting to work someone in every county in the US. Since many of the more than 3000 counties in the country don't have resident hams, the county hunters rely on mobile hams to make these counties available for contacts. If you detour slightly from the beaten path, the county hunters will reward you with a gigantic pileup.

Oh, and don't worry if you have no idea what county you are traveling through, because these hams take their county-hunting business seriously. If you can provide a road number and some landmark information, they will tell you where you are.

Keep in mind, the frequencies used by the County Hunter's nets are usually monitored by the Mobile Emergency Net. If you suffer the misfortune of having car trouble, someone may be available to summon aid for you.

There are literally hundreds of nets on the HF bands, each catering to a particular interest. RV mobile operators will want to check out the Good Sam RV Radio Network nets. Or you might want to check into one of the myriad of nets having nothing at all to do with mobile operation. The choice is yours.

Times and frequencies for nets can be found via the ARRL's online *Net Directory*, **www.arrl.org/FandES/field/nets/**. You can search for nets by category (Maritime, Local, Wide Coverage, etc.), by state, by band and by other criteria. Prepare a list before you travel; net operation can really make for an interesting trip.

10-Meter FM Repeaters

If your rig has FM capability, there's a world of DX with near-studio-quality audio awaiting you on the 10-meter FM repeater band. (propagation permitting, of course.) Although there are only four repeater outputs in the band plan (29.62, 29.64, 29.66 and 29.68 MHz), there are more than 100 repeaters listed in the

current *ARRL Repeater Directory*. Using these machines is simple.

Because a 100-kHz offset is employed, you'll need to program your rig to transmit 100 kHz below the output frequency of the repeater you want to work. (Various rigs accomplish this split frequency operation in different ways. Check your rig's instruction manual for setup information.) Once the rig is ready, operation is much the same as on VHF/UHF repeaters. Just key up and give a call. But don't be surprised if you hear hams answering you on several different repeaters. It's a common occurrence when the band is open. Usually, FM *capture effect* will permit you to hear only the repeater with the strongest signal, but rapidly changing band conditions can shuffle signals in and out like a deck of playing cards.

Nonetheless, 10-meter repeaters can be a lot of fun. Give them a try, and remember—since FM operation is a 100% duty cycle mode, you'll need to reduce your rig's RF output to prevent overheating. Follow the instructions in your rig's manual.

CW: Digital Operation for Your Digits

Think mobile CW operation is a dark and secret art, reserved for those who possess the nerve and manual dexterity of a brain surgeon? It's not true! If you operate CW from home, you can probably do it mobile. The only prerequisite is you must be able to copy in your head, since you obviously can't write what you hear and drive at the same time. But don't worry about it; copying CW without writing is a skill easily developed with a bit of practice.

Many first-time mobile CW ops find making their first few contacts while stationary is a great confidence builder before going out on the road. Once mobile, making your initial contacts with the keyers speed set somewhat below your capabilities will give you a leisurely opportunity to develop your CW "wings."

Operating style is a matter of preference. Many mobile CW operators find tuning to a clear frequency and calling CQ is most productive.

For example: let's assume you've done some listening and find a nice clear spot at 7.035 MHz. First, send "QRL?" to verify the frequency is not in use. Assuming it is clear, make a 3 × 3 call, "CQ CQ CQ DE [repeat your call 3 times]/M K". (You could have sent the "/M" after your call each time, but it isn't necessary.) As soon as you've finished calling, turn your rig's RIT so as to sweep about 2 kHz each side of your transmit frequency—a necessity to hear replies that might be outside the

passband of your rig's narrow CW filter. This filter is a must-have for mobile CW. And another tip: whenever you answer a CQ, engage the dial lock to ensure against unintended frequency hops.

No matter how you establish contacts, you'll find lots of hams anxious to talk to you and ask questions about your mobile setup. Of course, if you'd really like to be the center of attention, drive to a really rare county and check in on the County Hunter's net on 14.056 MHz. And you'll also find there are lots of general interest nets on the CW subbands. Refer to ARRL's online *Net Directory*, **www.arrl.org/FandES/field/nets/,** for more information.

Although it may be challenging at first, mobile CW can quickly become a pleasurable obsession. So consider yourself warned. Some hams have been known to deviate many miles from their planned routes just to extend their operating time.

KEEP IT LEGAL

In addition to the usual FCC regulations, when we take our radios out on the road we come under the jurisdiction of additional laws that can affect how and when we operate. Before you go mobile, determine what laws apply to your operation.

You might score big points with your non-ham passengers by using an earphone or headset, but make sure the device you use leaves you with one ear available to monitor for warning sounds from other vehicles. Not only is covering both ears with headphones or a headset very dangerous, in some states it is very illegal.

If you will be using a VHF/UHF rig, capable of receiving Public Service frequencies (such as police and fire), be careful you don't run afoul of a *scanner law*. Intended to prevent mobile criminals from using scanning receivers to monitor police activities, scanner laws have been an unwelcome (and sometimes unexpected) constraint for many hams. Although most states exempt licensed Amateur Radio operators from their scanner laws, and the FCC has gone on record as supporting a nationwide exemption, make sure you know the particular laws for the state or states where you'll be operating. In addition, it is a good idea to know the revised statute number for the applicable law. It's pretty frustrating to be detained for a long period of time while a policeman checks to determine the existence of a law you claim exists, but for which he isn't aware. It may also help to have a copy of your amateur license with you, to prove you are a ham.

Some radios with extended receive coverage present an additional hazard because they are capable of transmitting outside the amateur bands. Fortunately, modern amateur equipment will not transmit outside the ham bands. Whenever using an older transceiver, or one with an unknown (and possibly modified) history, it is a good idea to verify the radio will not transmit outside the ham bands before using it. Why is this a hazard? First, your Amateur Radio license does not grant privileges outside the ham bands. And second, Amateur Radio gear doesn't carry FCC Certification (approval) for transmitting outside the ham bands, so it is illegal for this purpose.

Okay, so you have no plans to ever use your *opened-up* rig to transmit anywhere but the ham bands. But what if you do it accidentally? Many mobile hams have experienced the embarrassment of sitting on the mike and pressing the PTT switch, allowing anyone monitoring the frequency to hear what (or who) they talked about when they thought no one was listening. Having your local police department hear your Candid Camera conversations could easily qualify as the ultimate embarrassment and more. Don't jeopardize your license. If you are contemplating the mobile use of a modified rig, you may want to consider reversing the modification that allowed out-of-band transmit. Or, when you program your rig with the Public Service frequencies you plan to monitor, program in a transmit frequency inside the amateur bands. Then when your mike PTT sticks for a half hour as you drive down the freeway, your pride will be the only thing that suffers.

REPORTING AN EMERGENCY

One of the best reasons in the world to have a mobile station is for emergency communication. If you should happen upon the scene of an accident, you might find you are the only person there with the means to summon emergency assistance. In situations involving serious injuries time is of the essence. Your first priority is to get EMS on the way immediately.

As soon as you determine injuries are involved, establish your radio communication with EMS dispatch. If you are using a familiar repeater, state you are making am emergency autopatch and dial the access code and emergency number.

If you don't know how to access the autopatch, give your call and announce you have emergency traffic and need immediate assistance. Someone can then bring up the patch for you or make the call from their home phone. If the repeater is in use, don't be shy. You have priority on the frequency.

If HF is all you are able to use, then use it. In most cases it will be quicker than driving somewhere to a phone.

When the call is answered, tell the dispatcher you are reporting an accident with injuries. Speak in short sentences and release the push-to-talk button between sentences to allow the dispatcher to break in with any question they may have. Give them the opportunity to tell you they already have the information, rather than tying up their emergency line with a three-minute fact-filled monologue.

You can tell them you are a ham, but typically this is not necessary. Some dispatchers will request your name before hanging up. Give the location of the accident. Include the highway number, nearest mile marker or exit number if you know it, and the direction of travel if you are on a divided highway. Be absolutely certain of your information. Routing the EMS to the wrong location is a potentially tragic mistake you don't want to make.

If you know the number of victims give this information to the dispatcher. Tell them if there is a hazardous substance involved, or any fuel leakage or fire. Don't waste valuable time or risk causing confusion by including nonessential information, such as your call or repeater ID. Stay on the line until the dispatcher indicates it is okay to hang up. If possible, remain at the scene to render whatever assistance you are qualified to give and to provide additional communications if necessary.

Interference and Automobiles

Terence Rybak, W8TR, and Mark Steffka, WW8MS, wrote this section about interference. Terence Rybak has been an Automotive EMC Engineer for 27 years. His previous experience was with the electrical utility industry. He is also a NARTE Certified EMC Engineer and was first licensed in 1963.

Mark Steffka, WW8MS, has been in the automotive industry since 1991 and has been an Automotive EMC Engineer since 1997. His previous experience was with military and aerospace electronics. He is also an adjunct faculty member in the Electrical and Computer Engineering (ECE) department at the University of Michigan. He was first licensed in 1975.

This section will give you a basic description of how modern technologies can influence the performance of a typical Amateur-Radio mobile installation. Because of the complexity of the newer systems, this section is not intended to provide specific details on various fixes that might be utilized. Those are best left to automotive dealers or mobile radio installation facilities. Readers who want to investigate those topics in detail can check out other publications, including recent editions of *The ARRL Handbook* and ARRL's *The RFI Book*.

This section will provide the reader with a fundamental background in vehicle system operation and its effect upon mobile radio installation and operation. We will review and describe various system and component operations that may generate, or receive, frequencies used by amateurs and how issues may be identified. Finally, we will discuss how some of the more common interactions of the systems and components are addressed and/or eliminated.

Manufacturer's Recommended Installation Requirements

The ultimate authority on whether a radio transmitter can be installed in cars is, naturally, the vehicle manufacturer. Surprisingly, the vehicle manufacturers' policies about all types of aftermarket equipment, especially radio transceivers, vary widely. An article published in the September 1994 issue of *QST*, detailed manufacturers' policies about installing transmitters in their vehicles. Some manufacturers indicated that they had published installation guidelines (how to obtain these is at the end of this section). Others, however, made it very clear that problems or damage caused by the installation of any aftermarket equipment would not be covered under warranty. One vehicle manufacturer even told us that the answers to the questions we asked were "proprietary"!

Interference: Starting with Shopping for A Vehicle

When shopping for a vehicle, it is useful to take along some portable (preferably battery operated) receivers or scanners and have a friend tune through your intended operating frequencies while you drive the vehicle. Tuning across various frequencies will help identify any *radiated* noise issues associated with that model vehicle. Since radiated noise can be more difficult to resolve than *conducted* noise, this process will give you an early indication of what to expect from that particular model. If you intend to make a permanent transceiver installation, give some consideration to how you will route the power and/or antenna cables. While you are looking for way to run the wiring, keep in mind that some newer cars have the battery located in the trunk or under the rear seat, and that may make power wire routing easier.

Transceiver Power Requirements

Because of the significant current draw of many amateur transceivers, the power connections should be made directly to the battery and fused as close to the battery as possible, as described earlier under Installation. The battery acts like an electrical filter to bypass noise. Since all the modules that are powered from the vehicle power supply may be a source of conducted noise, avoid using cigar lighter or "Power Point" receptacles as power sources. The power leads should be twisted together (or run side-by-side) from the back of the rig all the way to the battery. In addition: Follow the manufacturer's instruction about whether or not to use the vehicle chassis as a power return. The radio's power and antenna leads should be routed along the body structure, away from vehicle wiring harnesses and vehicle electronics. Any wires connected to the battery should be fused at the battery using fuses appropriate for the required current.

WIRING IMPACT

Sometimes, the length of the power and power return lines can also have an impact on radio/vehicle interactions. In general, the length of the lines needs to be much less than one-quarter wavelength of the frequency(ies) being used.

Inline Filters Can Help Reduce Conducted Noise

Filters are available for installation at a convenient point in the power leads, between the radio and the battery. These filters can be obtained at many electronics parts stores, automotive parts stores, or stores that specialize in automotive-entertainment-system installation. Place the filter near the radio to reduce the probability of unfiltered leads picking up unwanted signals.

Antenna Location

Before permanently drilling holes to mount an antenna, try a magnet mount antenna in the desired location. If undesired effects occur, they may be resolved by moving the antenna to another location.

Troubleshooting

If vehicle-radio interaction develops following installation, the source of the problem should be identified prior to further operation of the vehicle. Following these installation guidelines can eliminate most interaction problems. If any vehicle-radio interaction problems exist after following these guidelines, contact your car dealer for additional assistance.

Interference and Noise

INTERFERENCE FROM AUTOMOBILE SYSTEMS TO AMATEUR RADIO

There are many different types of systems in vehicles that may cause vehicle RFI issues, some of the systems include:

- The high voltage system in the spark-ignited engine (traditional in the United States).
- The systems used in compression-ignited engines (commonly existing in diesel engines).
- Fuel injectors on both spark-ignited and compression-ignited engines.
- Specialized systems in the newer hybrid vehicles that use both high-power electric motors and internal-combustion engines.
- The microprocessor-based engine-control systems.
- Chassis control systems.
- Driver information display systems (instrument panel).
- Adaptive cruise control
- RADAR obstacle detection
- Remote keyless entry
- Key-fob recognition systems

As a result, the application of many distributed electrical and electronic systems in vehicles can affect mobile-based Amateur Radio operators. Active electronic components have replaced many of the functions that used to be controlled by mechanical linkages and systems. Not all noise is from the traditional sources such as ignition. Some may be from various instruments and gauges or from pulse-width-modulated lighting systems on the vehicles. Most electronics modules these days contain a large number of active devices, including digital and analog components. These components and assemblies may emit RF noise and they need to be immune to external electromagnetic fields. Now add back to the mix the traditional noise sources, such as the numerous electric motors in today's vehicles and the interference potential becomes much greater than even a few years ago.

Noise Types

Most radio installations should result in no problems to either the vehicle systems or any issues with the transceiver. In those situations where issues do occur, the vast majority of issues are with interference to the receivers from the vehicle on-board sources of energy that are creating emissions within the frequency band used by the receiver. To resolve these issues, it is useful to know about the nature of the noise types. The traditional way to describe the noise is to characterize it as one of two types:

- Broadband
- Narrowband

Broadband Noise

Broadband noise is defined as noise having a bandwidth much greater than the affected receiver's bandwidth, and is reasonably uniform across a wide frequency range. For practical purposes, ignition, or similar pulse-type noises and brush-type motor noises can be considered broadband. References to these systems are often based on their operation—many of them generate an electrical arc or a spark and are called "arc-and-spark" devices. Not all broadband sources are arc-and-spark, since today many devices and on-vehicle data communication signals may also consist of pulse-width-modulated (PWM) signals. PWM signals exhibit noise interference potential similar to broadband noise.

Narrowband Noise

Narrowband noise is defined as noise having a bandwidth less than the affected receiver's bandwidth. Narrowband noise consists of noise present on specific, discrete frequencies or groups of frequencies. Microprocessor clock harmonics are an example of narrowband noise. Unlike today's PC clock speeds above the gigahertz range, automotive microprocessor clock speeds are still at or below 50 MHz. These speeds were chosen because of engineering issues involved in designing a computer that operates reliably over a wide temperature range and when exposed to other harsh environmental factors. With these clock speeds and their harmonics, it's easy to see why there can be RFI across several amateur bands.

Is it Broadband or Narrowband?

The first step to identify the source of a noise is to determine its characteristics, as described in the previous discussion. A straightforward and very effective way to do this is to use a battery operated portable receiver. As it turns out, an older non-digital receiver actually works best for this test, for a couple of reasons. The first is that since they have a continuous tuning capability they will tune to frequencies between the allocated broadcast frequencies, and the second is that they do not have auto-mute circuitry or noise-reduction circuits. If there is no noise heard with such a receiver, then it is almost a guarantee that any amateur transceiver will work very well in the vehicle. The process is:

1. Operate the vehicle systems and listen for noise on the radio audio. If the RFI is detected, by tuning across the band you can tell if it's narrowband or broadband.
2. If it is *not* possible to tune out of the noise, no matter where the receiver dial is set, then it is probably broadband noise.
3. If only one or two stations are affected by the noise, then it is probably narrowband noise.

These steps can also be performed with handheld transceivers for any of the amateur bands, which will tell you specifically about the RFI that may be expected for actual installations.

Examples of Broadband Noise

Ignition Systems Overview

A review of typical ignition systems and how they operate reveals how the various suppression schemes work. **Fig 1-33** depicts systems present in most vehicles. The first system, Fig 1-33A, is used in older vehicles and a few new ones with large-displacement engines. Newer electronic ignition systems omit the distributor in favor of one coil per cylinder (Fig 1-33B) or one coil for a pair of cylinders. During operation, a sensor in the distributor, or a crankshaft position sensor, sends a trigger signal to the engine computer and to an ignition module (essentially a high-power dc amplifier), which then drives a high-current pulse through the coil, T1. The secondary voltage can be 5 to 40 kV (depending on the engine operating conditions.) The secondary current travels through the plug wire to the spark plug. For the configura-tion in Fig 1-33A, the secondary current also travels through the distributor rotor and its gap in the distributor cap before reaching the plug wire.

The fast rise time of the pulses discharging across air gaps (distributor and spark plug) creates broadband ignition noise. The theoretical "ideal" models (zero rise time) of such pulses are called impulse functions in the time domain. When viewed in the frequency domain, the yield is a constant spectral energy level starting at 0 Hz and extending up in frequency. In practice, real ignition pulses have a finite rise time, so the spectral-energy envelope decreases as the frequency increases.

It turns out that both the noise generated by the ignition spark and that generated by the fuel injector is manifested as a regular, periodic *ticking* in the receiver audio output, which varies with engine RPM. If an oscilloscope were connected to the audio output, a series of distinct, separate pulses would appear. Since ignition noise is usually radiated noise, it should disappear when the antenna element is disconnected from the antenna mount. The radiation may be either from the secondary parts of the system, or may couple from the secondary to the primary of the coil and be conducted for some distance along the primary wiring of the vehicle.

Parameters in Secondary Ignition Systems

It turns out that the peak voltages and currents in the secondary circuits can reach up to tens of Amps and tens of kilovolts (fortunately only for a fraction of a

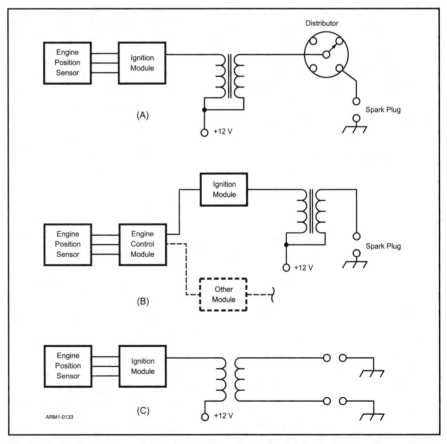

Fig 1-33—System diagrams representative of most vehicle ignition systems. At A, a distributor ignition system, which is common on older vehicles and some large-displacement contemporary engineers. B shows an electronic-ignition system, found on most newer automobiles. C shows an electronic-ignition system with two sparkplugs per coil.

second). Not only can radiation occur from the wiring, the dimensions of the hood and sheet metal of the vehicles can result in quarter-wave radiating elements for certain frequencies. As a result, the sources of the noise on a vehicle can be from the following:

- Ignition systems
- Generator systems
- Alternator systems
- Small powerful electric motors
- Switches
- Static discharges between chassis and body components.

Radiated emissions occur typically above 30 MHz, and conducted emissions at lower frequencies, because of the length of the path along the wire leads and cables connected to the various components. For the most part, because of the effect of the shielding that takes place by the body panels, adjacent vehicles or houses along the roadside are usually unaffected. On systems such as motorcycles that do not have an engine compartment to provide shielding, the spark plug may have a shield over it. The purpose of ground straps is to eliminate the potential across the gap. Otherwise, the gap may behave like a slot antenna. For years, there was also the common practice of adding ground straps from various vehicle body components to eliminate the voltage potential across the gap between the components. (This is *not* recommended, as it can defeat the rust prevention protection of those components.)

Specifics on the Source of RFI on Spark-Ignited Engines

The source of the RFI in spark-ignited engines is the point of the breakdown of the mixture into combustion initiation. Ignition noise is caused by the fast rise and fall times of the high-voltage ignition pulses. Most vehicles have low RFI emissions in the amateur bands, since there are industry standards that are used to control the levels of the emissions, primarily from 30 MHz to 1 GHz. The amplitude and the frequency of the oscillations that take place in the secondary circuit can be reduced by the circuit resistance (which affects the amplitude) and the inductance (which affects the frequency).

Resistance wiring is the only practical way to control the damping—it is not feasible to add it to an ignition system using discrete resistors unless they are manufactured into the wiring. For traditional inductive discharge systems, a value of about 5 kΩ in the spark plug provides effective suppression and, with this value, there is no detectable engine operation degradation. Capacitor discharge systems are required to have very low impedance on the order of tens of ohms in order to not affect the spark energy delivery amount. Most spark plug resistances are designed to operate at several kilovolts, so a low voltage ohmmeter may not give proper resistance measurement results. Actually, the term *resistor wire* is somewhat misleading. High-voltage ignition wires usually contain both resistance and inductance. The resistance is usually built into suppressor spark plugs and wires, while there is some inductance with resistance in wires, rotors and connectors. The elements can be either distributed or lumped, depending on the brand, and each technique has its own merit. A side benefit of resistance in the spark plug is improved electrode wear.

If Ignition Noise Gets Worse Over Time

Replace worn ignition components, spark plugs, wires and distributor (if present).

Other Sources of Noise

Watch out, too, for the "ignition noise" that is not caused by the ignition system. Many hams have been fooled into thinking that a noisy electric fuel pump, noisy pulsed fuel injectors or some other source was an ignition-noise problem. The ARRL Lab has received reports of ignition noise from diesel engines, which use no electrical ignition system at all.

In many cases, the problem extends beyond ignition noise. Fuel injectors, also controlled by the engine controller, can mimic ignition noise. They produce a *pop-pop* sound at low speeds and change to a whine as speed increases. Today's alternators are often required to deliver higher current than their older counterparts. This also increases their noise-generating potential.

Identifying the Noise Source Is Crucial to Fixing It

As previously discussed, ignition noise manifests itself as a regular, periodic ticking sound that varies with engine RPM. At higher speeds, it sounds somewhat musical, like alternator whine, but has a harsher note. There is one distinguishing feature of ignition noise in that it increases in amplitude under heavy acceleration. This results from the increase in the required firing voltage with higher cylinder pressure. A good test can be conducted while driving on a flat road. Accelerate to approximately 50 mph and provide just enough throttle in high gear to maintain a steady vehicle speed. There should be no road grade so as to provide a light engine loading condition. Apply throttle increase and evaluate noise just below downshift condition. Repeat at three AM broadcast band frequencies at the low, middle and upper end of the band. Exact frequencies are not critical. Ignition noise will typically increase across the AM broadcast band with the increase in engine load. Also, ignition spark noise usually disappears when the antenna is disconnected (because it is usually radiated noise). By comparison, the high-current pulse in the ignition coil primarily generates a noise that is conducted on the wiring harness. Unfortunately, what most hams call "ignition noise" often originates from one or more of several sources in a modern automobile.

ELECTRIC MOTOR NOISE

The challenge in using electric motors on vehicles is that the motors can result in both radiated *and* conducted noise, depending on the frequency of the noise. See **Fig 1-34**. Noise must be eliminated with filters at the source, before it has a chance to be conducted or radiated by long leads that act as antennas.

FUEL PUMP NOISE

Electric fuel-pump noise often exhibits a characteristic time pattern. When the vehicle ignition switch is first turned on, without engaging the starter, the fuel pump will run for a few seconds, and then shut off when the fuel system is pressurized. The noise will follow this pattern.

RESOLVING NOISE PROBLEMS

Work with the Dealer

The dealer should be the first point of contact because the dealer should have access to information and factory help that may solve your problem. Much of this chapter is about how to work within that process, and how to make the process work when it breaks down. The manufacturer may have already found a fix for your problem (this applies to all problems, not just ignition noise) and may be able to save your mechanic a lot of time (saving you money in the process). If the process works properly, the dealer/customer-service network can be of help.

Old Time Fixes Are Not Always Good Today

In amateur publications, many old-time

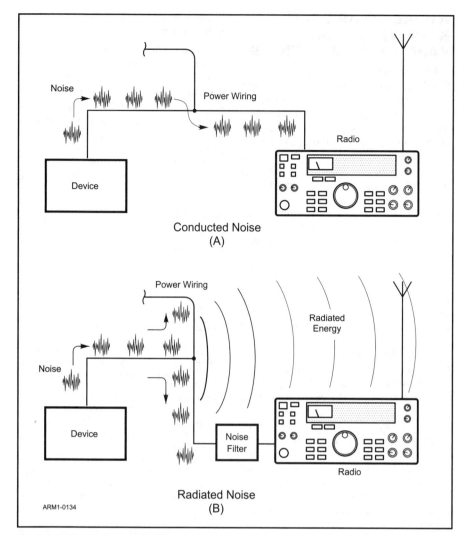

Fig 1-34—Noise from a device such as a motor can be conducted (at A) or radiated (at B) to a nearby radio. The noise filter shown at B is ineffective because noise energy has already been radiated by the quasi-antenna consisting of the long wire leads going to the filter.

RFI fixes have involved bonding various body parts together, with the intent of improving the grounding. With today's vehicles, adding ground straps between exhaust systems and chassis, or between body panels, may defeat the electrical isolation designed in for corrosion control.

General Troubleshooting Techniques

Receivers may allow conducted harness noise to enter the RF, IF, or audio sections (usually through the power leads), interfering with desired signals. To determine if this is occurring, check whether the interference is still present with the receiver powered from a battery or power supply instead of from the vehicle. If the interference is no longer present with the receiver operating from a battery or external supply, the interference is conducted via the radio power lead. Power line filters installed at the radio may resolve this problem.

Filters for DC Motors

If the motor is a conventional brush- or commutator-type dc motor, the following cures are those generally used, although as always, the mechanic should consult the vehicle manufacturer. To diagnose motor noise, obtain an AM or SSB receiver to check the frequency or band of interest. Switch on the receiver and then activate the electric motors one at a time. When a noisy motor is switched on the background noise increases.

A note concerning fuel pumps: Virtually every vehicle made since 1982 has an electric fuel pump, fed with long wires. It may be located inside the fuel tank. This motor is responsible for many RFI complaints, so keep it in mind as you look for broadband noise sources. This is one area where it is essential to consult the manufacturer. It is downright dangerous to work on fuel pumps and the wrong cure can result in an explosion!

Next to light bulbs, dc motors are probably the most common electrical devices in automobiles. A typical vehicle has many, ranging from small servos and stepper motors to high current ones such as 200-A starter motors. Most of these motors use brushes and commutators, but some are brushless. Electronically commutated brushless units are used on many vehicles.

Each style of motor produces a different noise spectrum. The armature current in brushless motors is controlled by switching transistors, as opposed to mechanical contacts for brush-commutator motors. Brushless motors are fairly quiet, but they may generate noise similar to that produced by an alternator.

The pulses of current drawn by a brush-commutator motor generate broadband RFI that is similar to ignition noise. However, the receiver audio sounds more like bacon frying rather than popping. With an oscilloscope connected across the receiver audio, the noise appears as a series of pulses with no space between the pulses. Such broadband noise generally has a more pronounced effect on AM receivers than on FM, because the noise consists of amplitude variations. Unfortunately, the pulses may affect FM receivers by impacting the background noise level and will reduce perceived receiver sensitivity due to the degraded signal-to-noise ratio.

Alternator and Generator Noise

A listening test may verify alternator noise, but if an oscilloscope is available, monitor the power line feeding the affected radio. Alternator whine appears as full-wave rectified ac, coupled with commutating pulses, superimposed on the dc level.

Alternators rely on the low impedance of the battery for filtering. Check the wiring from the alternator output to the battery for corroded contacts and loose connectors when alternator noise is a problem.

The resulting spark is primarily responsible for the *hash* noise associated with these devices. Hash noise appears as overlapping pulses on an oscilloscope connected to the receiver audio output. An alternator also has brushes, but they do not interrupt current. They ride on slip rings and supply a modest current, typically 4 A to the field winding. Hence, the hash noise produced by alternators is relatively quiet.

Generators use a relay regulator to

control field current, and thus output voltage. The voltage regulator's continuous sparking creates broadband noise pulses that do not overlap in time.

Alternator or generator noise may be conducted through the vehicle wiring to the power input of mobile receivers and transmitters, and may then be heard in the audio output.

Switches

Noisy switches may be present in vehicles. Unsuppressed switch contacts may generate fast, high-frequency voltage transients that travel through the vehicle wiring harness and radiate noise that sounds like *pops* in the receiver audio. These noise bursts may break receiver squelch and cause AGC-related dropouts. Problem switches may be easy to locate since many are readily accessible from the driver and passenger seats. In general, most hams don't bother curing these very transient pops. They may be an annoyance but they don't usually disrupt communications.

INTERFERENCE FROM AMATEUR RADIO TO AUTOMOBILE SYSTEMS

Immunity Is Ensured by OEMs

Just as vehicle electronic systems may interfere with mobile receivers, mobile transmitters may also affect the vehicle electronics if appropriate immunity measures are not in place. Electronic modules that radiate energy may also receive energy. This energy may come from on-board transmitters, nearby radio and TV stations, or any other device or event that generates an electric or magnetic field. Manufacturers use proven design techniques and run extensive tests to ensure RF immunity in their vehicles

WHERE TO ASK FOR HELP WITH RADIO/ VEHICLE INTERACTIONS

Every automobile manufacturer has an EMC engineering department, or contracts EMC testing to an external laboratory. There are regulations and standards that must be met, worldwide, and the engineering teams at manufacturers have to consider EMC issues as an overall part of the design. The EMC labs also get involved when problems are reported by helping to identify and characterize the problem, by identifying the components responsible and often by writing the technical service bulletins (TSBs) that will describe the problem and its cure to customer-service personnel and dealers.

If a customer has any problem with a car, including EMC problems, they need to start with the dealer. The dealer may have some experience, and if not, should be able to locate the appropriate TSBs, using the same search methods that would be used to locate a fix for an ailing transmission. If the dealer has a problem they cannot resolve, the dealer may enlist the services of the engineers at the regional or zone offices. If the zone staff cannot help, the zone office has the factory contacts that will ultimately be in contact with the EMC labs, setting a process into motion to identify the problem and find a cure. The EMC lab will then disseminate that cure to all of the dealers. This process is designed to ensure that the customer has the best available local service and that the work of the engineers at the EMC lab will be made available worldwide to help others with the same problem. When this process works, it works well.

The process sometimes breaks down, however. If all else fails, contact ARRL HQ. HQ can't always help directly, but is able to forward written inquiries to our contacts at the factory, mostly to serve as an indicator that the factory really may need to address something.

Mobile Radio Installation Guidelines

The bottom line: The first step is to ensure that your installation is in compliance with both the vehicle manufacturer's and the radio transceiver manufacturer's installation guidelines. Pay particular attention to the instructions about how to install radio transceivers' power leads. If you want the manufacturer to support your installation, do it exactly the way the installation guidelines tell you to do it.

Additional Resources

There are a number of web pages that the League has created or identified that may help you. See:

RFI—Automotive: **www.arrl.org/tis/info/ rficar.html**. The automotive RFI home page.

Other pages that may help are:

Automotive Electric Motor and Fuel Pump Noise: **www.arrl.org/tis/info/fuel.html**. How to diagnose and cure noise from fuel pumps and other electric motors. This includes a reprint of the information in the Ford TSB on the subject.

Automotive Interference Problems: What the Manufacturers Say: **www.arrl.org/tis/info/carproblems.html**. This page gives the RFI policy statements of vehicle manufacturers.

Lab Notes—Mobile Installations and Electromagnetic Compatibility: **www.arrl.org/tis/info/pdf/39574.pdf**. Some general guidelines on various automotive RFI problems.

Off-site web links:

NOISE and how to KILL it: **www.mindspring.com/~nx7u/mobile/noise.htm**. More information on automotive RFI and Ford fuel pump noise solutions.

Ford Explorer Radio Frequency Interference: **www.4x4central.com/tips.htm#rfi**. Still more information on Ford fuel-pump noise solutions.

Radio Interference to/from two-way radio receivers: **dodgeram.com/technical/tsb96/08_30_96.HTM**. Models: Dodge 1995-1997 BR Ram Truck.

CHAPTER 2

On the Go with Maritime Mobile

The author of this chapter, Steve Waterman, K4CJX, was first licensed at the age of 12 in 1955 as KNØAAE. He has been active in many aspects of Amateur Radio, including high-speed CW, DXing, providing phone patches for the US Military and the development of various digital modes. Since his retirement from the telecommunications software and voice-data network business in 1999, Steve been involved with maritime-radio installations and communications. He is currently the Worldwide Network Administrator of the *Winlink 2000* messaging system.

K4CJX was involved with emergency maritime operations for the three hurricanes in the summer of 2004 that plagued Florida and the Caribbean. He also lent a helping hand during the 2004/2005-tsunami disaster in South Asia.

A BIT OF HISTORY

US amateur operations in the US were brought to a close by the government during World War II. However, during the war many servicemen went to radio school to become proficient radiomen. Tales of spending many hours on the mill (typewriter) copying CW are still being told.

Once the war was over and Amateur Radio was again in full swing, many radiomen became avid hams, bringing their stories and expertise to others. Many former military radiomen also used their training to obtain various commercial licenses, and some became radio officers aboard commercial vessels. International trade was pushed by many countries trying to rebuild their economies.

Back then, only those lucky enough to be aboard larger commercial and military vessels could operate on the ham bands while at sea. Many land-based military operators stationed overseas also used military equipment on the ham bands when

they were allowed to do so.

In the 1940s, 1950s and even in the early to mid 1960s, most operation on the high seas consisted of CW, with some RTTY and single sideband toward the end of this period. I can remember staying up late at the age of 14 in 1957, waiting for hams on the high seas to get off duty so they could spin down to the 40-meter ham band to operate CW. There were some who regularly were available, and it was always a thrill to hear from them while they were underway.

After the single-sideband revolution, many military and commercial radio officers with amateur licenses used their ship's equipment to *phone patch* back to the US to serve the crew. Long lines were formed on military destroyers, aircraft carriers and supply ships by sailors who were waiting to take their turn in the radio room, where they could speak directly with their family and friends back home. I spent many hours stateside during the Vietnam War dialing collect to connect military crewmembers with their loved ones. It was a very heartwarming experience and one that I will never forget. See **Fig 2-1**.

Maritime Mobile Comes of Age

In the early years of maritime mobile radio, the sheer size and weight of the radio equipment precluded the use of Amateur Radio operations on all but the largest recreational vessels. It took a lot of room and a lot of electrical power to heat up all those tubes needed to generate a signal. However, with today's smaller and much more efficient solid-state transceivers, combined with effective automatic tuners, operating maritime mobile has become a joy!

The *dot Net* generation of early retirees, more affluent and with greater leisure time, has contributed to the number of people cruising coastal waters or enjoying

blue-water sailing. During the last decade there has been an explosion of those choosing the live-aboard cruising lifestyle. Entire families spend all their time moving around our globe, exploring all that they can possibly take in.

Amateur Radio has seen a jump in the numbers of hams who obtained their licenses just to use ham radio as their main source of communication with the rest of the world. Yes, there are satellite phones, which may be used for voice and data, but coverage can be spotty and they are expensive. Until recently, some of the major satellite carriers turned their transponders off while over unpopulated ocean areas to save battery power.

There are also commercial Common Carriers and private coastal associations that provide voice and digital services, but they are fee-based and have limited outlets and functionality. Those with worldwide resources, such as Globe Wireless, are geared toward commercial fleets rather than individual recreational vessels. This represents an opportunity for Amateur Radio to enhance safety and well-being at sea for recreational boats.

There are a number of SSB maritime networks now operating worldwide on the ham bands, fueled largely by the growth of e-mail service through HF digital communications. Amateur Radio has become the main link to the outside world for serious recreational sailors.

There are many types of activities in today's maritime environments, ranging all the way from lake bass fishing to cruising the high seas for months. Each has its own peculiarities when it comes to Amateur Radio installation and operation. However, in this chapter, the definition of *maritime mobile* only includes those motor-driven and sailing vessels capable of operating where land can be lost over the horizon. Operating on a 2-meter repeater

To Steve K4ØJX WITH THANKS FROM THE CREW OF THE USS SYLVANIA (AFS-2) OPERATING IN THE MEDITERRANEAN SEA.

KØPIV/MM
Jerry

K Ø J H
JERRY HALE

Jerry Hale
6334 EDWARD STREET
NORFOLK, VIRGINIA 23513
U.S.A.

Fig 2-1—At A, photo of USS Sylvania (AFS-2). At B, photo of radioman Jerry Hale, KØPIV/mm in the radio room of the Sylvania during the Vietnam War. At C, a 1981 QSL card from Jerry Hale, now KØJH, on a commercial vessel.

with a hand-held on an inland lake or river in a runabout will not be covered here.

LEGAL CONSIDERATIONS FOR HAM MARITIME-MOBILE OPERATION

Every time I read anything regarding the amateur regulatory environment, especially in an international context, I note the material is dated. The regulatory environment, like amateur equipment, changes rapidly. Rather than attempting to spit out specific rules and regulations for operating in other countries and offshore in international waters, it more appropriate to simply mention areas of interest and concern, refer to current practices and then provide the reader with references to find further information.

What is the definition of international waters or the *high seas*? Article 1 of the Geneva Convention of 1958 on the high seas states: "The term 'high seas' means all parts of the sea that are not included in the territorial sea or in the internal waters of a state." Article 2 states: "The high seas being open to all nations, no state may validly purport to subject any part of them to its sovereignty." The sovereignty of a state extends to the airspace above the territorial sea, as well as to the sea floor and the subsoil beneath it.

The ham licensed in the US and operating in international waters with a US registered vessel continues under FCC Part 97 rules, which he has an obligation to understand. Although not a regulatory body, the International Amateur Radio Union (IARU) has some guidelines, which do vary from region to region. When you're in international waters, you operate under the auspices of your FCC license, but you must be mindful of the frequencies assigned to other ITU Regions.

The world is divided into three Regions. You are bound by the privileges assigned to the ITU Region in which you are operating and by the privileges of your FCC license. If your station is operated in Europe, Africa or the adjoining waters, you're in *Region 1*. North and South America and the adjoining waters make up *Region 2*. The rest of the world, *Region 3*, is made up of the countries of Southern Asia (excluding the countries of the Arabian Peninsula) as well as the islands of the Pacific and Indian Oceans. A detailed map of the ITU Regions can be found in the ARRL *FCC Rule Book*.

You can obtain most of the information necessary to stay within legal guidelines on the ARRL Web site, **www.arrl.org/ FandES/field/regulations/io/ maritime.html**. My advice is to carefully explore this page, including the various reciprocal-operating authorizations. You should learn how to obtain authorization when you are anticipating transmitting in specific foreign ports. Of course, there is no substitute for contacting the countries under consideration for updated and accurate information. When in other countries you will be subject to their rules and regulations. [Communications by Amateur Radio for business purposes are always forbidden. That means, for example, that you shouldn't even consider trying to manage your investment portfolio by Amateur Radio, even if you are in the middle of the ocean.—*Ed.*]

A word of caution: Always ensure that you have the most current information from the countries you wish to visit to determine recent changes. Odds are that it will be to your advantage. At the time of this writing, due to the advancement of digital and other enabling technologies, there are many relaxed rules and many additional privileges now being either proposed or offered to the maritime amateur operator in many countries. One general rule that may be followed, especially in countries that have confusing regulations, is to operate only at the equivalent license class you were assigned when you received your US amateur license, especially if you have a US registered vessel. For example, in Mexico a "Receipt" for a temporary authorization certificate to operate may be easily and mistakenly issued to a US Technician class license to operate on the HF amateur bands, even though the Mexican government forbids such actions.

Where possible you should contact others who have had experience with this subject and who have successfully done whatever was necessary to obtain legal authorization to operate in your countries of interest. They may have information about the real situation and may be able to save you much time and frustration.

What About Operating Aboard a Commercial Vessel?

To those contemplating operating aboard a commercial vessel, such as a cruise ship or even a tramp freighter, this is always an adventure. I highly recommend that you first check with the shipping company, and if possible, the captain of the particular vessel you intend to board. Be prepared to spend a considerable amount of time preparing necessary documents— including possible CEPT licensing information, proof that you are a citizen in good standing and proof that you own your equipment. You want to do all this as early as possible. Depending on what you are attempting to do, it may take three to four

months to receive the proper permissions and authorizations to operate aboard a commercial vessel as a paying passenger. And unfortunately, no one but you will be in a hurry to obtain the proper documents with permissions and authorizations!

INSTALLATION PRACTICES FOR MARITIME-MOBILE OPERATIONS

Grounding

In today's maritime environment, not only is Amateur Radio equipment more sophisticated, but also other electronic devices aboard have become more numerous and sophisticated. This makes RF compatibility more troublesome in the maritime environment. In order for the various electronic devices used for communications, navigation, refrigeration, fresh-water production, automatic steering and other critical functions to coexist properly, you must take care in the installation of everything. One of the most important considerations is how grounding is handled on the vessel.

In a normal house on land, the problem of grounding is pretty simple. The green grounding wire in the ac wiring system is there to prevent shocks or electrocution. The ground connection is made by driving a long copper stake into the ground near the ac distribution panel.

On a boat things are considerably more complicated. In addition to the *ac* ground, you need a *dc* ground or return line, a *lightning* ground and an *RF* ground plane for the radio systems. Your first thought might be to simply make the ground connection to a metal through-hull propeller shaft or other underwater metal. This underwater metal will be grounded by connection to the seawater. Unfortunately, a connection between onboard electrical devices and underwater metal can—and probably will—give rise to serious electrolytic corrosion problems.

To keep your copper wires from turning into green powder, let's review the particular requirements of each system, resolve the contradictions between the systems and present a consistent and correct solution for a complete, integrated marine-grounding system.

DC Ground

The subject of dc grounding on your vessel is extensive, since it involves much more than your station ground and is beyond the scope of this chapter. DC grounding should be reviewed to ensure that the vessel under consideration is properly wired. If it is not, problems will probably arise for RF grounding also. Every on-

board electrical device must be wired with its own, separate dc return wire. Never use the mast, engine or other metal object as part of the dc return circuit. Each dc return for each branch circuit should be wired to the negative bus of the dc distribution panel. In turn, the negative bus of the dc distribution panel should be connected to the engine's negative terminal or its bus. The battery negative is also connected to the engine's negative terminal or its bus. The key factor here is that the vessel's electrical system is connected to seawater ground at one point only, via the engine negative terminal or its bus.

AC Ground

The best solution for ac grounding is the use of a relatively heavy and expensive isolation transformer. For most of us, an acceptable solution is to install a light and inexpensive galvanic isolator in the green wire, between the shore-side power cord socket on your boat and the connection to the boat's ac panel. A galvanic isolator is designed to protect your boat from galvanic erosion or metal loss by blocking dc currents. Connect the grounding conductor (green) of the ac panel directly to the engine negative terminal or its bus. This meets the American Boat and Yacht Council's (ABYC) recommendations for proper grounding. Visit their Web site: **www.abycinc.org/**.

When choosing galvanic isolators, make sure that you select one that has a continuous current rating that is at least 135% the current rating on the circuit breaker on your dock box. Certain galvanic isolators include large capacitors in parallel with the isolation diodes, which in certain situations theoretically provide better galvanic protection. Unfortunately, these units cost substantially more than conventional galvanic isolators. If you feel like spending real money on galvanic isolation, you might as well do it right and buy an isolation transformer. A good source for a galvanic isolator device is West Marine stores, which are located all over the world at **store.westmarine.com**.

It is also a good idea to use a Ground Fault Interrupter (GFI) in your ac wiring. GFIs will occasionally *nuisance trip* due to the humidity surrounding the wiring on boats, but the additional safety that they offer (particularly to nearby swimmers) in disconnecting power in the presence of ground currents is worth the nuisance. If your GFI starts to nuisance trip, it is a good idea to track down and clean up your damp wiring.

Lightning Grounds

If your boat is a sailboat, connect a

#4 AWG battery cable from the base of your aluminum mast to the nearest keel bolt from external ballast. If you have internal ballast, you should install a lightning ground plate. One square foot is recommended for use in salt water; fresh water requires much more. Do not rely on just a through-hull or a sintered-bronze radio ground (eg, *Dynaplate*) for use as a lightning ground.

For additional comfort, also run a #6 AWG wire from your keel bolt or ground plate to the upper shroud chainplates and to your headstay chainplate. Don't bother with the backstay if it is interrupted with antenna insulators. Have each of the cables that are used for lightning ground wires lead as directly as possible to the same keel bolt, with any necessary bends being smooth and gradual.

Given that you have grounded your mast solidly to the ocean, your mast will be at exactly the same electric potential as the ocean. There is no chance that you must dissipate the charge between the ocean and the atmosphere, so in my opinion you needn't bother with a static dissipater at the masthead. Wire *bottle-brush* static dissipaters may be useful to dissipate seagulls, but that is another subject beyond the scope of this chapter! When the mast has the same electrical potential as the ocean, you have done all that can be done to protect your vessel from lightning.

RF Ground

RF grounds are counterpoises for most HF antenna systems, and they reduce the impedance between the water and equipment being used. They assist in keeping the RF energy moving to the antenna, and not back into the vessel's electrical systems. This is a subject that will continually nag you if not handled properly.

VHF/UHF antenna installations do not need to use the ocean as a counterpoise. However, RF grounding of HF antennas must be considered or problems will develop. Careful deployment of proper RF grounds will be its own reward.

If you use an antenna tuner, as most do nowadays, mount it as close to the antenna as possible, preferably just under the after deck. Run copper ground strap from 2 to 6 inches wide (the wider the better) from the tuner to the stern pulpit/lifelines, to the engine and to a keel bolt. It is good practice to include the HF equipment itself in this network of ground strap, but remember that the tuner is the central point for all RF grounding. If the builder of your yacht had the foresight to bond into the hull a length of copper strap or an area of copper mesh, be sure to run a copper ground strap

to this as well, and say a blessing for builders such as these.

Sintered bronze ground plates (*Dynaplates*) can be used as radio grounds in situations where the ballast or engine is unavailable or awkward to connect. If the ballast, engine and lifelines are available, however, they generally make a high-performance ground. Be sure to keep the bronze, copper or aluminum ground strap away from water. Also do not use ground-mesh strap such as braid. It consists of smaller wires, which are more subject to corrosion.

If you have "RF in the shack," there are other actions you can take. Place an additional counterpoise wherever you have room. You could run a piece of coax under the bilge with the center conductor tied to the brad. Or you could use copper screen under the platform you use to operate and ground it too.

Always consider using a tuner external to your radio, preferably an automatic tuner (more properly termed a *coupler*.) Not only does an external tuner separate your radio from the RF ground point, it also affords great convenience when moving from band to band. An automatic tuner is certainly worth the additional expense, especially when you are concentrating on other priorities while underway.

RF Ground Rules

The RF ground needs to be a ground for RF signals only and therefore it does not need to conduct dc. You need to block the flow of direct current to prevent electrolytic corrosion problems. Find a dry secure place along each of the copper RF ground straps that are running to your engine and keel. Fasten the strap securely to an insulating piece of phenolic or to a terminal strip, cut a $^1/_{10}$-inch gap across the tape and solder several 0.15 µF ceramic capacitors across the gap. These capacitors will be transparent to the RF, which will be happily grounded by the ground strap system, but they will block any direct currents from running through the RF ground system. They will thus avoid any resulting susceptibility to *hot-marina* electrolytic corrosion. (Hot marinas are a subject that is too extensive to discuss in this chapter but something you should investigate thoroughly.) You should select the dc-blocking capacitors carefully, because they may carry a significant amount of RF current. See **Fig 2-2**.

As stated before in connection with dc grounding, make sure each electrical device on a mast has its own ground-return wire. If your steaming light, masthead light, tricolor, Windex light, etc are wired carefully and correctly there will be no

Fig 2-2—To avoid making another dc ground loop to the engine through the HF/SSB radio copper ground strip, fasten the copper strap securely to an insulating piece of phenolic or to a terminal strip. Leave a $^1/_{10}$-inch gap across the strap, and solder several 0.15 µF ceramic capacitors across the gap. These block dc flow in the ground strap. Suitable capacitors can be found at Digi-Key; tel 800-344-4539; type X7R monolithic ceramic capacitor, 0.15 µF, Digi-Key part number P4911-ND.

ground connection between their wiring and the mast itself. Make sure that this is the case.

This should also be true of your masthead instruments. An unintended dc connection between the mast and dc ground may result from a masthead VHF whip, one that connects the shield of the coax to the bracket connected to the mast. That shield also connects to the VHF radio, which is dc grounded by its own power connection. The easiest solution is to insert what is called an "inner-outer DC block" into the coax. This RF device puts a capacitor in series with the center conductor, and another capacitor in series with the shield. This device is transparent to the VHF RF signals in the center conductor and shield, but blocks any dc current in either the center conductor or shield.

You can make this block yourself or a good radio technician can make it. You can purchase one from radio supply houses, pre-fitted with any kind of coax connection on both ends. The commercial unit looks like a coax barrel connector. (PolyPhaser, Model IS-IE50LN-C1, also contains a lightning arrestor.) Once the dc connection from the mast to the VHF is broken, check for any other connections with an ohmmeter, and straighten out any other wiring errors or unintended connections. If your metal fuel tank is also bonded to the lightning ground system (per ABYC) then make sure that it does not have dc connections either to the engine via the fuel line or to the electrical system via the fuel level sensor. A piece of approved rubber fuel hose in the fuel lines to the engine solves that connection and a well-designed fuel-level sensor will not make electrical contact with the tank.

When you're done, there will be heavy

conductors running from the external keel or lightning ground plate to the mast, stays and to the metal fuel tank, but there will be no dc connections to the engine or to the yacht's electrical system.

Ground Summary

Grounding considerations are not only important to the successful operation of a maritime-mobile installation, but are critical to how the vessel will function with all electrical and electronic devices. By using capacitors to block dc connections in a few key areas, it is possible to have perfect ground systems for ac, dc, RF, lightning and to prevent electrolytic corrosion. You can have a boat that is immune to stray dc currents that are traveling through the water in hot marinas.

Note that in the old days, the technique of bonding everything together worked okay. In its defense, the bond-everything-together approach makes your boat less sensitive to electrolytic corrosion that can result from faulty wiring on your own boat. The problem is, the bond-everything approach leaves your boat totally defenseless to wiring errors in other nearby boats and shoreside facilities that cause stray dc currents to run through the water.

Today the technique of bonding everything together would still work fine if your boat spent all of its time on the high seas, in remote anchorages or in marinas that were wired perfectly and in which all of the nearby vessels were wired perfectly. Having underwater metal bonded together in crowded marinas today, however, is asking for expensive trouble. As outlined above, it is avoidable. With careful wiring and a few capacitors, you can have the best of all worlds.

Antennas for Maritime-Mobile Operations

Most maritime operations consist of not only Amateur-Radio operations, but also maritime operations in the marine bands on both HF and VHF. Let's explore optimal options for both.

VHF/UHF Antennas

Most 2-meter Amateur Radio antennas will work on the VHF marine bands (which operate in the 156 to 157 MHz range). However, it is always a good practice to use an SWR meter in the line to determine the best overall length for such antennas. Certain manufacturers do also make antennas especially designed for wideband VHF/UHF operations. One example of a wider-band dual-band VHF/UHF antenna is the COMET model CA-2X4SR, found at most Amateur Radio supply houses. For VHF only, try the Metz antenna with its

built-in lightning arrester and supplied right-angle bracket. For more information about the Metz maritime VHF antennas, try **www.metzcommunication.com**.

On VHF, antenna placement is not critical, although higher off the water is obviously better. On a sailboat the mast is always a good place to begin. If you wish to use the antenna on both the VHF amateur frequencies (146 to 148 MHz) and the VHF marine band (156 to 163 MHz, including the weather channels), make certain that the length of the antenna is a good compromise for both using an SWR meter.

Some vessels with a mast also use brushes to (supposedly) dissipate static charges. If you believe such a brush is useful, then placing a VHF/UHF antenna above probably defeats the purpose anyway. My personal view is to put the VHF antenna above the mast and forget about brushes, whose only useful purpose seems to be annoying sea birds. Placing the antenna down below the top of the mast will affect the antenna's SWR as well as its overall performance.

For vessels without a mast, placement is

more a matter of finding a place where it is higher than nearby metal objects, where it will perform better.

HF Antennas

There are many types of antenna configurations for maritime-mobile operation, depending on the type and size of vessel, particularly if you intend to use the same antenna for both marine and amateur bands, as most people do. See **Fig 2-3**, which illustrates the HF antenna types under consideration in this chapter.

As we have discussed, HF antennas require a good RF ground. You might think that with "water, water everywhere" RF grounding would not be much of an issue. But in reality, it is one of the most challenging aspects of a maritime-mobile installation. Remember, the vast majority of recreational vessels are made from wood or fiberglass. When considering antenna types for HF, keep in mind that the basic objective is obtain the smallest amount of inductance between your RF ground and the water.

Most sailing vessels use an insulated

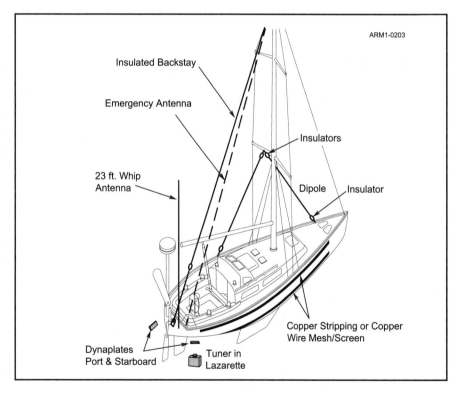

Fig 2-3—There are several ways to rig an HF antenna on a boat. For a sailboat, an insulated backstay is the most popular. A 23-foot whip on the transom, however, may be the simplest way to retrofit a sailboat with HF capability, and is usually the best way to get on HF on a powerboat without a mast. Many voyaging boats carry a spare wire antenna that can be rigged on a halyard or topping-lift should the main antenna fail. A dipole antenna rigged to a lower spreader (mostly, while at anchor) will provide excellent reception and transmission but unless multi-band coils or a balanced open-wire feed line is used it is very limited in frequency coverage.

backstay for their HF antenna system. Indeed, an insulated-backstay antenna may have already been professionally installed on your sailboat for marine SSB operations.

Recreational powerboats will usually have a professionally installed antenna in the form of a fiberglass whip about 23 feet long. Either the insulated-backstay or the 23-foot vertical whip is fed with an automatic coupler, often called antenna tuners. For the sake of simplicity I refer to couplers as *tuners* in this chapter. Automatic tuners are capable of tuning a random-length wire from 2 to 30 MHz, accommodating all marine and amateur HF frequencies.

Note that the insulated-backstay antenna is essentially an end-fed sloper, a form of vertical antenna. The challenge is to ensure that it has a good RF ground. Like any other type of vertical antenna, an insulated-backstay must have some sort of mirror image in the form of a counterpoise, regardless of frequencies used. Of course, what applies to a backstay antenna will also apply to a vertical whip antenna on a motorboat.

There are a lot of backstay configurations on sailboats. The simple single backstay used with a tuner at the bottom is an excellent HF single-wire antenna for marine and amateur frequencies. Some backstay configurations use two backstays. It is best to insulate both, since not insulating one may absorb some of the RF energy that would otherwise be radiated. When such a configuration exists, you can either feed both backstays as one, or even let the additional insulated backstay float. This is an individual choice and I have no specific recommendations regarding either method. If insulated, remember that the second backstay will re-radiate RF energy anyway. Also, make *certain* that you are using a Dacron topping lift on your sail. The topping lift (along the slanted angle of the sail) moves constantly, and if it is made of wire it can and will interact with the insulated backstay antenna. This will constantly change the tuning, which is definitely not good.

Sailboats with two masts, either ketch- or sloop-rigged, are more challenging for HF antenna design. My recommendation, shown in **Fig 2-4**, is to use the *triatic stay* (the stay between the two masts) in combination with the split backstay usually found on such sailboats. The triatic stay must be insulated, along with the split backstay, and the two must then be coupled together in an L-shaped configuration. None of the backstay configurations show any noticeable directional characteristics,

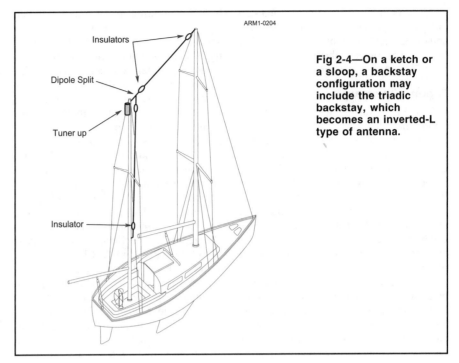

ARM1-0204

Fig 2-4—On a ketch or a sloop, a backstay configuration may include the triadic backstay, which becomes an inverted-L type of antenna.

Insulators
Dipole Split
Tuner up
Insulator

so this should not present any problem with directivity.

Regardless of the type of backstay antenna configuration on a sailboat, safety is a major consideration. Be sure to place at least seven feet of Teflon tubing over the first reachable portion of the insulated backstay to prevent RF burns when transmitting. See **Fig 2-5**. I recommend that the insulated backstay be installed professionally, but should you wish to tackle the task yourself check out Sta-lok Terminals, LTD, **www.stalok.com**, or Navtec Corporation at **www.navtec.net/home/index.cfm** for insulators and other related hardware.

Vertical Whip Antenna

A vertical whip such as the Shakespeare 390-series fiberglass antenna (**www.shakespeare-marine.com/antennas/ssb.htm**) uses no center-loading coil, unlike earlier whips that were designed for the 2-MHz frequency band. A modern marine whip is usually mounted on a swivel base at the bottom for lowering the antenna when going under bridges. Such a whip requires a top insulating bracket for stability. It works at input power up to 1 kW and is designed specifically for marine use, with fiberglass sealing the radiating wire from the elements.

Fiberglass whips are generally not as sturdy as a backstay, and they are not as long electrically for the lower HF frequencies as a typical sailboat backstay. In fact, fiberglass whips are not really meant for extremely rough seas. Nonetheless, a ver-

tical whip on a boat without a mast is a reasonable overall option when used with an automatic tuner.

There are a number of other types of antennas that can be found on boats, although some of them are not as effective on the marine bands. And some are more fragile and more subject to the elements at sea, such as shocking jolts, sea water and high winds. Depending on the nature of your maritime activity and the size of your boat, a 102-inch whip or a 30-meter Hamstick type whip could be sufficient for both maritime and amateur HF operations. When using an automatic tuners, you should configure the antenna so that it is *not* resonant on any band being used. Antenna tuners such as the ICOM AH-4, (100 W), ICOM AT-140 (150 W) or the SGC-230 (100 W) are excellent choices for any single-wire backstay or vertical whip antenna.

Dipoles and inverted-V wire antennas may be used where support such as a mast is available. However, these antennas are narrow in bandwidth unless used with an open-wire line, and they are flimsy compared to an insulated backstay or a marine whip antenna. Flattop dipoles require multiple supports and must be tightened so they won't get tangled in other rigging on the boat. Because of the greater efficiency and stability of marine HF vertical whips and insulated backstays, my own personal preference is to keep a wire dipole or inverted-V antenna rolled up with open-wire balanced transmission line for

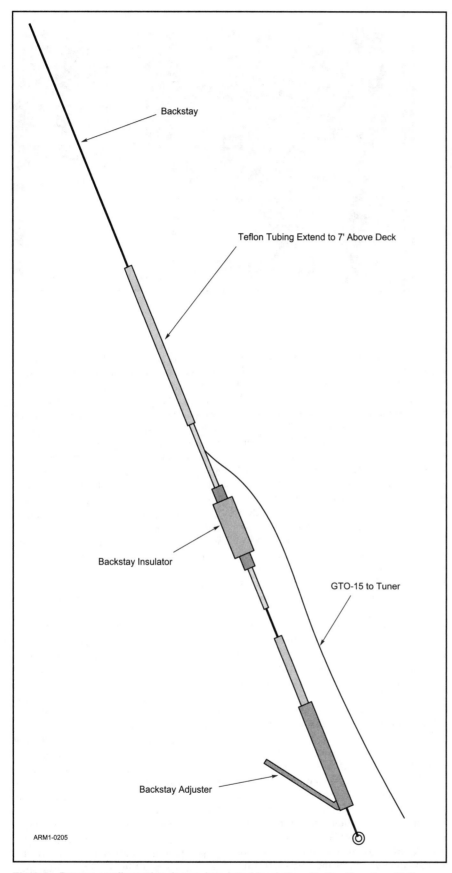

Backstay

Teflon Tubing Extend to 7' Above Deck

Backstay Insulator

GTO-15 to Tuner

Backstay Adjuster

ARM1-0205

Fig 2-5—Proper configuration for an insulated backstay. Notice the clear Teflon tubing on the first 7 feet. This will protect anyone who accidentally grabs the backstay during a transmission. Note that the backstay adjuster may be a turnbuckle. Unless you are certain that you can install this yourself, I recommend letting a professional do the work. The backstay is critical to your sailboat!

possible use as an emergency antenna.

Maritime-Mobile Radio Installations

Equipment choices are important, since offshore travel is always safer when marine and ham frequencies are available together. However, it is not as simple as modifying your Amateur Radio equipment for use on non-amateur frequencies—at least not legally if you are a US licensed amateur! The vast majority of amateur equipment is not designed to meet the stringent specifications of the marine-electronic service, and they are thus not *type accepted* per FCC Part 80, a subject that is too complex for this chapter. However, some manufacturers of type-accepted marine equipment, such as ICOM, do include the amateur bands in their marine radios.

Transceiver Selection

Transceiver selection is important, especially when considering a single unit for both marine and amateur operations. When compared to the flexibility and feature richness of the typical amateur transceiver, type-accepted marine transceivers are cumbersome to use on the ham bands. To put it another way, for those who are accustomed to the flexibility of amateur equipment, using a type-accepted marine transceiver is like trying to swim in a life jacket. By law, they contain no VFO, which tracks the transmit frequency with the receiver. In other words, they are *channelized* so that the end user can't easily transmit on an non-authorized frequency.

Type-accepted gear is designed to be configured by a technician and to be used by individuals with no knowledge of the parameters that have been set. The choices for a dual-purpose marine/ham transceiver are limited.

There are positive aspects, however, of using marine type-accepted radios. Most marine HF transceivers output 150 W and contain a temperature-controlled crystal oscillator that is more stable than off-the-shelf amateur gear. In addition, marine type-accepted transceivers contain an alarm system that is activated for emergency use on 2182 kHz. Keep in mind, however, that such a benefit is no longer critical since the Global Maritime Distress and Safety System (GMDSS) has removed the necessity of monitoring 2182 kHz for most offshore vessels. A modern Emergency Position Indicating Radio Beacon (EPIRB), which is about the size of a lunchbox and which operates on 121.5/243 MHz, when activated (usually dropped in the water) provides automatic

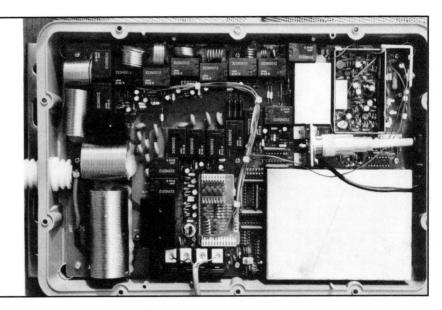

Fig 2-6—The ICOM IC-802 is a marine 150-W type-accepted transceiver that may also be used on the amateur bands. At A, the remotable head unit on top of the stowable transceiver box. At B, inside an ICOM AT-130 automatic tuner. (*Tuner photo courtesy Gordon West, WB6NOA.*)

distress alerting and locating in cases where a radio operator doesn't have time to send an SOS or MAYDAY call. Information about the GMDSS system may be found on the US Coast Guard Web site at **www.navcen.uscg.gov/marcomms/ gmdss/**.

My recommendation for an HF transceiver that fulfills all the marine-band radio requirements while also allowing operation on the amateur frequencies would be the ICOM IC-802 shown in **Fig 2-6A**. This unit uses DSP filtering and can be configured with the proper ICOM configuration software. It is much more versatile than previous marine type-accepted HF transceivers, but again it will not supply the normal functionality you are accustomed to enjoying with a ham transceiver.

However, if your primary operations are limited to amateur maritime SSB nets and digital e-mail, then the marine type-accepted version is viable. On the other hand, if you do plan to do a lot of casual CW or SSB hamming, then you must sacrifice somewhere. Or else you must carry two separate radios, a type-accepted one for the marine frequencies and another for the amateur bands. This is not a bad idea since it can also provide you with a backup on the amateur bands.

In addition, if you are *not* under the type-accepted constraints imposed by most countries, or if you wish to modify your amateur transceiver for expanded transmit in the marine bands should your type-accepted marine transceiver become inoperable during an emergency, then there are several Internet sites that provide you with modification information. Two such sites

are **www.qrz.com** and **www.mods.dk**. However, please remember that vessels licensed by most countries do not permit amateur equipment to be used for normal, non-emergency HF marine operations.

As previously mentioned in the section about antennas, you should definitely consider an automatic antenna tuner, especially if you intend to operate a wide variety of frequencies in the HF spectrum. The use of an automatic tuner is not only convenient but also an important safety consideration. There is something to be said about not having to stop and tune every time you change amateur or marine bands, especially when your main concentration should be elsewhere. Think about having to jump from one band to another while underway and having to stop piloting to manually adjust your tuner. This is especially true when your entire antenna system is bouncing around, causing changes in the way your tuner makes an impedance match. Being able to just push-and-talk while your tuner does all the work seems far more than reasonable and it certainly justifies the additional expenditure for an auto-tuner.

Remember that the tuner should be the central point for RF grounding, and it should be near the bottom of the backstay (or whatever antenna is connected to the tuner), rather than being located next to the radio. An excellent tuner for the marine radios up to 150 W is the ICOM AT-180. Another excellent automatic tuner for just about any radio is the SGC-230 auto-tuner. The difference in the two is that the SGC-230 requires nothing but RF to tune while the ICOM AT-180 requires a special control line from the radio to the tuner.

This is not an issue as long as the radio is an ICOM and has input for the control line.

Both of these automatic tuners, and others, may be purchased at major amateur equipment stores. I recommend that you only consider tuners that are weatherproof, regardless of where you intend to place it on your boat. Another important consideration is to ensure that the output power of your transmitter does not exceed the maximum output power recommended for a particular automatic tuner. For example, ICOM marine radios can operate at 150 W, while their AH-4 Amateur Radio automatic tuner has a 120-W maximum power limit.

Digital equipment, such as a computer, a modem for radio e-mail or GPS for position reporting, require a clean installation since they are more subject to RF than would be a transceiver, microphone, electronic keyer and tuner. Digital equipment will be discussed in its own section. But just for the sake of doing things right, let's assume that one or more of these digital items will be used in your shipboard station. Even if you don't currently intend to use such digital equipment, it will still be a good idea to plan ahead. You want to be assured you have done everything you can to take care of RF and grounding issues that can otherwise plague maritime-mobile operation.

Installing the station in the optimal spot will depend on your vessel type, but the principles for installation are mostly the same as seen in **Fig 2-7**. The most important installation rule is to ensure you have a good RF ground for all your electronics. But, where do you start?

Placement of the station equipment is

not critical as long as you can isolate the antenna tuner from the rest of the station's equipment. Ideally, you want to locate your tuner near the antenna, and you want a good quality coax running to your radio from the tuner. Be sure to use the tuner as your grounding point as shown in **Fig 2-8**, which also illustrates a perfect scenario. **Fig 2-9** is more representative of a real-life installation, where radiation from individual ground loops can cause negative effects not only on your own radio equipment but also on other electronic devices aboard.

So, after you've done what you can with the ground wiring itself, the next job is to make the alternative ground paths less attractive for antenna currents. This is started by adding RF impedance to the coax feed line, in the form of common-mode current chokes. These can be "line isolators" made with a large ferrite choke, or the choke may consist of multiple clip-on ferrites. These choke off common-mode currents that otherwise would use the coax's shield as an RF path. The transmitter RF output going to the tuner flows only inside the coax and is not affected by the choking action on the outside of the coax. **Fig 2-10** illustrates the methods to successfully deal with stray RF radiation.

A typical line isolator has about 20 turns of RG-8X wound around a ferrite rod, mounted inside a plastic housing with a female coax connector on each end. There are several commercially made models available, depending on the frequencies and RF power you use. I highly recommend the use of these line isolators for any HF maritime-mobile operation. The most complete source for these devices, along with weatherproofing your transmission line connections, may be found in the RadioWorks catalog on the Internet at **www.radioworks.com**. My favorite is the Radio Works model T-4 (ungrounded).

Like the line isolator, clip-on ferrites add common-mode impedance to choke off stray RF from entering other equipment on interconnecting cables. Clip-ons are made up of split ferrite cylinders, about $3/4$-inches in diameter, $1 1/4$-inches long, with a $1/4$ or $1/2$-inch hole through the middle. Clip-on ferrites should be used on audio or microphone lines. They should also be used on PC external devices, such as a mouse, keyboard, TNC (controller modem) or power supply. In other words, ferrite clip-ons should be use on any cable that enters or leaves an electronic device. The signals inside the cable are not affected, only the ground currents trying to use the cable shield as a "sneak" path.

Clip-on ferrites are hinged, and they snap together around the cable. My prefer-

Fig 2-7—Photo showing the nav-station of *Heart of Gold*, a custom-built 51-foot sailboat owned by Jim Cornman, KE6RK, author of *Airmail*. *(Photo courtesy KE6RK.)*

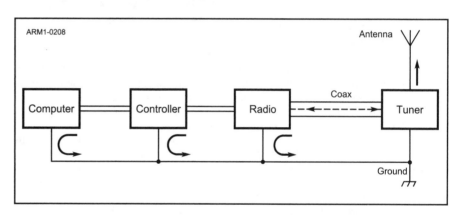

Fig 2-8—The ideal path of a radio signal going to the antenna and ground. All of the antenna current flows between the antenna and the primary ground connection. There is no ground current on the coax shield or through the secondary ground wires. Note the tuner is the optimal ground point, not the radio.

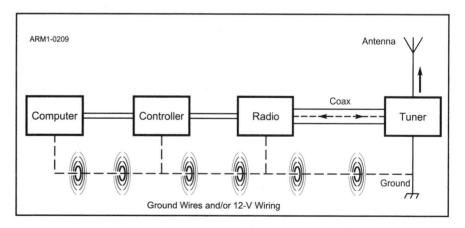

Fig 2-9—Here is a more realistic scenario. There are additional ground paths from the coax shield to the radio and to other equipment, as well as secondary ground paths. Antenna ground currents follow and radiate from all of these available paths.

ence employs large ferrites. I tightly loop the cable through the center of the clip-on, as many times as there is room before I snap the halves together. The common-mode impedance generated by looping the cable is squared each time it loops through, whereas the impedance obtained by placing several clip-on ferrites in series on the wire only increases linearly.

The ferrite halves must meet properly to be most effective. If in doubt, remove the ferrite halves from their plastic housing and secure the haves together with tape and/or tie-wraps. Also make sure that the cables themselves are properly shielded, with the shield connected to the connector shell (and equipment chassis) at both ends. This can be verified with an ohmmeter. If the metal shells of the DIN or DB-style rectangular connector at each end are connected, then the shield is terminated properly.

Steel or aluminum boats don't have a problem with ground systems, but aluminum boats usually have isolated 12-V dc neutral wiring to protect against electrolysis and can be subject to significant interference problems. In some cases the problem seems to be much worse than with a fiberglass or wood boat, probably because any stray RF energy is trapped inside a shielded box (the hull), reminding me of a fox locked inside a henhouse.

The process of grounding outlined above should be equally effective with metal boats. A line isolator should choke off stray RF at the source and would be the logical first step. If additional help is needed, the 12-V negative connections to the computer, radio modem (TNC) and transceiver may be RF-grounded using blocking capacitors to provide a RF ground, as described earlier in this chapter. As usual, ceramic-disc capacitors are a good choice for this duty, and a dozen 0.01 µF line-voltage-filter type capacitors wired in parallel will provide an inexpensive and low-impedance path for HF.

Noise from external sources is another very real concern. The FCC Class-B approval (CFR 47 Part 15 sub-part B) for noise excludes transportation vehicles, including maritime vessels. So, you cannot depend on the FCC Part 15 noise limitations for maritime equipment. On most vessels there is always something causing your S-meter to jump beyond a comfortable level. Depending on your choice of refrigerators, inverters, wind generators, lighting and other on-board equipment, noise may or may not be disastrous.

Common sense will assist you as much as pure technical information. Ask other boaters before you make a purchase of a new piece of electronic equipment. Better

yet, take a portable HF radio with you and check see what happens in the showroom when you expose it to the device you are considering. The Kenwood handheld TH-F6A is an excellent tool to have with you when considering a purchase of electronic gear. Most experienced cruising users of such equipment have information regarding brands that are less susceptible to both producing interfering noise and to being effected by RF radiation.

If you already own something that is generating noise, turn possible offenders off, one at a time, until you find the culprit. Check with the manufacturer of the noisy device to see if they can supply you with information on how to suppress its noise. The extent to which you need to worry about such problems will obviously depend on how much such electronic equipment you have aboard. For various types of noise filters, try RadioShack at **www.radioshack.com**.

MARITIME-MOBILE OPERATIONS

Safety first!

Regardless of what type of Amateur Radio equipment you install on your boat, or how you may intend to use it, the first and foremost operating criteria should be communication for your own personal safety and well-being. The primary use by the Amateur Radio community of a maritime mobile station is keeping communications open between you and the rest of the world. This is accomplished with something as simple as periodically checking into one of the many maritime-mobile nets and providing your position. Or it may be as complex as using your maritime station for radio e-mail to anyone, anywhere and at anytime, retrieving critical weather information or posting periodic position reports so that you may be tracked by your family and friends. All of these options are used extensively by maritime hams.

Another very simple, but important thing to do is to obtain a marine license for HF radio. If you already have a marine license and if your vessel is registered in the US, make sure your FCC marine license is current. The FCC Restricted Radiotelephone Operator Permit is required for boaters having a marine HF radiotelephone, for boaters having a VHF transceiver and traveling in foreign waters, or where a marine radio is required by law (eg, on boats 20 meters long or larger). For domestic VHF marine radio use, no license is required. Otherwise, there is a fee for the lifetime permit, but no tests are re-

quired in applying for a marine-radio license. An application is made on FCC Form 753, available from local FCC Field Offices or by writing to the FCC, PO Box 1050, Gettysburg PA 17326.

You may operate your marine radio after you have mailed your application(s) to the FCC, so long as you fill out, detach and retain the temporary operating authority attached to the application form. The temporary operating authority is valid for 90 days after you mail your application to the FCC and should be kept with your station records until you receive your official license/permit through the mail.

Popular Maritime-Mobile Nets

Individual maritime networks are abundant on the HF amateur and marine bands. Appendix 1 at the end of this chapter is a compiled list of the better-known maritime networks, displayed by hour of the day. These networks range from casual cocktail-hour social gatherings to more formal maritime service networks. The latter are specifically designed to assist you with weather, phone patches, indirect message relays or posting your position on publicly accessible graphic maps. These may be viewed on the Internet by your family and friends. Let's take a look at a few of these popular Amateur Radio maritime networks.

Maritime Mobile Service Network

The Maritime Mobile Service Network (MMSN) was founded in January 1968 by Chaplain A. W. Robertson, KB5YX, USN Retired. The nets original purpose was to "serve those who serve" in the United States military during the Vietnam crisis. Since that time, the net's hours and format have evolved with operational requirements. The MMSN is composed of a dedicated group of radio amateurs who unselfishly volunteer their time, equipment and efforts to serve and assist those in need of communications from foreign countries and on the high seas.

The primary purpose of the net is for handling traffic from maritime mobiles, both pleasure and commercial, and armed-service personnel deployed overseas. It also assists missionaries and persons working abroad.

The Maritime Mobile Service Network is operational every day from 1200 Eastern Standard Time until 2100 Eastern Standard Time, and from 1200 Eastern Daylight Time until 2200 Eastern Daylight Time, on a frequency of 14300.0 kHz. The network also operates on the alternate frequency of 14313.0 kHz, should the primary frequency be occupied. The Maritime Mobile Service Network immediately follows The Intercontinental Traffic Net.

Over 60 volunteer net control stations from throughout North America and the Caribbean maintain the network. Further, the net control stations are assisted by other relay stations. This assures virtually total coverage of the entire Atlantic Ocean, Mediterranean Sea, the Caribbean Sea and the eastern Pacific Ocean.

The MMSN is recognized by the United States Coast Guard and has developed an excellent working relationship with them. The MMSN has been instrumental in handling hundreds of incidents involving vessels in distress, medical emergencies in remote locations and passing health-and-welfare traffic in and out of areas affected by natural disasters.

The MMSN can act as a weather beacon for vessels during periods of severe weather, and it regularly repeats high seas and tropical weather warnings and bulletins from the National Weather Service and the National Hurricane Center. The network is also part of the MAROB (MARine OBservation) program of the National Weather Service. It gathers live weather and sea conditions from maritime mobile stations and forwards that information directly to the weather service on the Internet via a facility provided by *Winlink 2000*. Live data assists meteorologists with upcoming maritime weather forecasts. All Amateur Radio operators are invited to join the network simply by checking in. More about the Maritime Mobile Service Network may be found at **www.mmsn.org**.

The Intercontinental Amateur Traffic Net

The Intercontinental Amateur Traffic Net (INTERCON) was formally founded on September 8, 1960, by seven stations. These seven stations decided to meet twice weekly under the NCS Naval Reserve Amateur Radio Station, K4NAA in Arlington, Virginia. The net has expanded over the years to the present, with seven days per week programming. INTERCON is a well-disciplined operation, offering its services to amateurs worldwide on 14300.0 kHz daily from 0700 ET to 1200 ET.

The Intercontinental Amateur Traffic Net was created to promote goodwill and friendly relations among radio operators everywhere. It handles third-party traffic and information between individuals in any country where such traffic handling is permitted by treaty or mutual agreement. It also provides a means of emergency communications to any location where the normal means are disrupted by local disas-

ter such as fire, earthquake, storms, floods and terrorist activity.

Checking into the net is really quite simple. Tune to 14300.0 kHz and listen to ensure that no emergency traffic is in progress. When the Net Control Station (NCS) asks for check-ins, announce your call sign clearly and phonetically. The NCS will acknowledge your call sign, and in the order that you were heard, ask you to go ahead with your traffic. At this point advise the NCS of your name, QTH (location) and any traffic you may have. If you have no traffic, announce that you are standing by to assist with any traffic. You are more than welcome to check-in without traffic. If you hear a station calling but not being acknowledged by the NCS, please tell the NCS that you have a *relay*. More detailed information about this network may be found on **www.interconnet.org**.

The European Maritime Mobile Service Network

The European Maritime Mobile Service Network is an association of Amateur Radio organizations in various European countries covering all of Europe. It meets daily at 1630 UTC on 14313.0 kHz. Piracy reports, phone patches, position reporting and relay-traffic handling are available. At the time of this writing, the network is undergoing a great transition and expansion. The best place to catch up on the current status of the European maritime network association is to visit their Web site at **www.eu-mmsn.org/pages/home.html**.

The Pacific Seafarer's Net

The Pacific Seafarer's Net is a network of volunteer Amateur Radio operators that handles radio and Internet e-mail communication traffic between sailing and motor vessels operating on all oceans with land-based parties. The Pacific Seafarer's Net operates from 0230 UTC to 0400 UTC, 7 days a week, 365 days a year on 14313.0 MHz, and alternatively on 14300.0 MHz. Between 0230 UTC and 0324 UTC, general traffic is handled, and then at 0325 UTC the roll call portion of the Pacific Seafarer's net begins.

The Net Control Stations (NCS) are in various locations throughout the Continental United States, Hawaii and New Zealand. Communications traffic consists of daily position, message handling via e-mail relays, health-and-welfare traffic, phone-patch services, search-and-rescue coordination and vessel equipment inventories for search-and-rescue operations. Life-threatening emergencies are handled from any vessel whether or not they have

ham-radio licenses. Net control stations keep computer databases on participating vessels and their movements throughout the oceans.

By taking daily position reports on vessels in-transit, this net is able to serve the public by providing health-and-welfare updates and vessel position charting accessible through **www.bitwrangler.com/yotreps** and **www.pangolin.co.nz**. Information from the reporting boats advises directly other vessels about weather conditions and the state of the ocean. Additionally, the information is used by the New Zealand Meteorological Services and the National Oceanic and Atmospheric Administration for weather conditions on the high seas.

The Pacific Seafarer's Net also inventories vessels reporting so that if there is emergency information to pass on to search-and-rescue operations such as the USCG. This type of information is useful for any net control of any maritime net to have in the event of a missing boat or a disabled vessel at sea. The inventory form is on the website: **www.pangolin.co.nz/pacseanet/login.php?net=0**. It is the fifth item down on the right-hand column. Vessel operators and owners may wish to e-mail their inventory prior to departure. They may do so using the above submission form, which will automatically e-mail itself to the inventory manager KB6USC, Dick McNish. More information regarding the Pacific Seafarer's Net may be found on **pacsea.net/frames.html**.

The Waterway Radio and Cruising Club

The *Waterway Radio and Cruising Club* (WRCC) offers much more than just a network. It is a dues-paying organization with a monthly newsletter, the *Scuttlebutt*, which contains much information about WRCC activities. The actual network was launched in the early sixties, making it one of the oldest continuously operating maritime nets in Amateur Radio. Today the WRCC has around 1300 dues-paying members and anyone holding an Amateur-Radio license of any class is eligible to join.

Membership in the WRCC is not a prerequisite for participation in the Waterway Net, but a general class or higher license is required. The WRCC net meets on the air every morning of the year for about an hour starting at 0745 Eastern Standard/Daylight Time on a frequency of 7268.0 kHz LSB. The main purpose of the Waterway net is to encourage Amateur Radio communications to and from boats, with an emphasis on safety and weather information. Position reports help keep live-aboards and cruising boats in touch with families and friends, and float plans provide a strong measure of safety for offshore passage makers.

In addition to the SSB net on 7068.0 kHz, there is an active daily WRCC CW net at 0700 Eastern Standard/Daylight time daily. The schedule varies, and there are slow-CW days and faster-speed days for those who are interested.

The Waterway network holds an annual picnic every November in Melbourne, Florida. If you are located in the Bahamas, Caribbean or the Eastern United States, membership in the WRCC will provide you with good information about maritime hamming as well as letting you get acquainted with other active maritime hams. More information may be found on **www.waterwayradio.net**.

The Caribbean Maritime Mobile Service Network

The *Caribbean Maritime Mobile Service Network* (CMMSN) operates daily on 7241.0 kHz from 1100 to 1200 UTC, assisting cruisers contact each other, getting in contact with land-based stations and enabling them to keep in touch with their home bases, family and others. It also provides good daily weather information, handles emergencies and more. The net weatherman checks in promptly at 1115 UTC daily with direct satellite and other weather reports for the entire Caribbean. Following these reports he moves to 7086.0 to give fills, and send weatherfax photos to those who have the capability of receiving them. The reports for the Caribbean are generated from a variety of Internet sources when the telephone lines are working. There is a ground satellite station to collect photographs and program charts directly from the GOES-8 geostationary weather satellite, as well as high-resolution photographs from various NOAA orbiting satellites. Further information may be found on: **members.isp.com/kv4jc@isp.com/index.html**.

The International Boat Watch Network

The International Boat Watch Network is made up of many of the major amateur and marine HF maritime nets. These have banded together to form a worldwide emergency contact system for the maritime ham. Founded and currently managed by Mike Pilgrim, K5MP, this collaboration of separately sponsored maritime and amateur maritime networks, along with the US Coast Guard and other similar agencies in other countries, provides assistance to the families and friends of the

maritime ham should there be any need to locate or contact a vessel for any reason. Vessels in distress or otherwise in need of assistance use this collaboration of maritime networks to obtain assistance with on-board situations that may require additional resources. Through the International Boat Watch Web site, **www.boatwatchnet.org/**, family and friends who wish to contact any vessel worldwide can fill out a "Boat Watch form." The network then sends e-mail to various places throughout the world.

The International Boat Watch Network has a remarkable history of providing assistance in maritime emergencies, and the heroic efforts of those who participate have not gone unnoticed. The US Coast Guard depends on the International Boat Watch Network for assistance locating recreational vessels that are missing, or that must be contacted for a critical information exchange.

Not all maritime networks directly participate in the Boat Watch network, but the better organized networks are included. Appendix 1 contains a list of the participating networks, both marine as well as amateur, as does Appendix 2 from WB6NOA. You should bookmark these Appendices and keep them readily available. These participating networks are not set up strictly for emergency communications and they welcome regular maritime check-ins. In fact, maritime check-ins are viewed as special and given priority treatment.

Important Activities of Maritime Networks

Position reports may be sent to the *Winlink 2000* Amateur Radio messaging network via an e-mail, or verbally sent by radio to several maritime networks. They then place the positions on publicly available graphic-mapping displays, such as the Automatic Position Reporting system (APRS,) YotReps and ShipTrak. For the *Winlink 2000* user, a template is provided for manually or automatically sending a position report.

For a HF radio user checking into a SSB maritime net, the ability to provide this service usually depends on the ability of the specific network manager to manually enter the data. A Web browser template is provided for this purpose by the *Winlink 2000* Amateur Radio messaging network development team for use by SSB maritime networks. When data is placed into the system, it automatically shows up on APRS, YotReps and ShipTrak. All three of these graphic position-reporting services may be viewed on **www.winlink.org/aprs/aprs.htm**.

Those wishing to see a position report must have the amateur call sign of the boat they are seeking. They can simply enter the call sign and then view the position on a map.

The APRS system is used extensively by maritime and land-mobile users for announcing their current position for family and friends. It does not track individual positions over a long time. To continually show your current position on the APRS Web site, you must give a position at least every ten days.

YotReps is another service for reporting maritime positions and weather. YotReps is sponsored out of New Zealand. Unlike APRS, YotReps is specific to bluewater cruising and does not just show a current position, but also tracks the position over time.

ShipTrak, sponsored by the Maritime Mobile Service Network, is also good for tracking vessels over a period of approximately 21 days. There is both a text-based table as well as a graphic view of the positions reported within the last 21 days.

E-mailing a request for a position report is also another option in the *Winlink 2000* system. An e-mail user may send a position request to **QTH@winlink.org** with "subject: Position Request" and in the message body you enter the call sign of the vessel being tracked, followed by a space and the number of position reports desired. If the user wants all the positions, "999" may be placed after the space. More than one position may be requested, so long as each is placed on a separate line. If the originator of the request wishes to send the results to another e-mail recipient or Winlink user, each is displayed properly in the example below:

To: QTH@winlink.org
Subject: Position Request

[The following is in the message body of the e-mail]

N7MRO 3
DH1MZ 999
Reply SMTP:judywt@comcast.net
Reply SMTP:davelp@aol.com

In addition to the actual report, the vessel reporting may put brief comments in a position report. These will show only in the e-mail return and on the APRS report. More information regarding the e-mail reporting and request process may be found on: **www.winlink.org/instructions.htm**.

The National Weather Service's MAROB (MARine OBservation) weather reporting system, **www.nws.noaa.gov/om/marine/marob.htm**, is also furnished to any network that chooses to use it. This reporting system for recreational vessels (under 16 tons) is provided by the *Winlink 2000* development team in the same Web-based template as is position reporting. This is used by the National Weather Service in its weather forecasting. The Maritime Mobile Service Network, the Pacific Seafarers Network, the Canadian Le Réseau Des Petits Bateaux (The small boater's net,) and the Waterway Radio and Cruising Club are among those networks that are usually available for entry of position reporting and MAROB data.

Phone patches are still available, although they are not as popular as they once were. Should you wish to attempt a phone patch from an offshore boat, some maritime networks can still handle telephone patches. They refer to them as *two-ways* or *phone patches*. They are usually handled through the network manager. Of course, their availability is also determined by when a network participant with a phone patch is available. You should check into a maritime network and request information from the operating network manager about the availability of a phone patch. If a particular network has members who are willing to provide a phone patch to you, he will most likely request you to be present during a time when he can arrange it for you.

The Maritime Mobile Service network does currently provide a phone patch service, and it usually has information about other networks that also provide phone patches. Obviously, your phone patch request will receive a higher priority from any of these if you are underway rather than docked in a harbor where other means of communication are available.

Worldwide Radio E-mail, Position Reporting and Extensive Weather Information

Radio e-mail, which provides you access to the worldwide Internet e-mail system, is by far the most reliable and direct way for you to stay in direct contact with the rest of the planet. It can provide you with graphic or text-based weather, position reports to your family and friends and many other services. You only need to add a specialized modem (a terminal node controller, or TNC) and a computer with special freeware end-user client software.

Like land-based e-mail, what is most useful about digital messaging over Amateur Radio is its store-and-forward ability. Even if the recipient is not available on the other end for a real-time one-to-one connection, you can take advantage of radio propagation to any participating HF station that can receive the digital data, rather than worrying about a specific pre-deter-

mined schedule. Third-party traffic is thus more reliable using digital messaging over Amateur Radio interfaced with the Internet e-mail system.

The Amateur-Radio *Winlink 2000* global-messaging network, **www.winlink.org**, is currently the largest and most sophisticated HF radio e-mail network available. It has stations located within reach of just about anywhere you might travel. As a voluntarily operated worldwide system, *Winlink 2000* currently uses the Internet Engineering Task Force (IETF) RFC 2821 de facto e-mail standard and pushes approximately 150,000 e-mail messages (or approximately 250,000 minutes per month) through its network system to either Internet recipients or other amateur *Winlink* radio users.

As of this writing, there are over 6600 maritime hams using the system monthly to approximately 82,000 e-mail recipients. *Winlink 2000* is the primary means of communication used by the maritime ham. As the network administrator, I highly recommend it. Some of the advantages of deploying *Winlink 2000* on your own boat are:

- The use of a powerful end-user client program called *Airmail*, which resembles a standard e-mail program and which is provided to you as freeware if used on the amateur bands.
- The worldwide ability to send and receive multiple-recipient e-mails with reasonably sized binary attachments.
- The ability to request and retrieve over 700 graphic- and text-based weather products from virtually every source available to the public, via an *Airmail* built-in catalog.
- The ability to send position reports to the Internet for public view by family and friends, including the Automatic Position Reporting System (APRS), ShipTrak and YotReps or an e-mail request.
- The ability to subscribe to various list-servers.
- The ability to be privately tracked each time you check into a *Winlink 2000* participating network station. This requires you to have a GPS device with an NMEA port available (most do) connected via a serial port to *Airmail*.
- The ability to change your own user-definable parameters.
- The ability to locate the 30 nearest *Winlink 2000* or APRS users in an emergency, or if you just want to chat.
- The ability to send National Weather Service weather reports via *Airmail*.
- As with SSB, the ability to use *Airmail*, radio modem and marine station equip-

ment for certain maritime coastal (commercial) e-mail services.

As you can see by this list of features, there are many reason why *Winlink 2000* and the SSB maritime network systems are so widely used by the maritime amateur community. However, if you have already operated maritime Amateur Radio, you will know that operating *any* mode as a maritime ham somehow brings a feeling of adventure and mystique to both sides of any QSO! Let's investigate the *Winlink 2000* radio messaging system in some detail.

Details About *Winlink 2000*

The *Winlink 2000* Amateur-Radio global-messaging system was born in January 2000, after much experimentation with previous *Winlink* systems. *Winlink 2000* consists of several Common Message Servers (CMS) acting as network *hubs*, where messaging is interfaced to the Internet. Internet e-mail messages and weather data enter the system at various CMSs. They are sorted and sent to Radio Message Server (RMS) node stations—the stations actually contacted by the end-users. Likewise, when messages bound for other hams, for the Internet, or for both hams and the Internet arrive from a *Winlink 2000* user into an RMS node, the CMS hub station does the routing to the final destination.

The 40-plus RMSs randomly seek out one of the CMSs so that even if one or two are out of service, automatic alternate routing can still take place. This *star-network* configuration may be thought of as a wheel with spokes. The Common Message Servers are in the hub of the wheel and at the end of each spoke is a Radio Message Server waiting to be polled by a *Winlink 2000* user. All functions are mirrored throughout the system. In other words, you can check into one RMS, pick up three out of five messages and then check into another RMS to retrieve the remaining two pending messages.

In addition to radio pathways, a *Winlink 2000* end-user has other options. When an end-user connects to the Internet directly *Airmail* can be used as though the end user were actually using a radio. This is a wonderful way to learn how to use *Airmail* before you venture out onto the water.

Airmail also allows password-protected Web browser access for text-based e-mail. This option is often used in Internet cafes when in-harbor or during periods when the maritime user is away from his computer and boat. These non-radio pathways are encouraged when their use is possible to allow more room in the Amateur-Radio spectrum for others who have no other option than to use radio.

Serious *Winlink 2000* HF radio users currently use PACTOR II or PACTOR III modes, an ARQ protocol developed by Special Communications Systems GmbH and Co (SCS), **www.scs-ptc.com**, the German inventors of all PACTOR modes. To date there is no other digital protocol with a 100% error-correcting protocol that can match the performance of these modes. However, the *Winlink 2000* system is not mode-bound by anything other than performance. Should a more efficient protocol come along, you can be assured that the *Winlink 2000* developers will incorporate it into the system.

The original PACTOR I protocol has a maximum speed of 200 bits per second in a 500 Hz bandwidth. The more advanced PACTOR II protocol has a maximum speed of 800 bits per second in the same bandwidth, while the newest PACTOR III protocol has a maximum speed of 3600 bits per second in a 2.1-kHz bandwidth. Given the same amount of data, the efficiency of PACTOR III is far superior to the other two PACTOR protocols and it is more robust under all conditions. The use of PACTOR I is now discouraged on the *Winlink 2000* system, mainly because of the large number of users attempting to connect to each RMS in a 24-hour period. Most of the RMS system operators have elected to eliminate PACTOR I, but there are still some RMS stations that allow the use of PACTOR I. These are noted in the *Airmail* PMBO frequency list, which we will discuss later in this chapter.

The SCS radio modems (TNC) are used mainly for PACTOR II and PACTOR III. These modems, which are actually microprocessor driven with a robust CPU and fast bus speed, use their computing muscle and the additional power of DSP processing, to copy signals as much as 15 dB *below* the noise level. They also have the ability to put the transceiver's audio frequency tones on the center frequency of the modem should the actual carrier frequency be off as much as ±40 Hz.

When using the SCS PTC II Pro radio modem, a single serial port on your computer controls the modem while also controlling the radio by passing *Airmail* radio commands through the modem to the radio's TTL or RS-232 remote control port. Otherwise the less-expensive SCS PTC-IIex requires an additional serial port so that *Airmail* can control the radio's frequency and other parameters.

Many modern laptop computers lack a serial port, but you can use USB-to-serial adaptors. I have tried several manufacturers of these multi- and single-port USB-

to-serial adaptors, such as Digi EdgePort, Keyspan and StarTech and all were trouble free.

There are other manufacturers of these devices, but there have been some problems with other brands. Knowing that those listed above do work properly, I personally cannot recommend other brands. Appropriate adaptors may be found at just about any discount computer outlet on the Internet. I have always found a good selection of these adaptors at Computer Discount Warehouse, **www.cdw.com**.

USB-to-serial adaptors may also be used to supply *Airmail* with position information from the NMEA serial port on most GPS receivers for constant real-time position reporting over *Winlink 2000*. Be sure to leave the GPS off while booting your computer since Windows sometimes confuses a GPS for a serial mouse!

For ICOM users, the OPC-478U interface changes serial data to TTL logic levels, thereby eliminating the need for an additional, expensive ICOM CI-V interface. The OPC-478U is available where ICOM products are sold. Be sure to replace the 3.5-mm stereo phone plug that has the data on the ring (not the tip) with a 3.5-mm mono plug to mate with the connector on the back of the ICOM HF radios. Why ICOM uses the 3.5-mm stereo phone plug is confusing to me. I believe it is used with their VHF radios. However, a mono plug is required for all ICOM HF radios.

At the time of this writing, Farallon Electronics provides the interface with the proper plug on the end. Farallon is also the dealer for the SCS modem and can be reached at **www.farallon.us**.

THE *AIRMAIL* PROGRAM

The *Airmail* program is also used by some commercial maritime e-mail carriers for a fee, but it is offered at no charge to an Amateur-Radio user when used on the amateur bands. When used with *Winlink 2000*, it also contains much functionality not available through any other marine e-mail service, commercial or amateur. *Airmail* was written by Jim Corenman, KE6RK, who has cruised for the last several years in a custom-built 51-foot sailboat with his wife Sue, building and testing *Airmail* while seeing the world. As with the *Winlink 2000* network system it serves, *Airmail* is constantly being enhanced to provide expanded functionality on an ever-growing list of equipment.

An option mentioned on the *Winlink 2000* Web site for those boaters using *Airmail* who do not have an Amateur Radio license is to join SailMail, a non-profit association. SailMail provides radio e-mail service, but

on the marine bands rather than the ham bands. See **www.sailmail.com** for more information. [There are other commercial maritime e-mail services also. Do a Google search on "HF maritime e-mail service."— *Ed.*]

I recommend that you download *Airmail* and follow along on your computer as the functions and features of this remarkable program are discussed here. Installing *Airmail* is simple. There are three components of *Airmail* to consider:

1. The main *Airmail* program.
2. The *ICEPAC* propagation engine program.
3. The *GRIB* (gridded-binary) weather file viewer.

These may all be found on the following Web URL: **www.siriuscyber.net/ham/**. Each of these components may be easily installed from this Web site or downloaded and installed later. Install the main *Airmail* program first, the *ICEPAC* program second and the *GRIB* file viewer last. Follow the install wizard instructions for all three modules.

Once you install *Airmail*, you must answer some basic questions.

- What modem do you use?
- Do you want to install the built-in automatic control that *Airmail* features to control your transceiver so that you do not have to set all the parameters every time you want to use it?
- Do you want your GPS NMEA serial port connected to your computer so that *Airmail* will have real-time coordinate information?
- Lastly, when you make your first successful radio connection to a *Winlink 2000* participating station, do you then want to configure *Airmail* so you also use it over the Internet using the *Airmail* Telnet module?

Let's examine each of these questions. Your choice of a radio modem, which is also referred to as a *TNC* (Terminal Node Controller), depends on the seriousness of your intent to use *Airmail* with *Winlink 2000* to communicate with others, to provide a means for others to track your position and to gather weather information while on your boat.

If you are just cruising in the sight of land, your options are clearly dependent on your interest, balanced by what is in your budget for such items. However, if you are in the middle of a blue-water cruising experience, need critical weather information and want others to know where you are, your priorities may be different.

For example, the Kantronics modem, which only uses the Pactor I protocol, may be adequate to connect to those participating stations that allow Pactor I and to send and receive text-based e-mail. However, if you are expecting to request and receive 30,000 to 100,000 bytes of weather data after sending and receiving a few e-mail messages, you will want Pactor II or Pactor III.

An easy way to remember the difference between these protocol varieties is that when using Pactor I, it takes 80 minutes to send what can be sent on Pactor III in seven minutes. Obviously, the differences are great and there is a difference in cost to be considered. Consider also that you are one of many attempting to access a *Winlink 2000* participating station. You may have to wait five minutes for a Pactor III user, while for the same amount of data transferred you may wait forty-five minutes to an hour for a Pactor I station to finish. Try to visualize *you* as the one waiting! The choice becomes clearer this way.

Another important factor to consider when looking for the right radio modem is *Airmail*'s ability to control the radio remotely. All HF Pactor radio modems (TNC) supported by *Airmail* may be used while the program controls the radio, and there are two ways this may be accomplished. For the older SCS PTC II (no longer sold) and the SCS PTC II Pro, the radio is controlled through the modem port from the computer. In this way, only one serial comport is necessary for control of the modem and the radio. However, a second serial communications port must be used with less-expensive SCS modems, such as the PTC-IIe (no longer sold) or the PTC-IIex, as well as the PACTOR-1-only modems such as the PK-232 (firmware version 7.2), or the several KAM series PACTOR modems. As I have previously mentioned, there should be no lack of communications ports if you use the USB-to-serial adaptors.

Because there are several modem options to obtain PACTOR II and PACTOR III speeds, I do not recommend that the serious user consider a PACTOR-1-only modem for offshore use. It is always a good idea to look on the *Winlink 2000* Web site or ask a current *Winlink 2000* user prior to purchasing a radio modem. Models and protocols are subject to change and you never know when something more efficient may be available, perhaps even at a lower cost.

Regardless what type of transceiver you may have chosen, it is important to make certain that it can communicate with the modem, your computer and with *Airmail*. Depending on the model of your transceiver and modem, you may be faced with converting a TTL interface with an RS-232 interface. So determine if there is a need

for conversion. With modem or radio control or for any other connectivity, don't look for the standard expensive TTL-to-serial conversion products supplied by the radio manufacturer. Instead, consider finding such converters over the Internet. One exception described above is the ICOM OPC-478U, which converts the USB-to-serial-to-TTL all in one cable and for a reasonable sum.

Configuring *Airmail*

Configuring *Airmail* is a snap once you have chosen a radio modem and transceiver. When you first initialize *Airmail*, a configuration wizard will set up most of the parameters necessary to identify you and your hardware configuration. If you do not know the settings requested, they may also be set up later from the **Tools**, **Options**, **Connection** menu.

Once you have determined the proper settings for your radio modem and transceiver configured in the *Airmail* menu item **Tools**, **Options**, **Connection** and your personal data set up from **Tools, Options, Settings** you have only one more step before testing the configuration with your modem and transceiver.

Optional Modules

To have module options readily available in the main **Module** menu list, you must choose which of these modules you wish to use. You do this by going into **Tools, Options, Modules**. For a typical maritime operation, you might wish to set up the **Terminal Window** and **Position Reports**. Should you operate any VHF packet while near land, you may wish to choose and configure the **Packet Client** as well. Also, for those times when you have direct Internet connectivity to the computer hosting *Airmail*, you may also wish to set up and configure the **Telnet Client** module. Each of these modules is very intuitive to configure and use. Now you may proceed to the main *Airmail* **Modules** menu to find each module waiting for your use.

Setting the Levels

Configuring *Airmail* for transmitting a signal is not difficult if you follow a basic rule: Use just enough gain so that the ALC reach 95% of your radio's transmitting power under a single tone (FSK.) For those using the SCS radio modems, a provision in the *Airmail* Terminal Window is specifically designed to assist you with this determination. In the **Control** menu on the Terminal Window form, you can drop down to **PTC II Pro Amplitude**. There you will see two slide bars, one for FSK and another for PSK.

To accurately determine the amount of audio to send from the radio modem to the radio, I suggest you use FSK rather than PSK. PSK, like SSB, uses variable amplitudes and is not as accurate for setting output power parameters. In addition, make certain that you lower your transceiver data *input* audio levels toward their lower level (some transceivers do not have this option). The objective is to pump as much audio as you can out of the modem to avoid possible interference from some noise source on your boat.

Adjustments to the transmitter FSK level control will also show in the main menu **Tools**, **Options**, **Connections**. If you use the **Track** feature, my recommendation is to put both FSK and PSK at the same level and leave them that way.

For those who do not have a SCS radio modem, read the manual that accompanies your radio modem. You will either end up with a screwdriver in hand for a potentiometer adjustment inside the modem, or you will be asked to configure the audio output level from the radio modem with a software command. If this capability is available, *Airmail* has a *dumb terminal* mode from the main menu **Tools, Dumb Terminal**. Remember, the objective is to pump more audio from the modem to the radio to reduce the effect of any stray radiation from nearby noise sources, especially your own radiated signal. Raising the output of the radio modem's audio level while reducing the input level of the transceiver audio level input, will assist in reducing the possibility of stray radiation getting into the audio line between your transmitter and the radio modem.

Setting up the *Airmail* Address List

You are now ready to explore the feature set within *Airmail*. Let's look at how to automatically load the *Airmail* address book. If you have an address book from the Microsoft *Outlook Express* or *Outlook* e-mail programs *Airmail* has an import function for this purpose. First export the **Contacts** folder from your e-mail program into a file using **Comma Separated Values (Windows)** and save it as "Contacts.CSV." Then go to *Airmail*'s menu **Tools, Address Book, Import Addresses** and import the addresses into the *Airmail* address book. You could generate this file on another computer if you like, perhaps bringing it in on a floppy or CD.

Sending a Message With *Airmail*

To send a message using *Airmail*, you must compose, address and then post the message, just like any other e-mail message. Click on the white rectangular icon

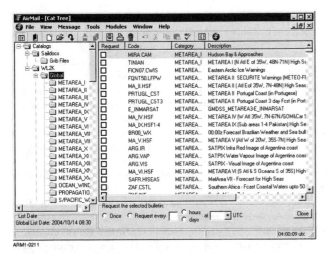

ARM1-0211

Fig 2-11—*Airmail* client, end-user freeware used with the *Winlink 2000* global messaging system, offers a powerful set of tools for the maritime ham. Shown is the built-in catalog of over 700 different weather products, worldwide.

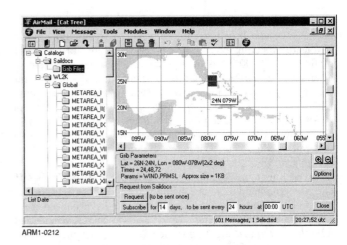

ARM1-0212

Fig 2-12—A built-in GRIB file viewer allows the user to define an area for a NOAA virtual Weather FAX (GRIB) request.

with the yellow star at the top right. This brings up the **Address Book** form. If you already have the address you need in the address book, then click on the proper address. Otherwise, you can either add an address manually or just cancel the address book form and the message form will automatically appear for manual addressing. Type in your text in the message form, just like you would any other e-mail program. Should you wish to attach a file, click on **File**, **Attach File** and like any other e-mail program, find the file and click **OK**. (See below for setup information for this feature.)

Now you must "post" your message for delivery. On the message form, at the upper right you will see **Post Via:**—For *Winlink 2000* message delivery, always post your messages to **WL2K**. The default checkbox states **Always use this path**. Now just click on the little red mailbox icon in the icon menu and you have posted your message for delivery. To ensure that your message is posted, look in the **Outbox** and you will see your message with a mailbox icon next to the message. When your message is sent, there will be a check icon next to the same message.

Attaching a File

Before you can successfully send a file attachment, you must first set your **File Attachment Limit** using the **Options Message** form. These options once set are automatically sent to *Winlink 2000* the next time you connect to the system and will be available to you system wide. To find the **Options Message** form, click **Window**, **Winlink 2000**, **Options Message**. Notice all the options that are available to you, such as adding a call-sign

prefix or suffix, necessary in some country harbors; the ability to turn incoming BCC (blind carbon copy) off or on; and a box to enter your file size limit. Once you have completed specifying your options, push **Send** on the form. The next time you log into a *Winlink 2000* participating station, the information will be sent to all participating stations, system wide.

If you attach a file to a message, you must consider the conditions between you and the participating *Winlink 2000* station and the Pactor mode you are using. For example, even under good conditions, sending an 80,000-byte GIF file that does not compress from its already compressed state will take over 70 minutes minimum on PACTOR I. This same 80,000-byte file will take less than 10 minutes using PACTOR III. Also remember that DOC, XLS, PDF and TXT files will compress up to 75% with the "B2F" binary format used by the *Winlink 2000* system.

Weather Products

Requesting and receiving weather products via *Winlink 2000* is a major benefit of the system. There are over 700 graphic and text-based weather products and help files available from a built-in catalog in the *Airmail* program. In addition, e-mail subscriptions are also available from NOAA and others, which can be delivered to you in a timely manner.

For the maritime user, the *Winlink 2000* Amateur Radio messaging system gives more opportunities for gathering important weather information while offshore than any other method available from any other source. Installing and using *Winlink 2000* for this purpose alone has brought thousands of maritime users into

Amateur Radio. **Fig 2-11** provides a small peek into the various weather options that are available for download. The total weather coverage is worldwide, with satellite images, weather faxes, real and virtual buoy reports, tide reports and text-based weather.

Simply use *Airmail* **Window**, **Catalogs** and find the desired category in the **WL2K**, **Global** weather folder in the folder tree located in the left frame, pick your particular area on the right hand side of the form, and check the appropriate box for whatever weather product you need. You may click multiple boxes and send multiple requests. However, remember that you will also receive a lot of traffic back into your system from your requests. Please make only serious requests.

In addition to the normal weather products from many sources, there is also a very specialized weather fax viewer built into *Airmail*. Through the use of *Saildocs*, a free service offered by the author of *Airmail*, NOAA virtual wind speed and pressure weather fax information may be downloaded in the form of a small text "GRIB" file for any reasonably sized geographic area, anywhere on the planet. **Fig 2-12** shows a user-defined area picked for requested NOAA virtual weather fax information.

Fig 2-13 shows the results of the request. This virtual computer-generated information from NOAA, which usually consists of a 2500-byte file, comes with a NOAA accuracy disclaimer, of course. But it also has proven to be accurate and is a lot smaller download then the 20,000- to 30,000-byte NOAA real weather fax. Not only is the download smaller and faster, it is much more readable than a black-and-white weather fax.

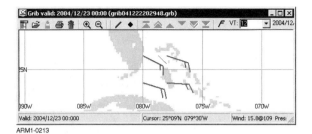

ARM1-0213

Fig 2-13—The resulting information after the system responds to the user-definable virtual Weather FAX area in Fig 2-12, showing details of wind speed and barometric pressure.

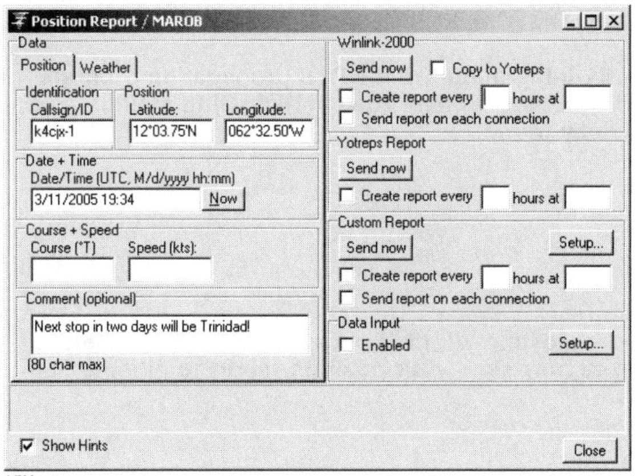

ARM1-0214

Fig 2-14—The option to send a simple position report, as well as a complex NOAA MAROB weather report, is available to the maritime ham when using *Winlink 2000* with *Airmail*.

You can also interrogate the virtual information on the GRIB file map with your mouse curser by longitude and latitude to gain very detailed information within the area you have chosen. These downloads may be requested once or you can subscribe for any user-definable period.

Updating Catalogs

To update your catalog items when you receive an all-user message that new or updated items are available, click on **Tools**, **Update Catalog**. An update will add what is new, replacing the current catalog file. You should update your current list rather than ask for a new one, since a new list is approximately 70,000 bytes and takes a while to download.

Each participating station (RMS) may have local bulletins specific to that station. To obtain these specific bulletins you click on **Windows**, **Catalog**, **WL2K** and look for the catalog of the specific station in the folder tree.

In addition to the *Airmail* catalog options, there are many list-servers that will periodically send weather information per your e-mailed request. You should exercise caution when choosing such an option since continual reports can mount up to a lot of transfers should you not be available to receive these automated incoming messages on a timely basis. Nevertheless, for some situations the option may be attractive. Saildocs and NOAA have a host of such automated information sources. There is a certain comfort in knowing you

have the ability to obtain accurate graphic and text weather information upon request using *Winlink 2000*.

Position Recording

Your position may be automatically recorded whenever you log into any *Winlink 2000* participating station if you have your grid square or longitude and latitude filled in on the **Tools**, **Options**, **Settings** form and have **Send QTH** checked. This is handled automatically each time you check into a participating station, wherever you may be located if you have your GPS turned on and connected to your computer via the computer's serial port. You do not have to do anything to be recorded. *Airmail* will handle everything if it is configured for this function. Should it be necessary to determine your location for whatever reason, the *Winlink 2000* sysops will place you where you were when you last connected to the system.

If you wish to publicly post your position reports on the Automatic Position Reporting System (APRS), or on one of the graphic position tracking systems, you may do so manually, or in the form of an automated message when you log into a *Winlink 2000* station. Or you can post at predetermined time intervals. Your family and friends can then view your route on various maps provided by the graphic-tracking systems as you move. They can even request and receive up to 21 days of position information by sending an e-mail request.

Nearby Boats

Also, within the *Airmail* **Global** catalog is the ability to download the 30 nearest users listed on APRS. Why might this be important? Let's assume you have left the Panama Canal and are headed for the Marquesas Islands. You expect to be at sea for at least 21 days. When you are 1700 miles out, you develop a problem with your rudder, or perhaps your rigging or any number of other significant problems. You are not in immediate danger, but you need help. Knowing that you started about the same time as several other vessels, you log into a *Winlink 2000* participating station and request the **Nearby User** list. Upon receiving it, you discover that there are three other vessels within 90 miles of your current position who are making the same voyage, and they too are using *Winlink 2000* for communications and weather.

You could send them an e-mail requesting assistance or you might try contacting them directly via VHF marine radio. Or you might check into one of the many SSB maritime networks to request help from the boats whose positions are known to you. Obviously, most of us will not have such an extreme need but the information is there regardless of where you are or under what circumstances you request the information. You could also just be curious about other maritime amateurs that you have met along the way, but have not been able to contact and wish to e-mail.

You may also request the position of any Amateur Radio operator who is also using the *Winlink 2000* position-reporting system. There is an easy to use template within *Airmail* for sending your own position report as well as for receiving the position of others who are also reporting on a periodic basis. Click on the *Airmail* menu **Window**, **Position Reports** to bring up the information in **Fig 2-14** or **Fig 2-15**.

Finding *Winlink 2000* Stations

Now that you have some basic information about how to send e-mail, request and receive weather information, update or replace the catalog, send and receive position reports and set yourself up to be tracked automatically by the *Winlink 2000* system, it is time to discuss how you find the *Winlink 2000* participating shoreside stations necessary to make all this happen and how to update the *Winlink 2000* participating station (RMS) list. Then you need to know how to use the system when you have an Internet connection available.

Finding participating *Winlink 2000* Radio Message Server stations while cruising on your boat is just like finding any other station at a specific location. The difference is

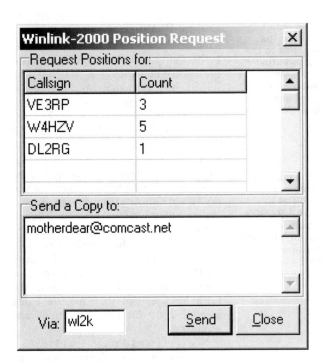

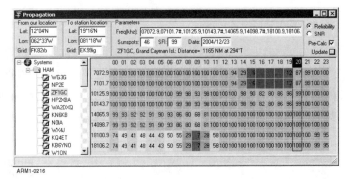

Fig 2-16—By integrating the *ICEPAC* propagation program engine into *Airmail*, *Winlink 2000* participating stations can be more easily found.

Fig 2-15—Requesting your own position or that of others is easy and you may send your position report to other e-mail recipients.

that there are many different RMS location, and each is scanning a variety of frequencies. On **www.winlink.org/stations.htm,** you will find an up-to-date, detailed list maintained by ZS5S. This is the same list that may be updated upon request to *Airmail* over the air. *Airmail* also contains a list of participating stations and it also lists other HF BBS stations that have nothing to do with *Winlink 2000*. Check the *Airmail* menu **View, Frequency List, Text Format** to see the station list. If is not the same date as the one viewed on the ZS5S link you should update it.

Go to the menu **Window, Catalogs, WL2K, Global** and within that folder, find **PMBO** (ZS5S_BULLS category). Check the checkbox, and the request will be sent to a participating station (RMS) the next time you check into that station. Once you have retrieved the list from any participating station, you should go into your **Inbox** and open the message with the list in it. With the list opened or the message highlighted, go to menu item **Tools, Make Frequency List**. You should see the same list come up in the frequency-list form. Update and save the list by clicking on the **Update** and **Save** buttons at the bottom of the frequency-list form. Now that the list is current, you may use it with some assurance that the station you are calling is actually scanning the frequency that you have chosen.

But how do you know what frequencies to use? Most of us rely on our own experience and intuition about how HF works to choose the station and frequency. But

there is a very powerful and helpful aide that is available for finding available *Winlink 2000* stations.

Propagation Predicting

The built-in *ICEPAC* propagation engine driving the *Airmail* propagation module can be useful regardless of the amount of HF experience you might have. Let's explore this powerful tool. **Fig 2-16** shows the *Airmail* propagation module. Since all listed *Winlink 2000* RMS stations have their grid squares (and longitude and latitude) known, the only other variable needed is your own location. Once this is established, the chart is updated in several ways:

• By your clicking the **Update** command button when you find a *Winlink 2000* participating RMS station you wish to use.

• Once you have connected to any of these stations, the current 10-cm Solar Flux Index (SFI) will be automatically loaded into the *Airmail* propagation module.

• *Airmail* also displays the time and date of the last SFI download so that you can judge the initial value of the number that is present when you first use the program and before your first connection to a participating station.

To select a station for the propagation module, go to the *Airmail* menu **View, Propagation**, or press function key F8. Click on **Ham** to bring up all listed stations, which *Airmail* will automatically

sort in order by increasing distance from your position.

In Fig 2-16 notice that the boat's location [Maiden grid FK82rb] is near Trinidad and the chosen *Winlink 2000* station, ZF1GC, is located in the Cayman Islands. This appears to be a great choice with many options for connection. Obviously, the first priority is "just look for the green" but there are some subtleties that can add reliability and speed. Notice the time 2000 UTC is highlighted on all bands. That is the current time in this example. Also, notice that 7, 10, 14 and 18 MHz frequencies are all available for PACTOR II or PACTOR III connections (PACTOR III is indicated by "#" after the frequency). And they will all probably work perfectly, assuming that another station isn't using the chosen frequency.

Of the available options what should be the first choice? Experience dictates that the higher you can go toward the Maximum Usable Frequency (MUF), the better signal-to-noise ratio you will experience. This usually results in a speedier connection for a given strength of signal. In addition, *multipath distortion* from signals arriving at your receiver on different paths through the ionosphere is likely to be less on the higher frequencies.

At the time I am writing this, I find that 17 meters (18 MHz) seems to favor Northeast-Southwest pathways. This is subjective data and it may not be substantiated at any other moment in time. But over the last several months, this has been the case at *Winlink 2000* stations in the US, Central America and Caribbean. The point is that my own experience, combined with what I can obtain from the *ICEPAC* propagation engine, is going to make it much easier to continually connect to desired stations than if I relied only on my own intuition.

But what if the PACTOR III frequency is busy or does not call up the chosen station? My advice is to stay with PACTOR III and move down to the next lower band if PACTOR III is available there. If not, find another PACTOR III PMBO that has a clear frequency near the MUF and call. Remember that PACTOR III has an average rate of over 2400 bps, and on a good connection, can reach 3600 bps. PACTOR II has a maximum rate of 800 bps. With PACTOR III, lower frequencies and lesser signal strength may still easily override a stronger PACTOR II signal. Again, PACTOR III capability in the frequency list is shown with "#" after the listed frequency. Always depend on the list at the top of the propagation module form for actual frequencies since some stations have more than one frequency per band, but only one frequency per band is listed in the chart.

Those with little-to-no experience with HF propagation will most likely use this module more often than those who have HF operating experience. Therefore, in my opinion, those with less experience will likely have better connectivity and data throughput rates than those who like to guess!

Telenet

The Telnet function of *Airmail* is most convenient for times when you're not near your target shore station, but wish to operate as though you were using the radio. Actually, however, you are using the Internet, and operating at Internet speeds. The Internet is becoming increasingly available to coastal regions through cell-phone connectivity, as well as wireless Internet (802.11b) that may be found in many marinas throughout the world.

The Telnet module is easily to configure, and has the look and feel of *Airmail* when it used over radio. In order to be accepted as telnet client, you must first have been logged into a *Winlink 2000* participating radio station over the air. When you do log in over the air, you are immediately self-registered, your call is checked and is approved, if valid. Thereafter you may log into the system via the *Airmail* Telnet client.

To find the Telnet client module, go to menu **Tools**, **Options**, **Modules** and check the appropriate boxes. Thereafter, you will find the module available on the main menu

under **Modules**. Specific instructions are available in the *Airmail* **Catalog** under the Global Catalog's **WL2K_HELP** category that will give you the configuration information necessary to log into the various participating stations Telnet servers. Of course, you must check into a *Winlink 2000* participating station over the radio and make the request before you receive the information. Remember, this is first and foremost a *radio* e-mail system, and not a substitute for a normal ISP!

This brief tour of the functions and features of *Winlink 2000* through an *Airmail* perspective is by no means meant to completely prepare you for any sort of offshore operation. Hopefully, however, it has provided you enough information to get on the air and give it a whirl.

Web Browser Password-Protected Access to *Winlink 2000*

The Web Browser password protected Access to *Winlink 2000* radio e-mail is limited to text, and meant for those who are in locations where they cannot use their own computers to gain access to their messages. It does not allow attachments, nor can you request weather bulletins. It is usually used in Internet cafes or other instances when you are not on your boat and have no Internet connectivity to your own computer.

OPERATING MARITIME HAM RADIO JUST FOR FUN

Just about every ham gets a thrill out of communicating with someone who is operating from a boat underway. Depending on where you are, you may create a "pileup" of stations calling you, or you may just be able to relax and have a great QSO. When conditions are not optimal, PSK-31 and CW are usually the best modes for making contact rather than SSB. Of course, this depends on the same factors that effect any amateur operations.

While underway, however, SSB may be easier to manage since other modes do take a bit more concentration when attention may be needed elsewhere. A word of caution: When your vessel is in or around harbors, yacht clubs and other areas where vessels are concentrated, be certain to first inquire about "radio quiet times." These are times when a group of vessels may all be waiting to check into some service network, or times when weather or other marine broadcasts may be scheduled. You don't want to blast them with your fun-time transmissions when they are listening carefully for something else altogether!

Operating CW

When operating CW the "/MM" after a call always attracts attention. People seem to put up with a weaker signal to talk to someone adventuresome enough to be on the high seas. If you are a CW aficionado you will want to consider having a true amateur transceiver rather than a marine transceiver, since you will not have the flexibility to adjust the CW signal compared to an amateur transceiver. By reviewing Appendix 1, you will note that there are several CW networks for those who wish to participate.

Other Digital Modes

PSK-31 and other such real-time conversational digital modes may be used, but may be more subject to QSB (fading) and multipath distortion than either SSB or CW, which do not seem to have an effect on the human ear. Even so, there should be no real issue with all but really marginal conditions. With today's soundcard technology and digital protocols, these modes may be excellent for having a QSO while operating maritime mobile.

PSK-31 is by far the most popular and easiest to find. There are several excellent programs available for operating multiple digital modes, including CW, and many of these programs may be found in *QST* articles from the ARRL Web site at **www.arrl.org**, as well as the *ARRL HF Digital Handbook*.

A single-sideband conversation while underway is obviously easier to manage than typing away at your keyboard on a digital mode. SSB is a marvelous way to keep up with those friends that are on land as well as fellow cruisers. All maritime networks on HF use USB, while the amateur maritime nets are on the sideband that is normally used on any particular amateur band.

As with CW, other people seem to be drawn to those who are operating SSB from a vessel underway or who are anchored in some exotic location. They will put up with a marginal signal just to be able to talk with a maritime-mobile station.

Many hours of enjoyment are available to those who are familiar with the marine and amateur SSB maritime networks. As you move around, you will certainly find that the people you meet will be checking into the same networks. They will be happy to move off and have a one-to-one QSO with you. Talking to those who have been where you are going, or who can benefit by your own information from places you have been, offer hours of enjoyment, and will provide the comfort of knowing that there are others out there sharing the same exciting adventures.

VHF/UHF FM

VHF/UHF FM activities are primarily used in two places: between nearby vessels underway, and in coastal areas where amateur repeaters are available. Marine VHF frequencies are used more often than amateur VHF/UHF frequencies, especially between vessels, and when communicating with harbors and other boats. So having the capability to use both amateur and marine VHF frequencies helps keep communications open is important.

LET'S GET STARTED!

The old saying *"Thank goodness I installed my burglar alarm before I got robbed!"* is also applicable when it comes to any installation and operation offshore. It always amazes me to find the number of people seeking help with their installation and operation after they get underway, rather than before. It is certainly wise to install and learn how to operate your station before you get going. Before you set sail you have time to adjust grounding, radio equipment, take care of interference problems and other station parameters. Being familiar with your station's operation also allows you to concentrate on piloting the boat.

Each year for the last six years, the *Winlink 2000* Development Team travels to Melbourne, Florida, at the invitation of the Seven Seas Cruising Association to participate in a two-day how-to lab session for members experienced in all types of maritime communications, as well as new members interested in cruising as maritime Amateur Radio operators. Much of the material covered in the SSCA communications lab sessions has been presented in this chapter. The material is based on the highest-priority concerns of those experienced amateurs regarding good maritime mobile installations and operations. It certainly is not meant to be an all encompassing how-to guide, but should help get you kick started.

Other Sources of Information

For non-amateur VHF/UHF, as well as HF Marine frequencies and other such valuable information, visit the US Coast Guard Navigation Center Marine Telecommunications site, **www.navcen.uscg.gov/marcomms/**. There is a wealth of navigation information at this site, and I highly recommend that you become familiar with the information presented there.

For non-amateur maritime weather-related networks and broadcasts, the National Oceanic and Atmospheric Administration (NOAA) Web site, **www.nws.noaa.gov/om/marine/commar.htm**, also offers valuable information.

Making Friends

Most maritime amateurs are interested in meeting other maritime amateurs. In fact, there are several maritime associations that are quite active with Amateur-Radio activities, including cocktail hour nets and other such activities that bring maritime hams together. Depending on where you are harbored, you may find that many of those owning vessels are also hams. The more you associate with those who spend most of their time offshore, the more you will find hams aboard. Full-time boaters, who actually live on their vessels have increased in numbers since the ".dot net" generation started retiring. These people are interested in socializing with others who do similar activities. Let's explore several organizations.

United States Power Squadrons

The United States Power Squadrons is the world's largest non-profit, recreational boating organization. Established in 1914, USPS has some 55,000 members organized into 450 squadrons across the United States and in U.S. territories. About 12% of its members are licensed amateur radio operators, and many are members of ARRL. Most members of this organization are near-shore rather than bluewater cruisers, since their emphasis is on power boating. However, there are a number who do go offshore, and they must take maritime communications as seriously as those who circumnavigate the world in a sailboat.

In conjunction with the Canadian Power Squadron, USPS operates a net on 14.287 MHz every Saturday at 1700 UTC. This net is primarily a conversational net; however, it is structured to handle message traffic or emergencies also. Maritime-mobile stations are encouraged to check into the net to file float plans, pass traffic or join in a discussion of marine related topics. There is also a once-a-month 10-meter net that meets on the first Saturday of each month at 1800 UTC on 28.357 MHz. In June 2005, USPS and ARRL will sign a Memorandum of Agreement to encourage and promote the related activities within both organizations.

Seven Seas Cruising Association

The Seven Seas Cruising Association (SSCA) is an international organization formed in March 1952, at Coronado, California, by a group of cruisers living aboard six seagoing sailboats. Their purpose was to share cruising experiences and information through the medium of a monthly bulletin.

Many years later, the SSCA still provides information about all facets of cruising to its members on a monthly basis in a publication called the *Commodores' Bulletin*. The *Bulletin* does not contain advertising, nor does it have stories by professional writers. The stories in the *Bulletin* are real accounts of the adventures of SSCA members in their own words. The stories include practical information on provisioning in foreign ports, procedures for clearing in and out, recommendations for cruising guides and charts of an area, depth and bottom type to facilitate anchoring, weather tips and any other facts that can help turn your cruising dream into reality. Several of the founding members still belong to SSCA and share their knowledge, accumulated over a lifetime, with current members.

Currently the SSCA has more than 10,000 members. This forms a caring, supportive family that reaches out with international friendship and goodwill. Members are committed to leaving a *clean wake* by treating others and our environment with profound respect. There is an incredible camaraderie engendered by being one of those who brave the elements of the oceans in small boats. As an SSCA member, you have an option to contribute your time, energy and knowledge to the association.

All members join as Associates, pay the same dues, receive the 50-page monthly *Commodore's Bulletin* and partake in all of the organization's functions. Those members who have lived aboard their cruising sailboat for at least a year, who have fulfilled a sailing-distance requirement and who have been sponsored by two current Commodores, may be elected as voting-member Commodores. If a Commodore no longer resides on his boat for more than three months he becomes a Rear Commodore and retains the voting privileges.

SSCA members fly a distinctive burgee to identify themselves to others coming into an anchorage. To see an SSCA flag flying is to know that fellowship is at hand. Members help members and also organize great parties!

The SSCA holds several events each year, one in Annapolis, Maryland, and another in Melbourne, Florida. These gatherings draw a sizable crowd and are an excellent place to meet new friends, share tales and explore the cruising lifestyle. The SSCA is very pro-ham radio, and about 30% of the members who attend the gatherings in Melbourne, Florida, have obtained their ham licenses. In fact, SSCA has a VEC available for obtaining a ham license at their events, as well as an all-day lab how-to session for those interested in deploying *Winlink 2000*. Learn more by visiting **www.ssca.org**.

Waterway Radio and Cruising Club

The Waterway Radio and Cruising Club is another association that supports several nets, both formal and informal. If you live in the eastern half of the country, you may wish to consider joining the WRCC. They publish a monthly newsletter named the *Scuttlebutt*.

They hold a daily Waterway Net, meeting at 0745 ET on 7268.0 kHz LSB. WRCC also sponsors several gatherings, including an annual picnic the day immediately preceding the SSCA gathering in Melbourne, Florida. Check out WRCC at: **www.waterwayradio.net/**.

West Coast Nets

On the US West Coast, you may wish to check into various Amateur Radio SSB nets to get a feeling for what is going in the cruising community there. There are several interesting informal networks:

- The *Manana Net* is an Amateur Radio service net serving maritime mobiles in the Sea of Cortez, the coast of mainland Mexico, South America, Alaska and as far out into the Pacific as possible. The net operates on 14340.0 kHz. Early-bird session of the net starts at 1830 UTC with the net opening at 1900 UTC, Monday through Saturday (but not on Sunday). The early-bird session is there for you to meet and greet other maritime amateurs. Further information may be found by visiting **www.geocities.com/TheTropics/3989/**.
- The *Sonrisa Net* at 1400 UTC on 3968.0 kHz.
- The *Chubasco Net* at 1530 UTC during Standard Time, 1445 UTC during Daylight Saving Time, at 7294.0 kHz.
- The *Baja Net* at 1530 UTC on 7238.0 kHz.

Fig 2-17—Photo showing a typical group in Georgetown, Bahamas, gathering on Monument Beach for their daily ham study group. *(Photo courtesy Mary Jean Tedford, KE4WHA.)*

- The *Happy-Hour Net* at 0000 UTC on 3968 kHz.

More Than Hamming on the Water

Operating a maritime mobile station usually involves offshore sailing or motor yachting, which in turn indicates a particular lifestyle. There are thousands of individuals out there at any moment, boating and hamming for their own safety and pleasure. Amateur Radio has become a part of the cruising lifestyle. In many areas of the world where you find serious boaters, you will find ham classes conducted by mariners for mariners. **Fig 2-17** show a typical group in Georgetown, Bahamas, gathering on Monument Beach for their daily ham study group. Similar activities take place continually on the island of Trinidad, on the West Coast of Mexico in various harbors, and in other such places where mariners gather.

As a mariner, you will find that Amateur Radio is yet another common interest you can share with others as you travel. There is a special bond between mariners. Amateur Radio plays an important role not only in strengthening this bond, but also contributes greatly to the safety and well-being of those who travel, while adding comfort to their family and friends.

Amateur Radio Maritime Mobile Networks

Time (UTC, Z)	Freq (kHz)	Name	Days	Area	Notes	Contact
1600-0200	14300.0	Mar. Mobile Service Net	D	CAR,PAC		KE4EDX
0100	3925.0	Gulf Coast Hurricane Net	D	GC	wx, tfc	WD5CRR
0130-0300	28313.0	10M MM Net	D	EPAC,Haw		N6URW
0130	7126.0	MM Fast CW Net				
0200	21402.0	Gerry's Happy Hour	M-F	PAC, Baja	soc	
0200	14334.0	Brazil/East Coast Net	D	EC, ATL	wx, tfc	K3UWJ
0200	3932.0	Great Lakes Emergency	D	GL	wx, tfc	WD8ROK
0200	7126.0	MM Slow CW Net				
0200-0400	14300.0	Seafarer's Net	D	PAC, WC	MM	WA6ZEL
0300	14313.0/14300.0	PAC Maritime Net Warmup	D	PAC	WU	KH6UY
0330	14313.0/14300.0	PAC Maritime Net	D	PAC	MM, rc	KH6UY
0330	7294.0	Sandia Net	D	WC, Baja	soc	KA6HFG
0400	14115.0	Canadian DDD Net	D	PAC	also 1730Z	VE7DB
0400	14318.0	Arnold's Net	D	SPAC	MM, wx	ZK1DB
0500	21200.0	UK/NZ/African Net	D	PAC, IOC	MM	VK3PA
0630	14316.0/14105.0	S. African Maritime Net	D	SATL, IOC	also 1130Z	ZS5GC
0630	14313.0	International MM Net	D	ATL, MED, CAR	also 1700Z	DKØCM
0700	7235.0	HHH Net				
0700-0000	7085.0	MED, SEA Cruisers Net	D	MED, SEA	MM	
0715	3820.0	Bay of Islands Net	D	NZ, AUS, PAC	MM	ZL1BKD
0800-0830	14315.0	Pacific Inter-Island Net	D	SPAC, SEA	tfc	KX6QU
0800	14303.0	UK Maritime Net	D	ATL, MED	also 1800Z	G4FRN, Bill
0900	14313.0	MED, SEA MM Net	D	MED	MM	5B4MM
1000	14320.0	South China Sea Net	D	SPAC		
1030	3815.0	Caribbean Weather Net	D	CAR	1045, 2230Z	VP2AYL
1045	7163.0	Caribbean WX Net	D	CAR	1030, 2230Z	
1100	3770.0	Mar Provinces WX Net	M-Sa	NE Canada	wx	VE1AAC
1100-1200	7237.0/7241.0	Caribbean MM Net	D	CAR	MM	KV4JC
1100	14300.0/14313.0	Intercon Net	D	NA, CA, SA	tfc, & 2200Z	K4PT
1100	14283.0	Carribus Traffic Net	D	EC, CAR	tfc	KA2CPA
1110	3930.0	Puerto Rico WX Net	D	PR/VI	wx & 2310Z	KP4AET
1115	14316.0	Indian Ocean MM Net	D	WPAC, IOC		VK6HH
1130	14316.0	South African MM Net	D	SATL, IOC	0030, 0630Z	ZS5MU
1130	21325.0	South Atlantic Roundtable	D	SATL	tfc & 2330Z	PY1ZAK

Time (UTC, Z)	Freq (kHz)	Name	Days	Area	Notes	Contact
1200	14121.0	Mississauga Net	D	E Canada	ATL wx	VE3NBL
1200	14320.0	Southeast Asia Net	D	SEA, SPAC		WB8JDR
1230	7185.0	Barbados Info Net	D	CAR	tfc	8R6DH
1245	7268.0	East Coast Waterway Net	D	EC, CAR	MM	NU4P
1300-1400	21400.0	Trans Atlantic MM Net	D	NATL, CAR	rc, wx	8P6QM
1300	7083.0	Cen. Am. Breakfast Club	D	CA	MM, soc	AB6HR
1415	3968.0	Sonrisa Net	D	Baja, So. Calif.	MM	XE2VKB
1400-1600	7263.0/7268.0	Rocky Mtn RV Net	D	Mid West	tfc	K5DGZ
1430	7294.0	Chubasco Net	D	Baja, So. Calif.	MM	XE2/N6OAH
1600	7238.0	Baja Calif. MM Net	D	Baja, So. Calif.	MM	N6ADJ
1600	7200.0/7268.0	Taco Net	D	Baja		
1600-2200	14300.0/14313.0	Maritime Mobile Service Net	D	ATL, CAR, PAC	also 2400Z	KA8O
1600-1800	7263.0/7268.0	Pacific RV Service Net	M-F	Western US	tfc	K6BYP
1630	14303.0	Swedish Maritime Net	D	MM	0530, 2030	
1630	14313.0	INTERMAR eV, German Maritime Mobile Net	D	MM		DLØIMA
1700-1800	14308.0	RV Service Net	D	US	tfc	KB1Z
1700	14323.0	US/Canada Power Sqdn Net	Sa	US/Canada	tfc	W7LOE
1700	14340.0	Cal Hawaii Net	D	EPAC	tfc	K6VDV
1700	7240.0	Bejuka Net	M-F	CA	tfc	HP3XWB
1700-1900	14280.0	Inter Mission RA Net	M-Sa	CA, SA	tfc	WA2KUX
1730	14115.0	Canadian DDD Net	M-F	PAC	MM 0400Z	VE7CEM
1800	7076.0	South Pacific Cruising Net	D	SPAC	MM, wx	informal
1830	14340.0	Manana Net Warmup	M-Sa	WC, EPAC	MM, WU	KB5HA
1900	14297.5	Italian Amateur Radio Maritime Mobile Net	D	MM		ON6BG, AA3GZ
1900	14285.0	Kaffee Klatch Un-Net	MWSa	Hawaii, Tahiti	MM, news	KH6S
1900	14305.0	Confusion Net	M-F	PAC	tfc	W7GYR
1900	7285.0	Hawaii AM Net	D	Hawaii	tfc, wx	KH6BF
1900-2000	21390.0	Halo Net	D	NA, SA	tfc	WA4FXR
1900	14329.0	Bay of Islands Net	D	NZ, SPAC	MM	ZL1BKD
2000	7080.0	New Zealand WX Net	D	NZ	MM, wx	ZL1BTQ
2000	7095.0	Harry's Net	D	WPAC, SPAC	MM	KL7MZ
2000	14260.0	Partyline Net	D	CAR, CA		N3QVW
2000-2200	21390.0	Inter-American Traffic Net	D	NA, CA, SA	tfc	WD4HAY
2030	14303.0	Swedish Maritime Net	D	ATL	0530, 1630Z	
2030	14315.0	Tony's Net Warmup	D	NZ, SPAC	MM, WU	ZL1ATE

Time (UTC, Z)	Freq (kHz)	Name	Days	Area	Notes	Contact
2100	14261.0	Ben's Friends MM Un-Net		EC	MM, soc	K3BC
2100	14315.0	Tony's Net	D	NZ, SPAC	wx, 2130Z	ZL1ATE
2100	14113.0	Mickey Mouse Connection	D	SATL, SPAC	MM	CX9ABE
2130	14290.0	EC Waterway Net		EC, US	MM	
2200-2230	3963.0	EC Recreational Vehicle Net	D	EC, US	also 1300Z	KB1Z
2200	21402.0	Pacific Maritime Net	M-F	PAC, Baja	MM	KB7DHQ
2200	21412.0	Maritime Mobile Service Net	M-F	PAC	MM	KA6GWZ
2200-2400	14300.0/14313.0	Intercon Net	D	NA, CA, SA	also 1100Z	K4PT
2230	3815.0	Caribbean WX Net		CAR	wx & 1030Z	VP2AYL
2310	3930.0	Puerto Rico WX Net	D	PR, VI	wx & 1110Z	KP4AET
2330	21325.0	South Atlantic Roundtable		SATL	tfc, soc	
0000-0200	14300.0/14313.0	Maritime Mobile Service Net	D	CAR, Baja, PAC	MM	
0000	3968.0	Happy Hour Net	D	Baja, W. Mexico	MM, soc	
0000	14320.0	Southeast Asia MM Net	D	PAC, SEA	MM	
variable	14325.0/14275.0 /14175.0	Hurricane Watch Net		ATL, CAR, PAC	hurr. track	KØIND

Abbreviations:

ATL - Atlantic, AUS - Australia, Baja - Baja California, CA - Central America, CAR - Caribbean, D - Daily, EC - East Coast, ECAR - Eastern Caribbean, EPAC - Eastern Pacific, GC - Gulf Coast, GL - Great Lakes, IOC - Indian Ocean, MED - Mediterranean, MM - Maritime Mobile, NA - North America, NATL - North Atlantic, NZ - New Zealand, PAC - Pacific, PR - Puerto Rico, rc - roll call, SA - South America, SATL - South Atlantic, SEA - Southeast Asia, soc - socializing, SPAC - South Pacific, tfc - traffic, VI - Virgin Islands, WC - West Coast, WU - Warm-up session, check-in, wx - weather, Z -Z (Zulu zone)

International Boat Watch Network

Organization	Frequency kHz	Time (UTC, Z)	Coverage	Web URL
Winlink 2000 (e-mail)	Multiple Gateways	24/7	World Wide Coverage	www.winlink.org
Sail the World French World Sailing Assoc.	Multiple Gateways		3900+ members, 850 boats; 450 boats currently sailing	www.stw.fr
Mediterranean M/M Net	7085.0	0700Z	Med. Coverage	
U.K. Maritime Net	14303.0	0800/1800Z	Eastern Atlantic	
Trans-Atlantic MM Net	21400.0	1300Z	Atlantic	
S.A. M/M Net	14316.0/ 7045.0	0630/1130Z	Indian Ocean / S. Atl.	
Caribbean Emerg. & Weather Net	7162.0/ 3815.0	1030/2230Z	E. Caribbean / Atl.	
Caribbean M/M Net, St. Croix, US V.I.	7241.0	1100Z	Caribbean	www.viacess.net/~kv4jc
Mississauga Net	14121.0	1245Z	Canada/Bermuda	
Intercon. Traffic Net	14300.0	0700-1200 ET	U.S./Carib./Cen. Amer./W. Atl.	www.interconnect.org
Maritime Mobile Service Net	14300.0	1200-2200 ET	U.S./Carib./Cen. Amer./W. Atl	www.mmsn.org
Waterway Radio & Cruising Club	7268.0	0745 ET	U.S. East Coast, Carib.	www.waterwayradio.net
Pacific Seafarers Net	14313.0	0315Z	Pacific	www.pacsea.net
Maritime Emerg. Net	14310.0	0400/1800Z	Pacific	
Baja California MM Net	7238.0	0800 PDT	Coastal Baja / Calif	
Chubasco Net	7294.0	1530 PST 1430 PDST	Mexico West Coast	
Comedy Net	7087.0	2040Z	SW Pac., New Caledonia, Vanuatu, New Zealand, Australian E Coast	
Sonrisa Net	3968.0	1415Z	Baja/SW US/Mex	
Manana Net	14340.0	1830Z	E. Pac/AK to Panama/Hawaii	
Happy Hour Net	3969.0	0000Z	Baja / SW US / Mex.	
Mobile Maritime Net	14323.0	0025Z daily	S/E Asia	www.cruiser.co.za
Pacific Inter-Isl. Net	14315.0	0800Z	Micronesia to Hawaii	www.cruiser.co.za/radionet.asp
Pac. M/M Service Net	21412.0	0200Z	Pacific World Wide	

Organization	Frequency kHz	Time (UTC, Z)	Coverage	Web URL
Roy's Net, Perth	14320.0	1115Z	N&W Indian Ocean	
Robby's Net (Australia)	14315.0	1000/2300Z	South Pacific	
Tony's Net (NZ)	14315.0	2100Z	South Pacific/Australia	
John's Weather Net	14315.0	2140Z	SW Pacific	
Tony's Net (Kenya)	14316.0	0500Z	Indian Ocean/Red Sea	
S.E Asia M/M Net	14320.0	0630/1130Z	Indian Ocean / S. Atl.	
Le Réseau du Capitaine	14118.0	7am Montreal time, daily	Med, E Pac, Indian, Atl, Carib, Bermuda	http://lereseauducapitaine.qc.ca & va2af@videotron.ca
INTERMAR (Germany)	14313.0	1630Z daily	Europe	www.intermar-ev.de & www.eu.mmsn.org

Non-Ham Boat Watch Member Networks

Name	Freq.(kHz)	Time	Area	Web URL
Herb Hilgenberg's Southbound ll Net	12359.0 SSB	2000Z	Detailed Wx, Atlantic, Caribbean & Pacific	http://hometown.aol.com/hehilgen/ myhomepage/vacation.html
Cruiseheimer's Net	8152.0 SSB	0830 ET	U.S. East Coast to Eastern Caribbean	
Caribbean Wx Cntr Daily Net	8104.0 SSB	1230Z	Caribbean and SW N. Atlantic/Bahamas	www.caribwx.com
Caribbean Safety and Security Net	8104.0 SSB	1215Z	Caribbean and SWN. Atlan.	boatmillie@aol.com
N.W. Caribbean Marine SSB Net	8188.0 SSB	1400Z	Western Carib. to Panama & Gulf of Mexico	
Panama Canal Connection Net	8107.0 kHz	1330Z	Eastern Pacific & SW Caribbean	
Russel Radio, New Zealand	12359.0/ 12353.0	0830/1630/ 0915/1600 NZT	Bora Bora to Australia	
BASRA (Bahamas Air-Sea Rescue Association)	4003.0 USB VHF CH 72 7096.0	0700/0715 Nassau/ 0720 ET	Wx and Emergency traffic thru Bahamas	www.basra.org see also http://caribbeansearchandrescue. freeservers.com

* Owned and Managed by Mike Pilgrim, Amateur Radio K5MP

Last updated: December 2004

Top 100 Marine/Ham Channels/Frequencies, by Gordon West (WB6NOA)

Category	Frequency (kHz) Rx/Tx	Mode	Information
DISTRESS	2182.0	USB	International Distress and Calling
HAM	3963.0	LSB	RV Net, daybreak
	3968.0	LSB	Sonrisa Mexico Net, early AM
	3964.0	LSB	East Coast Waterway Net, early AM
	7238.0	LSB	Baja California Net, 0800 PxT (alternate: 7260.0)
	7241.0	LSB	Caribbean Net, early AM
	7268.0	LSB	East Coast Waterway Net, 1145Z
	7294.0	LSB	Chubasco Mexico Net, 0730 PxT
	7268.5	LSB	RV Net, 0800 PxT
	14300.0	USB	Intercontinental Marine Net, 24 hours
	14313.0	USB	Daily Marine, evening Pacific Maritime Net, 0400Z
	14340.0	USB	Manana Net, 1830Z, M-Sa
	14263.0	USB	Noontime RV Net
	14302.5	USB	18 Wheelers, friendly daytime radio checks
	14325.0	USB	Hurricane nets, when needed
	21390.0	USB	East Coast Halo Net, days
	21402.0	USB	Pacific Maritime Net, 2200Z
	21412.0	USB	Pacific Maritime Mobile Service Net, 0200Z
	21415.0	USB	Evening Happy Hour maritime mobile net
	28400.0	USB	Center of 10-meter ham band
	7294.0	LSB	Sandia Net, evenings
	3856.0	LSB	Taco Net, evenings
	3815.0	LSB	South Pacific net, evenings
	7250.0	LSB	Gordo West, 0845 PxT, most mornings
TIME	2500.0	AM	WWV Time/WX: 13 min. , propagation:18 min. after hour
	5000.0	AM	WWV Time/WX: 13 min. , propagation:18 min. after hour
	10000.0	AM	WWV Time/WX: 13 min. , propagation:18 min. after hour
	15000.0	AM	WWV Time/WX: 13 min. , propagation:18 min. after hour
	20000.0	AM	WWV Time/WX: 13 min. , propagation:18 min. after hour
FAX	4344.1	USB	Pacific Coast weather facsimile
	8680.1	USB	Pacific Coast weather facsimile
	12588.6	USB	Pacific Coast weather facsimile
	17149.3	USB	Pacific Coast weather facsimile
	9980.6	USB	Honolulu weather facsimile
	11088.1	USB	Honolulu weather facsimile
	16133.1	USB	Honolulu weather facsimile
	12751.1	USB	Canada West Coast weather facsimile
	10534.1	USB	Canada East Coast weather facsimile
	13508.1	USB	Canada East Coast weather facsimile
	6338.6	USB	Boston weather facsimile
	9108.1	USB	Boston weather facsimile
	12748.1	USB	Boston weather facsimile
	4316.0	USB	New Orleans Gulf weather facsimile
	8502.0	USB	New Orleans Gulf weather facsimile
	12788.0	USB	New Orleans Gulf weather facsimile
	11028.0	USB	Melbourne Australia weather facsimile
	13549.0	USB	Aukland New Zealand weather facsimile
	10553.1	USB	Darwin Australia weather facsimile
USCG	2182.0	USB	International Distress & Calling
	2670.0	USB	Coast Guard SSB, short range
	4125.0	USB	HF Channel 4 Safety, mon. by USCG, Canad. weather, 400 mi.
	6215.0	USB	HF Channel 6 Safety, mon. by USCG, 1000 mi.
	8291.0	USB	HF Channel 8 Safety, mon. by USCG, 2000 mi.
	12290.0	USB	HF Channel 12 Safety, mon. by USCG, 4000 mi.
	16420.0	USB	HF Channel 16 Safety, on-request to USCG, 4000 mi.

Category	Frequency (kHz) Rx/Tx	Mode	Information
SHIP-TO-AIR	8843.0	USB	Ship-to-Aircraft, Distress Only
SHIP-TO-SHIP	2638.0	USB	Ship-to-Ship, short range
	4146.0	USB	HF Channel 4A, Ship-to-Ship
	4149.0	USB	HF Channel 4B, Ship-to-Ship
	4417.0	USB	HF Channel 4C, Ship-to-Ship
	6224.0	USB	HF Channel 6A, Ship-to-Ship
	6227.0	USB	HF Channel 6B, Ship-to-Ship
	6203.0	USB	HF Channel 6C, Ship-to-Ship
	8294.0	USB	HF Channel 8A, Ship-to-Ship
	8297.0	USB	HF Channel 8B, Ship-to-Ship
	12353.0	USB	HF Channel 12A, Ship-to-Ship
	12356.0	USB	HF Channel 12B, Ship-to-Ship
	12359.0	USB	HF Channel 12C, Ship-to-Ship
	16528.0	USB	HF Channel 16A, Ship-to-Ship
	16531.0	USB	HF Channel 16B, Ship-to-Ship
	16534.0	USB	HF Channel 16C, Ship-to-Ship
MEXICAN NETS	4060.0	USB	Baja southbound evening net
	4051.0	USB	Baja southbound evening net
	8104.0	USB	Baja southbound early morning net
	6212.0	USB	Picante Net, 6 AM (local Mexico)
	8116.0	USB	Amiga Net, 7 AM (local Mexico)
	4054.0	USB	Southbound, 6 PM (local Mexico)
	6209.0	USB	Bluewater, 7 PM (local Mexico)
TELEPHONE	4396.0/4104.0	USB	WLO Telephone Channel 414
	8788.0/8264.0	USB	WLO Telephone Channel 824
	8803.0/8279.0	USB	WLO Telephone Channel 829
	8806.0/8282.0	USB	WLO Telephone Channel 830
	13110.0/12263.0	USB	WLO Telephone Channel 1212
	13149.0/12302.0	USB	WLO Telephone Channel 1225
	13152.0/12305.0	USB	WLO Telephone Channel 1226
	17260.0/16378.0	USB	WLO Telephone Channel 1607
	17362.0/16480.0	USB	WLO Telephone Channel 1641
SWL	5935.0	AM	Shortwave News, tune around for BBC, VOA, etc
	9955.0	AM	Shortwave News, tune around for BBC, VOA, etc
	12172.0	AM	Shortwave News, tune around for BBC, VOA, etc
	13760.0	AM	Shortwave News, tune around for BBC, VOA, etc
	15685.0	AM	Shortwave News, tune around for BBC, VOA, etc
OTHER	27125.0	AM	CB Channel 14, Mexican shore boats
	13282.0	USB	Air tower weather, East Coast
	13270.0	USB	Air tower weather, West Coast
	1070.0	AM	Broadcast band -- tune dial is active on ICOM radios

Aeronautical Mobile

CHAPTER 3

This chapter was written by Dave Martin, W6KOW, who has been a ham since 1953. Dave is a longtime pilot who has had many aeronautical Amateur Radio adventures, some of which he describes here. During a 21-year U.S. Navy career, he occasionally operated both aeronautical and maritime mobile from Navy vessels. After Navy retirement, he returned to journalism and edited *KITPLANES* magazine for 17 years. This subscription and newsstand monthly is for builders and pilots of Experimental-category aircraft. He now operates aero mobile from his two-seat, open-cockpit Spacewalker II homebuilt airplane. (See **Fig 3-1.**)

In the history of Amateur Radio in motion, aeronautical mobile is certainly among the most obscure. Two reasons are apparent: lack of easy access to suitable airborne platforms and few obvious practical benefits. Nearly everyone has a car or truck into which a mobile rig will fit, and—at least until the advent of the ubiquitous cell phone—hams on wheels routinely provided first alert for emergency response to traffic problems and accidents. Thousands of high-seas sailors have relied on marine Amateur Radio for safety-related daily position reports as well as regular contact with family and friends. Contrasting these types of ham operations, aeronautical mobile seems to lack these attractions.

At first glance, aero mobiling seems to fit into the categories of Amateur Radio that may be a corollary of the mountain climber's credo ("I climb it because it is there."). The ham equivalent: "I do it because I can." For amateurs, this category includes moonbounce (EME), meteorscatter DX and communications involving orbiting satellites. These activities are pursued primarily because they are fun.

Fig 3-1—Author Dave Martin, W6KOW, operates VHF aeronautical Amateur Radio from his Spacewalker II.

As in all aspects of the hobby, the practical side of these operations is knowledge gained through experimentation and the satisfaction when it works.

There is one significant and well-known exception to the lack of practical value for aeronautical Amateur Radio. It is the story of how the U.S. Air Force Strategic Air Command boss used airborne ham gear to investigate better long-range communications. We will revisit the story briefly here in this chapter. Several less-well-known aero ham stories with apparent practical value are also featured.

HELP FROM SOME EXPERTS

Many of the aero mobile hams who have contributed to this chapter have devised or built Amateur Radio equipment specifically for ham radio in flight, and most have dealt with antenna issues including efficiency and installation. Two of the contributors, Jim Weir, WX6RST, and Bob Hayos, K6CUK, have particularly strong connections to both aviation and airborne Amateur Radio, including dealing with technical and legal issues of installing and operating the equipment.

Jim Weir is a longtime ham, pilot and flight instructor (airplanes, gliders and instruments). His company, RST Engineering, produced low-cost aviation electronics (avionics), including kits for aircraft audio panels, headsets and other equipment. He was the first and only person to get FCC and FAA approval to produce aircraft-band transceivers in kit form. Starting with single-channel and then six-channel AM transceivers intended initially for use primarily in gliders, he eventually produced a kit for a 720-channel *navcom*, which combines communications and navigation equipment. The approval required the builder to return the completed kit to RST for checkout and calibration.

Jim's company also designed and donated the aircraft antenna systems for the successful around-the-world *Voyager* airplane designed by Burt Rutan (winner of the $10 million X-Prize for the first commercial spacecraft) and flown by Dick Rutan and Jeana Yeager in 1986. For more than 15 years Weir has written the "Aero 'Lectrics" column in *KITPLANES* magazine. He is an FAA-licensed Airframe and Power (A&P) aircraft mechanic and his A&P ticket includes the "IA" designation, which means he has *inspection authorization* to approve the work of other A&P mechanics. The registration number on his Cessna 182 is N73CQ.

Bob Hayos, K6CUK, is a retired USAF pilot and aerospace engineer with a long background in aeronautical mobile Amateur Radio, both in the military and in a series of airplanes he has owned and built. He has a private pilot's license and an FAA mechanic's rating with the IA designation. Bob has owned a large number of light airplanes, the latest of which is a Murphy Rebel, an all-metal two-seater he built from a kit that comes from Canada. See **Figs 3-2** and **3-3**. Bob has installed Amateur Radio equipment in all the aircraft he has owned. He is also a designated airworthiness representative (DAR) for the FAA. He inspects and initially licenses new Experimental-category aircraft before their first flights.

Bob's perception of some of the FAA-related legal issues is slightly different from Jim Weir's, and this difference illustrates an important point to note when considering the installation of ham gear in aircraft: To some degree, the legal attachment of Amateur Radio equipment is a matter of interpretation of FAA rules by (in ascending order) the pilot/owner, an A&P mechanic, an AI inspector and by local FAA airworthiness inspectors. The FAA always has the last word but Bob notes that FAA inspectors are reasonable and sometimes change their decisions. If the intended Amateur Radio installation in an aircraft appears to be in a gray area when considered against the advice and examples in this chapter, the ham/pilot/owner should consult with an A&P mechanic, preferably with an IA rating.

A BIT OF HISTORY

We begin with some aeronautical mobile background from Bob Hayos, K6CUK:

"The operation of portable and mobile ham radio goes back to nearly the beginnings of amateur radio. As low-priced automobiles became available, hams quickly conceived of ways to put their radios on wheels. The first practical installations appeared shortly before WWII. The need for compact packaging to fit within the often-confined spaces in vehicles became quickly apparent. Rapid technological advances during the course of the war quickly led to such needed components as miniature tubes, lightweight assemblies and many other design enhancements. Such advances became part-and-parcel of aircraft radio design after the war. Dynamotor high-voltage generators and efficient component layouts became the basis for the design of the well-known Command and Liaison transmitters on even small fighter aircraft of the period.

With the cessation of WWII, huge quantities of air- and ground-based radio equipment became available. Who among the old-timers can forget the numerous radio magazine articles dealing with the conversion of the Command series of transmitters and receivers such as the SCR274N and the VHF equivalent ARC5s and SCR522s? Many of these popular radios were easily tuned up on the ham bands while still installed in military aircraft by Amateur Radio licensed crewmembers. A CQ signed with an aeronautical mobile postscript never failed to attract a large number of postwar radio enthusiasts.

Aircrew members were frequently able to tune up an ART-13/BC348 combination on the HF ham bands. Such operation of course could only be accomplished during periods of little or no official coms.

Relatively long aircraft antennas necessitated the use of tuners and sometimes long trailing wires with weighted lead ends. Such antennas were reeled in and out until resonance was obtained by watching RF load meters. I remember an incident when a C-47 transport was making a long low-level approach into Chicago's downtown airport. A delay in reeling in the trailing wire antenna caused the embarrassing effect of causing the weighted end of the wire to smash through the garage roof of a startled Chicago citizen. To everyone's good luck, no one was injured...that we knew of.

The use of VHF was neither forgotten nor overlooked. Late WW II aircraft were often equipped with VHF radios, typical of which were the SCR522 and ARC-5 series. These sets were easily tuned to op-

Fig 3-2—Bob Hayos, K6CUK, built his latest aero-ham platform from a Canadian kit. The sheet-metal two-seater is a Murphy Rebel.

Fig 3-3—The Rebel's panel may have space for a built-in Amateur Radio transceiver, but Bob uses a 2-meter handheld rig in this airplane.

erate in the 144-148 MHz band with surprisingly good performance. The ability to operate using short resonant antennas further enhanced the proliferation of hams on these frequencies.

Military ham operators were not the only sources of airborne radio activity. Ham licensed pilots and other crewmembers of civilian commercial aircraft also found opportunities to load up HF antennas on their aircraft during quiet moments of flight. The radio equipment has generally been all-band SSB capable and was normally used for long-range communication during transoceanic flights. Antennas in most cases were relatively short, fixed length and tuned automatically with fast acting couplers or tuners.

An interesting note to be considered is that while ham marine and vehicle mobile installations and communications are authorized in most foreign countries, aeronautical mobile radio frequently is not. Careful attention to such restrictions is important before beginning operations."

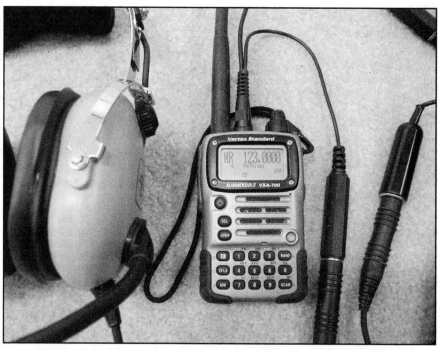

Fig 3-4—The full-feature, 5-W Vertex Standard VXA-700 2-meter FM transceiver also operates on the AM aircraft com frequencies and displays Omni navigation signals.

OFF-THE-SHELF AERO HAM GEAR

Until quite recently, no commercially available Amateur Radio equipment appeared to be dedicated to aeronautical ham mobile operation. That changed when Vertex Standard introduced the VXA-700 2-meter handheld transceiver (see **Fig 3-4**). This small radio incorporates all of the features expected in a state-of-the-art 2 meter handheld. In addition it offers 5 W using amplitude modulation on the VHF aircraft communications frequencies, plus aircraft navigation using VOR (known as *Omni* for its 360° capability) and localizer reception, NOAA weather broadcast and reception of the FM broadcast band. The VXA-700 comes with an adapter that screws into the top of the radio for connecting with standard aviation headset earphone and microphone plugs. The PTT switch on the transceiver may be used to key the transmitter, or a standard portable aviation PTT switch on its cord may be inserted between headset's mic plug and the VXA-700's interface cord.

With this radio, a pilot can communicate with airport towers and Air Traffic Control, navigate along airways or anywhere else with VOR stations, get area weather conditions and forecasts or listen to favorite music found on a nearby FM broadcast station. All of that is in addition to talking with hams on 2 meters.

A Vertex spokesman said recently that the company intends to offer the VXA-700 for the foreseeable future. Whether other manufacturers will follow the Vertex lead and offer products specifically for

the Amateur Radio aviator remains to be seen. However, the advent of small radios—many with detachable control panels—will entice amateurs with access to aircraft to join the fun of participating in aero mobiling.

Whether operating VHF or UHF through repeaters or long range on the lower bands, aero mobilers often find themselves the object of pileups, rather like running a special-events station or being on a DXpedition…without the need to be far from home. A few have even designed special QSL cards for their aero mobile operations.

In this chapter, you will learn about numerous types of aeronautical Amateur Radio operation from hams who have done it. Their descriptions make it obvious that this type of operation is unusual enough to make it fun for those airborne and for their ham contacts on the ground. You will learn about special considerations and FAA requirements. You will find out how to build efficient antennas and how to install equipment in an aircraft. But even if you never initiate an air-to-ground ham QSO, you will find that learning about aero mobile is fun.

Here are some of my own stories about aeronautical mobile ham operations, followed by those by several others.

FROM A GLIDER

One of my first aeronautical mobile

adventures was while flying a single-seat glider. In the early '80s, I had half interest in a Schweizer-1-26 sailplane, which I often trailered to Santa Ynez Airport, northwest of Santa Barbara, California.

The Santa Ynez Mountains run east and west as they parallel the Pacific coastline near Santa Barbara, and the occasional north wind provides excellent ridge lift for a glider flying from Santa Ynez Airport, which is a short distance from the north side of the mountain range. A glider pilot could buy a tow at the airport, catch the ridge-lift 5 minutes after takeoff and soar the ridges as long as wind and daylight remained. You needed to avoid the prohibited airspace at the west end of the ridge: President Reagan's vacation home, Rancho del Cielo (Ranch in the Sky), was there.

A simple 2-meter handheld transceiver with a spare battery pack provided more than an hour of QSOs, both simplex and through several Santa Barbara repeaters. Most of my contacts had never talked with a glider aero mobile station before and I still haven't done it myself.

For years hang glider pilots have used Amateur Radio in the mountains east of San Diego and probably elsewhere. Some pilots fly with their handhelds and provide real-time condition reports. They use 2-meter mountaintop repeaters within line of sight of some of the popular hang glider launch sites to communicate wind and lift conditions to other ham hang glider pilots considering driving to the sites.

AIRLINE-PASSENGER MOBILE

My first experience with aeronautical mobile was unplanned, unexpected and unlikely to be repeated, considering modern security requirements. I was in the Navy, working in Washington DC in the mid '70s. I often made day trips to Warminster, Pennsylvania, flying from Washington National Airport (now Reagan National) to North Philadelphia Airport. There I rented a car for the drive to Warminster for the day's work. I often carried my first 2-meter handheld (a crystal-controlled Wilson) in my briefcase, because a Philadelphia repeater matched one of the crystal pairs.

On this particular day, I was chatting with a ham through a Philadelphia repeater when another ham broke in after I mentioned that I was headed for a flight out of North Philadelphia Airport. "If you're going to North Philly," he said, "you're going to be flying in one my airliners."

"Oh, really?" I said in effect. "Yes," he said. "My name is Dawson. Tell the flight attendant (actually, I believe he said *stewardess*) I said it was OK."

The Eastern Airlines shuttle I would board was a turboprop operated by Ransome Airways. Coincidentally, I had seen a Pitts Special built by Dawson Ransome in the brand new Smithsonian Air & Space Museum two weeks before and had learned that he owned an airline company.

The head flight attendant raised an eyebrow when I told my story, but she took my Wilson handheld to the captain for his approval. She returned it in a moment and said I should let her know when I was going to transmit. Once we were in level flight, I made several brief contacts through repeaters and on a simplex calling frequency. Most attempts resulted in bringing up more than one repeater, which caused me to get off the air quickly.

It should be noted that this was on a clear day. The major concern about transmitting from aircraft has been the possibility of affecting flight instruments, as will be noted in greater detail later in the chapter. In cloudy conditions I would not have asked to operate. After landing, the flight attendant said that the flight crew saw no indication of my transmissions. Here's betting I will never get another invitation like that one!

Much more likely than my airborne airline experience is the possibility of working ham airline pilots while you are on the ground and they are at high altitude over international waters. We will hear from the cockpit end of QSOs like that later in the chapter.

Fig 3-5—Eight hams operated from the FujiFilm HKS-600 blimp. Their 4-hour flight included many amateur contacts through four onboard 2-meter handheld transceivers. Pilots in command were well-known hams John and Martha King, who produced ARRL Amateur Radio training videos. The HKS-600 requires a large ground crew to launch and recover it.

BLIMP-BORNE

"Hi, This is John King," the voice on our answering machine said. "On Tuesday afternoon, Martha and I are flying the blimp. Do you and Lois want to come along? Let us know."

Hams know John and Martha King from their ARRL amateur license training videotapes and CDs. Pilots know them as the only couple that has every FAA pilot license class and category rating…and for their aviation training videos. Beginning in the mid '90s, the Kings were volunteer pilots of the FujiFilm HKS-600, the world's largest blimp. We of course jumped at the chance to fly with John and Martha.

Nine people boarded the blimp for a 4-hour flight around San Diego on the last day of 1996. See **Fig 3-5**. Eight of us were hams and four brought along handheld transceivers. Our contacts on the ground seemed to have as much fun talking with hams in a blimp as we were having. But maybe not. Everybody aboard got a turn at the controls from the copilot's seat.

INSTALLATION REGULATIONS

The rules set by two government agencies come into play when you plan to operate Amateur Radio from an aircraft. The FCC's rules for running your ham station of course apply, but beyond that the FCC does not officially care how or where you operate your Amateur Radio station. The FAA, however, may have an interest in the installation of your rig and its antenna, depending on the licensing status of the aircraft.

For factory-built, FAA-certified aircraft (also called *certificated*), installing

things rely on the following general rule: If any device such as an antenna or a panel mount for a radio is *permanently mounted*, the installation may require an FAA Form 337 and all the paperwork and testing that goes along with it. In any case, an FAA-licensed A&P mechanic with inspection authorization (known as an *IA*)—needs to make or supervise the installation. The IA assures the electrical and mechanical soundness, recalculates the weight and balance of the aircraft with the additional equipment and signs the Form 337 (if he or she determines it is necessary), also indicating that the aircraft will not be affected negatively from an aerodynamic standpoint. The definition of *permanently mounted* would include use of screws, rivets, welding or adhesives, or anything requiring drilling holes or modifying any part of the certified aircraft. The Form 337 process is cumbersome and potentially expensive. Therefore, one practical approach is to find a way to mount both the radio and its antenna temporarily, avoiding the Form 337 issue entirely.

For example, if the radio can be powered from a cigar lighter and is light enough to be held in place safely with bungee cords or hook-and-loop fasteners such as Velcro, no official installation record may be required. Many handheld GPS receivers are operated this way with the external antenna attached to the inside of the windshield with suction cups that come with the antenna. The same technique may work with ham gear and antennas in the VHF region and above. Antenna patterns will, of course, be affected by the orientation and proximity to metallic aircraft structure.

As noted, the Form 337 installation may be required only for hard-wired or hard-mounted equipment in certified aircraft. In an aircraft that is not factory-built and FAA-certified, the owner/pilot may be able to install equipment permanently without official FAA interest, providing it is done safely and documented properly.

What civil aircraft are not FAA-certified? Experimental-category amateur-built aircraft (*homebuilts*), whether made from kits or plans, are in this category. See **Fig 3-6**. More than 20,000 homebuilts of all types are flying now and more new ones are launched every day. They include single- and multi-engine land and sea-planes, gliders, helicopters and gyroplanes. Even a few home-sewn hot-air balloons fall into this category. If one or more unpaid workers (typically the owner of the project) build the major portion of the aircraft, it qualifies for registration in the FAA's Experimental, amateur-built category. The original builder is offered a *repairman certificate* for this one aircraft and may—with only a few restrictions—maintain, modify slightly, and sign off the airplane for its annual condition inspection. The major-portion regulation is known as the *51% rule*. A new weight-and-balance calculation should be produced for any permanently mounted equipment subsequently installed in a homebuilt aircraft. For safety, powered equipment needs to be grounded and fused properly.

OTHER VIEWPOINTS ON THE LEGAL ISSUES

Here is Jim Weir's:

"In the first place, 'aeronautical mobile' is not mentioned at all in FCC's Amateur Radio regulation (Part 97), nor is it addressed at all in Part 87 (Aviation Services). You have to read Part 2 (Treaties...) to find the briefest of mentions, and there it is defined as a communication between an 'aeronautical station' and an 'aircraft station.' You go further to find that an 'aeronautical station' is a ground station in the aeronautical service, which is a reference to Part 87. Thus there is no official FCC term for an airborne radio station. Call yourself aeronautical mobile or aircraft mobile or airplane mobile and you will not be in violation of any of the FCC rule.

The FAA, on the other hand, has a significant official role in what is installed in aircraft. 'Wait a minute,' this agency says. 'You can't just bolt that thing onto the airplane and fly.' From a legal standpoint, the FAA is interested in having the airplane return to earth in a condition to be able to fly again. Specifically, FAA Parts 20 and 91 tell us what we have to do to be legal in an airplane.

I will cover FAA regulation Part 91 first. It is titled General operating and flight rules. Part 91 is quite specific in saying that the ultimate authority for the operation of the aircraft is the pilot in command (91.3). Part 91 is also specific in prohibiting any portable electronic device from being used on board an aircraft in flight... unless the pilot in command of the aircraft has determined that the device will not affect the aircraft communication or navigation radios. If the pilot says that it is OK to use your handheld on board, you may use your handheld on board. (Note that this discretion is not available to air carrier aircraft. The airline company itself—not the individual airline captain—must have a policy that allows you to use your handheld in flight. We know of no airline companies with such a policy.)

Misconceptions abound about legal and

Fig 3-6—Thousands of aircraft like this Kitfox are *homebuilts*, licensed in the FAA's amateur-built Experimental category. One advantage is that the builder may modify the aircraft slightly (including the installation of electronic equipment and antennas) without official FAA involvement. Aviation-quality workmanship and consideration of weight and balance and non-interference with control systems and aerodynamics are required.

feasible aero mobile operation. Here are four of them:

1. **You can't use a handheld radio while flying in the clouds (IFR)**. *At one time that was specifically prohibited, but that section of the rules has been re-written. While I'm not about to bore through clouds with somebody in the right seat chattering on the ham rig, there is no longer a rule prohibiting it.*

2. **I'll just use a magnetic-mount temporary antenna and run the coax in through the window**. *Two problems arise immediately. First, the airplane is all aluminum; and the magmount antenna base isn't going to stick. Second, airplane structures are strong but fragile. It's not a car window made out of glass with a steel frame, so trying to close the lightplane window on coax is going to bend the thin sheet of clear acrylic and its light aluminum frame. Don't do it.*

3. **You have to check every frequency of the handheld against every frequency of every com and nav radio in the airplane to be sure there is no interference**. *Nonsense. There are 760 channels of com radio on two radios plus 200 channels of nav radio on two radios plus two frequencies of transponder plus the GPS plus the loran plus the ADF plus the 40 channels of glideslope receiver plus the marker beacon receiver times the 800 channels of a 2-meter Amateur Radio rig*

on 5 kHz spacing. That adds up to almost 4 million tests before you can go flying. At one test every 10 seconds or so, this gives about 75 years of testing for 24 hours every day. That isn't reasonable. What is reasonable is that if the pilot thinks that there is interference with any aircraft radio, you will be asked to shut the rig off until the pilot is sure that it isn't the ham radio causing the problem.

4. **You can't wire the radio into the aircraft system**. *Let's explore this one.*

FAA Part 21 (and by inference Parts 23 and 43) regulate what can and cannot be attached to an airplane, and who can do the attaching, and who can approve the attached part. We will not go into a long description of the whole mechanism for working on airplanes, but here are three citations that may help to understand the process.

The FAA published a paper (Advisory Circular 20-98) that gives clear guidance to how 'auxiliary two-way radios' can be installed in aircraft. This is a rather old reference (May 1977) but it has not been updated, superseded or cancelled. AC 20-98 is out of print, but my website (www.rstengineering.com) has it scanned and posted for download. This FAA Advisory Circular gives clear guidance to the installing mechanic and to the inspector exactly what the requirements are for installing a ham radio permanently into an airplane.

The FAA Western Region also published a document in roughly the same time period called 'Plane Sense—Aircraft Alterations' in which the FAA tells the installing mechanic how to differentiate between a minor installation (easy logbook entry) or major installation (the paperwork blizzard including the famous Form 337).

The bottom line is that it is up to the mechanic who does the work and the inspector who signs it off as to whether or not an installation is major or minor. For example, I have a degree in electronics, I teach electronics and avionics at the college level, I've been working in avionics since I was 14 years old, and I have a company that manufactures avionics. For me, installing anybody's 2-meter rig in an airplane is the equivalent of a walk to the corner store. To another mechanic who has spent his whole career overhauling engines, the effort may be like climbing a tall, rocky mountain. The point is that if you want to install a ham rig in your airplane, it is much to your advantage to find a mechanic with extensive avionics experience who is willing to view it as a minor installation with a logbook entry.

Finally, the FAA, in a display of brilliance, wrote Part 21.303(b)(2), which says that the owner of an airplane may manufacture parts for his or her own aircraft. That is, you don't have to buy a bracket from the company that built your airplane to mount your radio in it. And you don't have to use an "approved" antenna. You can make it yourself. The caveat, of course, is that the installing mechanic has to agree that it 'looks like an airplane part to me' and that the final inspector has to agree with the mechanic. In general, it is best to find a mechanic who can also be the final inspector and who has the appropriate avionics background."

Bob Hayos checks in on the subject with a slightly different perspective. Regarding the temporary electrical connection of radio equipment such as through a cigar lighter as mentioned above, he advises checking with a local A&P mechanic with an IA rating before doing it. He knows of no regulation prohibiting it, but some FAA inspectors, he says, claim that nothing except the lighter can be powered from the cigar lighter socket. That, of course, makes no sense because the lighter element draws more current than most small radios, and this wiring is fused or has a circuit breaker. Beyond that, he suggests that we divide the subject into two activities: choosing the gear and planning the installation.

Finding radio transceivers and associated equipment for suitable mounting in aircraft is one of the most pleasant parts of such a project, he says. The transistor, followed by integrated circuits and computer technology, has changed everything. Miniaturization and the associated reduction in weight and power requirements have given us tiny radios that make aeronautical mobile feasible in even the smallest aircraft. As capability has increased, the relative cost has declined. What could be better? All the required accessories are included: antennas, tuners and virtually everything else to provide a neat, compact installation.

The simplest possible setup in a small aircraft is of course a compact handheld unit on a VHF or UHF ham band. No mechanical or electrical attachment to the aircraft itself is required.

Amateur Radio installations in Experimental aircraft have much the same technical considerations and constraints as those discussed for certificated airplanes. The documentation requirements, however, are much simpler. An airframe logbook entry by the builder describing the installation is all that is really necessary.

The physical installation of HF equipment is another matter and requires careful consideration of mounting possibilities. Brackets from the radio package to the aircraft structure have to provide rigidity and strength. The physical vibration modes of the equipment as well as structural resistance to common acceleration must be evaluated. A suddenly detached radio in the proximity of the pilot during extreme flight buffeting could produce disastrous results.

For a better and more efficient installation, a permanently mounted antenna is desirable. Such an antenna should be located on either the top or bottom surface of the fuselage so that an adequate groundplane is obtained. The skin of the aircraft (if metal) requires sufficient backing to prevent possible buckling or "oil canning." If the aircraft skin is fabric, a metallic groundplane under the skin must be provided. Good bonding to structure must, of course, be maintained.

HF installation presents additional problems because of much more complex antenna requirements. Resonant antennas on a small aircraft may involve the need for an on-board tuner, either automatically or manually operated, especially if the plan is to operate on more than one band. Fortunately, equipment manufacturers offer compact fast-tuning tuners capable of resonating short wire antennas. A wire antenna mounted between a wingtip and a vertical stabilizer can be made to radiate reasonably well even on 75-meter frequencies. However, narrow bandwidth is a penalty that may have to be accepted.

Amateur Radio equipment that can be installed in both air carrier and military aircraft is yet another matter, but few hams will be faced with this problem. Air carrier, corporate and military aircraft are usually equipped with HF equipment that can be used, as you will read later, if the pilot in command agrees (but in an airliner, don't bother asking). Yet in the cases where ham gear is to be installed professionally for some special purpose, the aircraft manufacturer, as well as the operating authorities, must be consulted. Virtually all that has been discussed for simpler aircraft applies, but there are many caveats. Aircraft in these categories frequently have pressurized cabins whose outer skins must maintain pressurization integrity. As such, any antenna penetration or mounting must be evaluated.

Electromagnetic interference (EMI) is another potential problem, and any amateur radio equipment has to evaluated for possible RF interference to both communication and navigation equipment. The Radio Technical Commission for Aeronautics (RTCA) Document DO-160R is a good reference for environmental conditions and test procedures involving airborne equipment.

The installation of amateur radio equipment aboard commercial-type aircraft poses some additional challenges, Bob notes. The first of these concerns government rules and regulations as defined by both the FAA and the FCC as noted above. In the case of the FAA, nearly every item of communications equipment, according to Federal Air Regulations, must be approved for aircraft operations. Such approval defines how the equipment will be installed, how it is approved and by whom, and how and under what conditions it may be used. It quickly becomes apparent that any attempts at "hay wiring" ham gear into either an airliner or a corporate aircraft cannot be accomplished easily. FAA Advisory Circulars AC 25-10 and AC 91.21-1A clearly define what can and cannot be installed into transport-category airplanes. Does this mean that it is impossible to bolt your pet Kenwood or Yaesu into a Boeing 747? The answer is no, it is not impossible, but it can be done only when alternate methods to the Advisory Circular limitations are provided and found acceptable by the FAA. The path to such action is difficult, costly and filled with frustrations for the average ham.

Now let's discuss operating aboard large aircraft. For operation as a passenger aboard a scheduled airline, you can all but rule it out. As noted above, some if not most airline companies have policies that preclude the captain from allowing passengers and in some cases the aircrew to transmit personal messages from the aircraft. In this case, even though FAA regulations permit such operations with approval of the pilot in command, that option is removed from the captain who wants to keep his or her job. (Seatback air telephone service installed in the airliner is, of course, exempt.)

For other passenger operations such as a charter flight or in a corporate aircraft, the answer is simple: Before the trip, contact the owner/pilot/operator of the aircraft for permission and guidance on air operations. Then take along a VHF/UHF handheld radio with a self-contained power supply and antenna and operate according to the company guidance.

Ham-radio operations aboard military aircraft are several orders of magnitude easier, Bob points out. In this case, strict FAA rules are not in effect. With carefully planned prior discussions, ham gear can frequently be taken aboard and installed. Of course, military regulations covering who may be aboard the aircraft and how they must function are clearly spelled out. This means that the aircrew and authorized passengers are usually military service people. Here is where reservists and guardsmen who are hams sometimes get to do something special. As an active USAF reservist, Bob Hayos notes that he was allowed to remove an inspection panel under a wing, mount a quarter-wave VHF antenna on a substitute plate, and operate 2 meters with great success. (Later in the chapter, Jim Weir, WX6RST, describes how to make an inspection plate antenna that could be installed on many light planes.) Bob was also authorized to temporarily retune an installed HF radio on 20 meters for some fantastic DX contacts. It can be done!

INTERFERENCE CONCERNS

As pointed out above, the main reason that the FAA and many airline company regulations prohibit transmitting from an airborne air carrier is the concern that navigation or communication systems will be affected. For the same reason, all types of receivers in the cabin have to remain off. At least theoretically, the radiation from a receiver's local oscillator could have an effect. (Above 10,000 feet, passengers may be authorized to fire up some types of electronics such as laptop computers, audio players and electronic games. These devices also have oscillators, but apparently no problem has been detected with them.)

Of primary concern to aircraft navigation are VHF Omnirange (VOR) signals and ILS localizers. These ground-based stations transmit continuously and are lo-

cated between to the top of the commercial FM band and just below the aviation voice VHF frequencies. The range is 108.0 to 117.95 MHz. Both VOR and localizers provide pilots with angular information from a ground station. VOR stations are located on hilltops or unobstructed clear ground for line-of-sight transmission. Using phase-detecting circuits, the aircraft VOR receiver provides the pilot with an indication of the radial on which the aircraft is located at that moment. For example, an aircraft east of the VOR station will show the pilot that it is on the 090° radial. The ground stations are oriented to magnetic rather than true north so they coincide with the magnetic compass in the cockpit. VOR stations are often located with UHF DME (distance-measuring equipment), which provides a cockpit indication of slant-range distance from the station in nautical miles. The combined system, named VORTAC, pinpoints the aircraft's location in two dimensions.

ILS is a three-dimensional, ground-based VHF/UHF system that guides the pilot to touchdown on a particular runway. The VHF localizer signal works like VOR but is useable only near the extended centerline of the runway. It is four times as sensitive as a standard VOR signal. On the cockpit indicator, a centered vertical needle or its electronic equivalent tells the pilot that the aircraft is on the extended centerline of the runway. The UHF part of the system indicates when the aircraft is on the glide slope (usually a 3° angle up from the runway touchdown point), above or below it. If the needle is horizontal, the aircraft is on the glideslope.

Pilots use the VOR and ILS systems to navigate and make instrument approaches in bad weather, although VOR is also used for navigation when the weather is clear. At a typical airport, ILS can be used on an approach to as low as 200 feet above the runway if the visibility is one-half statute mile or better. If the pilot does not see the runway clearly when reaching an altitude of 200 feet above the runway, full power is added for a go-around. Fuel to fly to an alternate airport forecast with better weather is required when you file a flight plan for this type of operation.

The reason to provide this level of detail is so that amateurs can understand the importance of not interfering with navigation systems. I have never heard of an accident resulting from Amateur Radio interference, but several National Transportation Safety Board accident reports have implicated other spurious transmissions in fatal aircraft accidents. In the 1990s the crash of a commuter airliner attempting an instrument approach in

Colorado was thought to be related to illegal cell phone use by a passenger.

This section will be of interest to Amateur Radio operators only because they won't be transmitting or receiving while flying in an airliner. The possible exception is Amateur Radio operation by the airline crew, and they won't be doing it while on a departure or approach or at low altitude.

MORE REAL-LIFE EXPERIENCES OF AMATEURS OPERATING IN THE AIR.

General LeMay's Airborne Shack

General Curtis LeMay, W6EZV, was an active ham who was in charge of the U.S. Air Force Strategic Air Command (SAC) in the late 1940s and into the 1950s. During this period, SAC transitioned from propeller B-29 and B-36 bombers to the jets—B-47s and B-52s. Communications between headquarters at Offutt Air Force Base in Nebraska and SAC's 21 US bases and 11 overseas locations was difficult enough that standard commercial and military circuits were sometimes supplemented by HF communications through USAF MARS (Military Affiliate Radio Service) stations on SAC bases. Licensed hams using ham gear on assigned frequencies near the ham bands operated the MARS stations.

Even more difficult than messages between the ground stations was reliable communication with SAC's aircraft. The new jets lacked the dedicated radio operators who had provided HF CW communications in the prop-driven bombers.

The base MARS station at Offutt was undermanned due to a lack of licensed hams. Hoping to encourage more ham and MARS activity, General "Curt" LeMay, W6EZV, and his deputy, Major General "Butch" Griswold, KØDWC, built some Heathkit equipment for the MARS station and, for the first time, heard SSB contacts on the 75-meter phone band. (The military was interested in SSB even before World War II, and after the war the Army Signal Corps had a contract with Collins Radio to investigate further.) LeMay borrowed some Collins SSB ham gear and found that it worked far better than either the AM HF Amateur Radio equipment or the USAF radios. But how well would it work from an aircraft at great range? He would find out.

In 1956, LeMay had a Collins KWS-1 SSB transmitter and a 75A-4 receiver installed in an Air Force C-97 transport. He dispatched Butch Griswold, KØDWC, Collins Radio President Art Collins, WØCXX, and Air Force Tech Sergeant Willie Wilson, W5DTA, to fly to

Okinawa, communicating on the ham bands while en route. They logged more than 1000 aero mobile ham contacts and followed up with QSL cards. (See **Figs 3-7** through **3-9**.) The result was spectacular, and the success caused the military to adopt SSB aboard SAC aircraft around the world. It also helped speed the transition from AM to SSB among amateurs.

An article by Charles A. Keene in the May 1997 issue of *QST* provided this story in more detail. The Collins Amateur Radio Club and Rockwell Collins also helped research this section.

Sherm's Cessna

Sherman Boeen, WØRHT, has been licensed to fly light aircraft since 1940 and he has been an active amateur since 1934. For more than 25 years he combined hosting a commercial television broadcast on WCCO-TV in Minneapolis with flying and making 16-mm movies of Minnesota and nearby aviation events. His weekly 15-minute show about general aviation in the Midwest used his Beechcraft Bonanza, his movie camera and his ease in front of a microphone to produce the station's longest-running local program.

Before that, Sherm was operating HF Amateur Radio mobile from his new Cessna 140. Here's how he did it:

"In 1948 I wanted a radio for my 1946 Cessna 140 (Fig 3-10). Aircraft VHF was new and expensive, so instead I built a low-frequency outfit. I found a cheap Harvey Wells radio range/broadcast receiver that worked well and built a transmitter for 3-4 MHz. I used a vibrator power supply that included a transformer and rectifier that produced 200 V dc for the tubes.

The 10-W transmitter had three tubes: a crystal oscillator, an amplifier and a Class A modulator. The transmitter had two crystals: 3105 kHz for aircraft communications and 3830 kHz for the 75-meter Amateur Radio band. The transmitter had a changeover relay and used an electric eye for tuning.

The ham band receiver was tunable. A battery-powered tube circuit converted onto the aircraft radio range-navigation frequencies. The engine was shielded so ignition noise was not a problem. The power pack consisted of 1.5 V dc and 67 V dc batteries, and the system worked well with the antenna extended.

The antenna was very flexible No. 16 braided copper wire that was wound on a reel mounted above my head. The wire was fed through a grommet in the fuselage and then to another grommet on the vertical tail. The wire was terminated with a small windsock that kept the trailing end from

being pulled through the tail grommet when it was wound in and kept the antenna stable as it was reeled out to operating length and back in.

The antenna covered 3-4 MHz depending on the length reeled out. With the antenna trailing 66 feet, I could work airport control towers on 278 kHz and talk with Flight Service Stations on 266 kHz with

excellent results, but range was limited to about 10 miles.

In daylight hours, I worked amateurs all over the Midwest with ranges of 60-100 miles. I tried nighttime operating but usually could not penetrate the QRM. However, I did get a brief report once from Waco, Texas.

The system worked well on both aircraft

Fig 3-7—In 1956, USAF General Curtis LeMay, W6EZV, equipped this C-97 transport with Collins gear to test the feasibility of using SSB for long-range contact with SAC bombers. The three hams aboard, from left-to-right Tech Sergeant "Willie" Wilson, W5DTA, Major General "Butch" Griswold, KØDWC, and Collins Radio President "Art" Collins, WØCXX, worked more than 1000 hams during flights to Okinawa. A three-part QSL card was sent to those who worked the C-97 in the air. [*QSL photos and information courtesy Rockwell Collins and the Collins Amateur Radio Club.*]

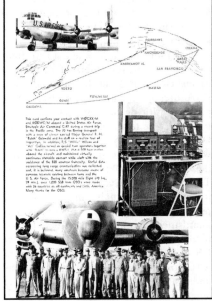

Fig 3-8—The QSL card included details of the mission including the C-97, its route of flight, a crew photo and a shot of the Collins KWS-1 SSB transmitter and 75A-4 receiver shack aboard the aircraft.

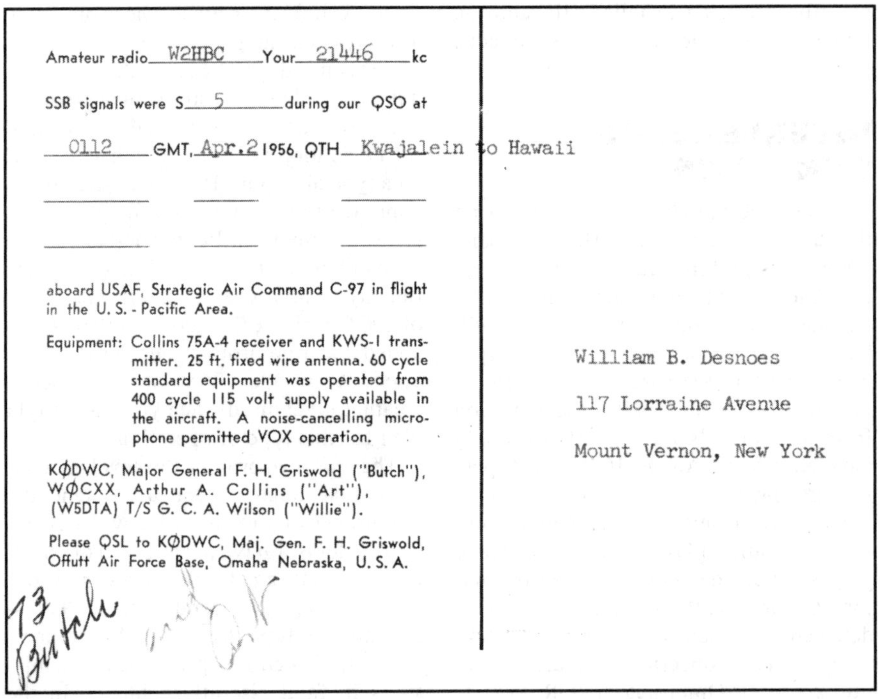

Amateur radio ___W2HBC___ Your ___21446___ kc

SSB signals were S___5___ during our QSO at

___0112___ GMT, ___Apr. 2___ 1956, QTH ___Kwajalein to Hawaii___

___ ___ ___

___ ___ ___

aboard USAF, Strategic Air Command C-97 in flight in the U. S. - Pacific Area.

Equipment: Collins 75A-4 receiver and KWS-I transmitter. 25 ft. fixed wire antenna. 60 cycle standard equipment was operated from 400 cycle 115 volt supply available in the aircraft. A noise-cancelling microphone permitted VOX operation.

KØDWC, Major General F. H. Griswold ("Butch"), WØCXX, Arthur A. Collins ("Art"), (W5DTA) T/S G. C. A. Wilson ("Willie").

Please QSL to KØDWC, Maj. Gen. F. H. Griswold, Offutt Air Force Base, Omaha Nebraska, U. S. A.

73 Butch

William B. Desnoes

117 Lorraine Avenue

Mount Vernon, New York

Fig 3-9—This QSL was made out to a ham in New York, W2HBC. It was signed by "Butch" Griswold and "Art" Collins.

and Amateur Radio bands with the antenna extended but there were a few problems. All the hams who heard me wanted to work an airborne station and sometimes I had to circle to allow time to answer them. The other problem was cockpit trouble: I forgot to reel the antenna in and lost it several times.

The HF experiment ended when I moved up to an aircraft VHF transceiver. But later, I also worked 2-meter Amateur Radio from my Bonanza. From 9500 feet I could work 200 miles with a 1-W Regency transceiver. One day I was flying on an instrument flight plan en route to Davenport, Iowa, monitoring Minneapolis Center. At the same time I was talking on 2 meters. Center came on frequency, "Bonanza 758 Bravo, this is Minneapolis Center, over." I was surprised and dropped the Amateur Radio mic, picked up the other mic, and said, "Center, this is WØRHT..." Center said, "Say again!" At that point I decided not to work 2 meters when alone and flying IFR (on instrument flight rules).

Amateur Radio in 1948 was fun and exciting, especially when it was spiced up by operating aeronautical mobile."

Homebuilt Aircraft/Homebuilt HF Antenna

Craig Cowles, WR7U, flies a Van's Aircraft RV-6 that he spent three years building from a kit (**Fig 3-11**). The RV-6 is a side-by-side two-seater and he decided to try aero mobile with the HF transceiver he uses regularly in his pickup truck.

The temporary installation of the fairly large older transceiver had the radio panel facing up and the transceiver with its back sitting on the floor on the front edge of the right seat pan. The arrangement used the seat belt and other secure fittings to assure that it stayed in place and did not interfere with the control stick, but it precluded a right-seat passenger. He previously designed and built his own aircraft nav and com antennas, which are mounted in the fiberglass wingtips at the ends of the sheet metal wings. The system works well. (As indicated elsewhere in this chapter, such installations are permissible by the builder of a homebuilt aircraft and would require an inspection and paperwork by an FAA-authorized mechanic for use in a certified aircraft.)

For his HF Amateur Radio antenna, Craig modified off-the-shelf items to make a simple, clean installation. The transceiver is grounded to the metal airframe and the feedline is RG-58 coax. He slightly modified a Radio Shack CB antenna mirror mount (RadioShack 21-937) that includes a beefy L-shape aluminum bracket and associated hardware, and a base for a Hustler mobile

Fig 3-10—Sherm Boeen, WØRHT, bought his 1946 Cessna 140 new for $2200 and equipped it with homebrew radios and antennas, including both ham and aviation communications. It was the first production Cessna 140 and it advertised the television station where Sherm's aviation program was broadcast.

Fig 3-11—Craig Cowles, WR7U, built this RV-6 from a kit and installed HF and VHF Amateur Radio gear in it.

ham antenna. The bracket is bolted to his tailwheel rod, which is nearly horizontal on the RV-6.

The nearly vertical part of the L-shape bracket was bent so that it is truly vertical despite the 20° angle of the tailwheel rod. The Hustler base is fitted to the vertical part of the L-bracket so that the antenna element is horizontal. (See **Fig 3-12**.) In Craig's case, a ³/₈-inch aluminum rod was cut to 12 inches in length and threaded with a ³/₈-24 die at each end. The forward end fits the Hustler mobile antenna base and the aft end is fitted with a standard Hustler RM-20 loading coil and its adjustable tip. In automotive installations the

antenna is vertical but Craig's application configures the antenna parallel to the ground and only about 6 inches from it. The distance from the forward end of the antenna base rod (which would be the bottom of the base in a vertical antenna) to the threaded coupler on the HM-20 element is 10.5 inches.

After the tail of the airplane was raised more than a foot above the concrete hangar floor, an antenna analyzer was used to tune the movable element of the horizontal RM-20 for the middle of the phone band on 20 meters. The length from the tip of the antenna to the loading coil was adjusted to 19 inches, resulting in an SWR

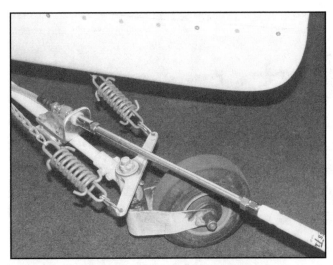

Fig 3-12—The 20-meter RV-6 antenna, attached to the tailwheel rod, was made mostly of off-the-shelf parts including a RadioShack antenna mount and a Hustler RM-20 loading coil and antenna top section. The bracket was bent to match the angle of the tailwheel rod, positioning the antenna parallel to the ground.

Fig 3-13—The 20-meter antenna worked so well that Craig bought a new Icom 706 for use in his airplane. The remote control head is attached to the panel with a cable to the transceiver, which is behind the pilot's seat.

on the ground of less than 1.4:1.

First flight reports in Oregon indicated excellent signals to and from a fixed-base amateur in Colorado and a mobile station in Louisiana. Craig was so impressed with the performance of this setup with his large old HF transceiver that he bought one of the new smaller, detachable-panel transceivers for use in his airplane. The bulk of the transceiver is attached firmly behind the pilot's seat, and the remotely cabled control head is fixed to a bracket fitted to his instrument panel (**Fig 3-13**).

Lacking a high-enough cloud ceiling for an airborne radio test with his new rig in place, Craig joined an early morning 20-meter net as he sat on the ramp at his home airport. Craig heard everyone and every station around the country reported perfectly usable signals from his horizontal, 3-foot-long, 6-inch-high antenna! The sheet-metal airplane works well as the ground plane for the antenna and the setup works even better with some altitude under it.

In addition to HF, Craig's new Icom 706 covers VHF and UHF ham frequencies. For testing on 2 meters he temporarily attached a bent-wire VHF antenna behind the RV-6 seats (**Fig 3-14**). The location is not ideal because the sheet-metal fuselage distorts the pattern but the antenna works well enough into line-of-sight repeaters and simplex stations.

Again it must be noted that from a practical standpoint, this particular type of installation and experimentation can be done legally only by the builder of an Experimental, amateur-built aircraft. However, later in this chapter Jim Weir, WX6RST,

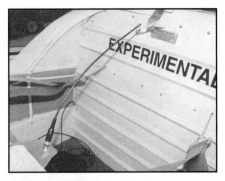

Fig 3-14—Craig attached a temporary antenna behind the seats to try VHF FM, which is included in his IC-706 transceiver.

provides details on an antenna-installation option that should work technically and legally on a factory-built sheet-metal airplane. In any case, mechanical and aerodynamic considerations must be taken into account with any antenna installation and a new weight-and-balance calculation must be done and available for display upon request from FAA personnel.

Surprise in the Air

Here's another airborne Amateur Radio story from Sherm Booen, WØRHT. You may have read this one before, as it was published as a "Strays" item in the May 1994 issue of *QST*.

"It was the summer of 1946. The war had taught me to fly, and those stressful days were over. It was time to enjoy my new love: flying. Now I could fly when and where I wanted, in grateful peace.

I had purchased a brand-new Cessna 140 and equipped the little silver plane with an Amateur Radio set so I could talk with my friends on the ground as I flew over the rich farmlands of the upper Midwest. On this day I was on my way to a meeting in Iowa and chatting on the radio as I approached the Iowa border. On the circuit was Marv, a pleasant fellow I had talked to many times from my ground station. When I told him I was flying down the highway to Mason City, he became excited and radioed, 'You'll fly almost over my station. Will you circle a few times? I'll tell you where it is.' I told him I'd check with him on the return trip.

Airborne on the way home, the first voice I heard was Marv's. He told me that he lived on a farm a mile west of the highway on the Iowa border. 'You can't miss it,' he radioed. 'It's the first farm-house on the north side of the road, a white house with red barns, surrounded by a grove of trees.'

He was right. It was easy to locate, and a few minutes later I was circling the farmhouse. I saw an open field on the east side of the farmyard, and I told Marv I'd fly low over the field and give him a good look at my new plane.

I lined up for a low pass, remembering old military maneuvers. I punched the throttle and roared along at 50 feet, not more than 50 yards from Marv's station. I could see his antennas. As I pulled up in a steep climb at the end of the field I radioed, 'How do you like my new plane, Marv?' There was a slight pause, and his voice came over the radio. 'I didn't see your plane, Sherm. I guess you didn't know. I'm blind. But the engine sounded beautiful.'

No, I didn't know. I had talked with him many times and I never knew. I leveled off on a heading for home, said 73 to Marv, and switched the radio off. I was lost in my thoughts, recalling the many sights I had seen and could still see...things my friend Marv would never see. 'But the engine sounded beautiful.'"

It is necessary to note that current regulations prohibit a low-altitude maneuver like Sherm's. The requirement is to stay at least 500 feet from any person, vehicle or structure. But Sherm is forgiven for his enthusiasm: Even if such a rule existed in 1946, the statute of limitations has passed.

Tips from a Corporate Pilot

Barry A. Brown, KØYLU, has flown for Rockwell Collins Flight Operations since 1977 and is currently flying company jets. Rockwell Collins makes HF transceivers for large aircraft to communicate with Air Traffic Control (ATC). For contacts with airborne hams flying these classes of aircraft he has some comments and suggestions:

"Airborne commercial HF transceivers normally tune in fixed increments—some as small as 1 kHz—but less expensive units may tune in larger increments. With their fixed frequencies or tuning steps, some air mobiles appear to be 'off frequency' to hams they are trying to contact. Please don't chide the airborne HF ham for being slightly off frequency as the transceiver may be as close as it can get.

These same units tune the limited-length antennas on aircraft when the operator initiates a tune cycle, usually by pressing the PTT switch. Once initiated, the tune cycle runs its course and may take 1 second or more to achieve the best match. Many antenna couplers use motorized rotating matching coils, sometimes requiring a long carrier time to find the best match. Any frequency change invalidates the tune status and another tune cycle will be required before the operator can transmit again. While flying we don't intend to tune up on the frequency of a DX station; we simply don't have the option to tune up nearby and then slide onto the active frequency without another tune cycle.

Some of the less expensive units make it difficult to change frequencies easily or quickly and they may have limited capabilities. The more basic units may have a limited number of remotely programmed 'channels' for the common HF frequencies used by the particular aircraft operator, with only one or two of the channels programmed for ham frequencies at the request of the amateur operator. If QRM

exists and you need to change frequency, ask first if he or she can go to the new frequency. If you just tell the airborne ham to 'go down 10' without waiting for a response you may find no one on the new frequency.

Because of the reduced efficiency of the short radiating elements on many aircraft, most air mobile operators will use bands above 40 meters. Also understand that the international standard for aviation HF communications is USB mode. Some transceivers don't even have LSB, AM or CW capability.

A good place to look for air mobile operators is on the International Association of Airline Hams net frequencies, 14.280 and 21.380 MHz. The group understands the operating conditions for air mobile stations, takes frequent standbys for check-in and welcomes newcomers. Check the website at **www.airlinehams.net** for net times and Net Control callsigns.

ATC communications, weather situations, amended clearances and passengers who have access to the cockpit crew in corporate and military aircraft may temporarily occupy the Amateur Radio operator's attention, thus causing a sudden absence when you stand by for them. Wait a while. They are usually still there and will call you again when they can.

For the same reasons, contacts need to be short unless the airborne operator prolongs the contact. Be considerate of the need to perform other functions and that the amateur contacts are secondary in priority.

Realize that the owner of the aircraft and the passengers may not object to the crew using the HF radio for ham contacts, but many are security and image-conscious and may not want their identity, the company identity or the flight itinerary revealed to the public. Corporate aircraft are frequently used for strategic purposes that they do not want their competitors to know. These tasks include classified projects, competitive bidding work and looking at companies to purchase. During a QSO, I usually share the aircraft type, altitude, speed, approximate location and sometimes where we are heading. I rarely share the company identity on the air unless the trip is routine or of casual interest to the contact and no security issues will be compromised. Airline and military flights frequently have similar considerations, so please don't press the aero ham for too much information. I often QSL the contacts I've made and will disclose the company identity and onboard equipment details then when the world isn't listening."

From the Flight Deck

Consider these stories by Lon

Cottingham, W5JV, a retired airline pilot:

"I went to work for Continental Air Services in Vientiane, Laos, in January, 1966 and I was given the Lao ham call XW8BT. There was a lot of hamming and CB going on in Vientiane, but I was the only FCC-licensed ham. Most of our C-47s and C-46s had AM HF radios in them. By this time, most ham HF was SSB. All of our HF sets were on AM. I worked several foreign stations but never did work the US on HF AM.

Air America had several brand new Caribou airplanes and they all had the latest Collins 618T HF sets. When the word got around that I was a licensed ham, I was allowed to use the HF sets in the Air America airplanes when they were idle. It was fairly common to have a queue of Air America and CAS pilots standing under the wing of a Caribou waiting to talk to their families back home in the US. Incidentally, I gave Conditional Class license tests to 12 people while I was in Laos.

Some time later, I transferred to Saigon. But the Air America ramp was on the far side of Tan Son Nhut Airport from our operations and this put an end to my air-mobile operation for a while.

I had a Heathkit station shipped out, put it together and then discovered that American hams could not operate from Viet Nam. But since I had a legal Lao license and call, I decided to bootleg from Saigon using my Lao call. This really worked out great...for a while. One evening, my bootleg 'CQ DX' from Savannakhet, Laos, was answered by a tremendously strong signal. It was Ambassador William Porter calling from the American Embassy about a mile away. Bill was a friend from my days in Laos. He said, 'Hi Lon. You are really strong to be in Savannakhet.' That ended my bootleg operation from Viet Nam.

My next air-mobile operation came in 1970-72 from the last DC-6B in scheduled airline service. Our DC-6 had only AM operation on its HF sets. But we occasionally borrowed a DC-6 from Evergreen (Air America), and it had SSB HF sets. When we used it, I worked 20 meter air-mobile quite a bit back to the states.

Our DC-10 fleet had two Collins HF sets. I started flying as copilot from Honolulu to Sidney through Fiji and American Samoa in the early '80s. To operate Amateur Radio from the airplane, two things were required. First, you had to have permission of the captain of the airplane, and second, the radio you used had to be independent of the radio that the aircraft used for its position reports.

Most of the captains were thrilled to ham from their aircraft but a few were afraid to do anything that was not 'in the book.' Several of the captains were also hams.

Later, when I became a captain on the DC-10, I hammed every time I got the chance. One day, coming back from Sydney to Honolulu, just after passing over American Samoa, I realized that it was about 0700 in Texas. My dad worked a group nearly every morning on 80 meters at about that time. I had never worked 80 meters from the airplane, so I decided to try it. Fortunately, those early HF sets worked LSB as well as USB. (The LSB mode was removed from many HF sets shortly after this.)

I tuned the HF set to 3.950, and to my amazement, I heard some familiar voices...not strong, but I could understand them. I made several calls before getting any of them to acknowledge me. When I finally got an acknowledgement, somebody said, 'Did that guy say he was in Samoa? Who does he think he's kidding?' I went right back to him and said that I was not kidding and that I was looking for my dad, K5AOY. I guess the call did the trick because one of them remembered that K5AOY had a son who was an airline pilot. That did it. They all wanted a signal report. We had several exchanges before the skip changed. Unfortunately, my dad was not on that morning.

I only flew the Boeing 777 regularly to Tokyo, London and Paris. Occasionally, I would trade a trip with George, W6OBZ, and fly the Newark to Tel Aviv trip.

At the time of day that we flew to Tokyo there was seldom any propagation anywhere. Coming back from Tokyo, as we passed abeam Anchorage, to the south, I often worked the 20-meter Trivaloney group (half trivia, half baloney) with excellent propagation to both the east and west coasts. The boys in Wisconsin and Minneapolis were also very strong. I could work them all the way to the West Coast or until they shut down. The biggest problem in working this group was getting Roy Neal, the net leader, to move the group to an even kHz so that we could understand each other. Roy did not like changes, and getting him to move the group from 14.238.5 MHz to 14.238.0 or 14.239.0 was sometimes a major effort. It must be remembered that the airliner HF sets only operate on even kHz frequencies; they would not work on the half-kilohertz frequency that the daily Trivaloney net uses.

The most fun I had with air mobile on the B-777 was on trips in the late afternoon and early evening between Houston and London/Paris. Kirby, K7EC, flying a B-767 to South America, would take off about 2 hours before I took off for Europe. We would meet on 14.238.5 and often find Tom, K4IC, a retired Marine Corps pilot, on the frequency. We always seemed to have excellent propagation between the three of us at that time of day and would often continue the QSO until Kirby had to start his descent to Buenos Aries or wherever he was. At this point I was somewhere around Gander, Newfoundland."

SOME TECHNICAL DETAILS

Jim Weir, WX6RST:

"Here are some things in no particular order that you might find of interest about aircraft radios and electrical systems.

1. Most light aircraft electrical systems are nothing more than an automobile 12-V system with all the heavy components eliminated or reduced in weight. Batteries, for example, are about half the weight (and half the amp-hour capacity) of a car system. Alternator fans may turn backward compared with the same make and model alternator in a car. Some alternators are geared off the front and some off the rear of the engine, and some are belt driven. Most have better filtering than you will find in an automobile. However, some aircraft have 24-V systems. Find out which system your aircraft has by measuring the bus voltage or ask a mechanic who has the specifications. Be cautious, however, as some manufacturers saved a dollar and used the same cigarette lighter on the 24-V system that was used on the 12-V systems. Don't presume. Measure. If you have a 24-V airplane in which you want to run a 12-V radio, use the circuit shown in drawing in **Fig 3-15** for anything drawing up to about 4 A. Above that current requirement, see any good reference on switching power supply design.

2. Aircraft VHF com and nav, marker beacon and glideslope are all AM and sensitive to high-power devices transmitting nearby. Use the absolute minimum power necessary for communications (which is not only good operating practice but an FCC regulation anyway).

3. An antenna above the ground has a radio horizon given by the equation $d = \sqrt{2h}$, where d is the radio horizon in miles and h is the aircraft's altitude in feet above the ground. For example, an aircraft flying 10,000 feet above the ground has a radio horizon of about 140 miles. This is another reason for using low power and avoiding repeaters—especially those without subaudible-tone (CTCSS) keying requirements—whenever possible.

4. Most light aircraft you will be flying in are either sheet aluminum or cloth cov-

ering over a wood or steel tube framework. In any event, using a handheld-type flex antenna inside the airplane may give you the same less-than-optimum results as using it inside a steel automobile.

5. On the other hand, bolting anything onto the airplane comes under the aforementioned discussion of legal installations and approvals. Having said that, removing an inspection plate and installing a temporary whip antenna onto the hole left behind is a rather trivial temporary modification and one that most mechanics would sign off as an owner-installed modification. That is, you can get it blessed once and then install and remove it yourself as you wish. See **Fig 3-16** for details on such a VHF or UHF whip mounted to an inspection plate.

6. I'm sure your mechanic/inspector will be quite interested in your installation technique, but there are two phrases that you need to drop from your vocabulary right now: duct tape and glue.

7. Along these same thoughts, you need to consider that anything you bolt onto, string through or weave around an airplane must not cause the airplane to tumble from the sky. I have inspected airplanes where the owner installed a ham rig cleaner than any avionics shop installation I've ever seen. But there were others where the owner draped the antenna coax over the aileron cables, through the rudder pulleys, and over the elevator chain. How that coax and that airplane arrived in condition to inspect is a mystery to me. If you've never worked on an airplane before, ask your mechanic the proper way of doing something if yours is to be a do-it-yourself installation."

HEADSETS AND MICROPHONES

Aircraft headphones and microphones are a throwback to the 1920s. When we first started to communicate air-to-ground, telephone parts were used to make our headsets. As the world changed, aviation people stayed with the technology that made Bell Labs famous. In particular, aircraft headphones still use telephone-style 300-Ω earpieces wired in parallel to give the current standard 150-Ω aircraft headphone impedance. Aircraft micro-phones simulate the carbon microphone that was all the rage when Babe Ruth was knocking the ball out of Yankee Stadium. In particular, microphones are biased with current and they give about 0.5 V p-p of audio out.

(Here is some aviation terminology: A

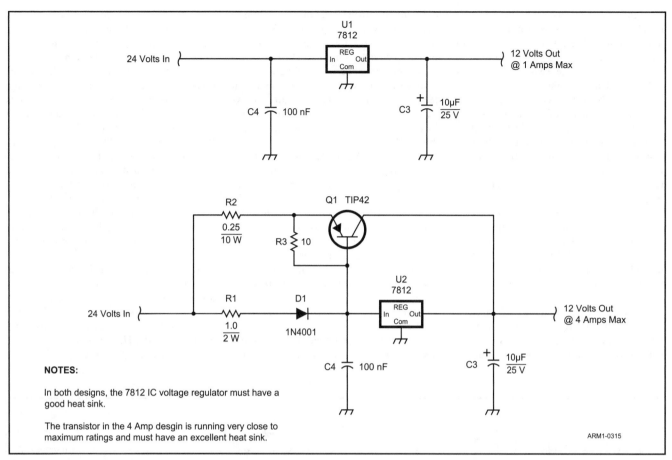

Fig 3-15—If your aircraft has a 24-V dc electrical system, you can build this converter for 12-V dc equipment.

headphone refers to the earcup devices. The combination of a headphone and a microphone is called a *headset.*)

The push-to-talk switch is a relatively simple device that grounds the aircraft key line to put the radio into transmit mode. There is no good reason why the aircraft PTT switch can't be used with a permanent ham radio installation. A portable PTT switch intended for use with a handheld aircraft radio may not work with ham handheld transceivers even if adapters are used for plug and jack compatibility.

The microphone element itself may well be a modern electret or dynamic element, but the headset manufacturers have to provide circuitry inside the headset to make that modern microphone element look to the radio like a carbon element. Designers and manufacturers of headsets use either transistor or integrated-circuit amplifiers, powered by what would have been the carbon element's bias current, to simulate a carbon element. In particular, most manufacturers use a relatively simple single-resistor bias to provide 10-20 mA of current, which will provide the required 1 V of audio.

How do you get 10 mA of current from

Fig 3-16—An inspection-plate VHF antenna, seen below the leading edge of the wing. For certified aircraft, this requires a signoff by a licensed aircraft mechanic. Inspection plates are found throughout sheet-metal aircraft.

a 12-V supply? Ohm's Law tells us that a single 1.2k Ω resistor will work well. Most designers use a slightly smaller resistor (470 Ω to 1k Ω is quite common) and a regulator to bring either the 12-V or the 24-V battery bus down to about 10 V. That

Fig 3-17—Aircraft made of fiberglass, such as these versions of the four-seat Velocity, are ideal for imbedded antennas, which are installed as the aircraft is built from a kit or plans. Quarter-wave antennas require an internal metallic ground plane.

way, if an emergency supply is desired or we want to check out the setup on the bench, we can use a common 9-V battery for testing in place of the 10-V regulated supply.

Now comes the problem. Most ham rigs use a relatively low-impedance, low-voltage dynamic microphone. Some rigs use a medium-impedance, medium-voltage electret microphone. No one solution will work in every instance. However, there is one constant technique to solving the problem, and that is a resistive voltage divider to take the 1 V of audio from the aircraft microphone and knock it down to the level and impedance that the ham radio likes to see. For example, the TM-221A that flies in "Snarly Charlie," the Cessna 182A Jim Weir flies (N73CQ), wants to see 10 mV p-p at an impedance of 600 Ω. He needed 1 V at an impedance a little lower than that, so he made a 100:1 voltage divider. Was that a problem? No, and the circuit in **Fig 3-18** accomplishes this. (One 560 Ω resistor and one 56 kΩ resistor make it work.)

A lot of amateur rigs intended for mobile use have only a speaker output. That turns out not to be a problem. A radio speaker output will drive a set of aviation headphones quite nicely. However, one helpful hint is that two radios wired in parallel cannot be used to drive one headphone. It doesn't work that way. You can't listen to the aircraft radio and the ham rig on the same set of headphones without some sort of isolation summing circuit. A

summing amplifier is relatively simple to make, however. The circuit in Fig 3-18, mentioned above for its mic impedance-matching device, contains an LM386 summing amplifier that allows several radios to drive as many headsets as you might need in the aircraft. This circuit is powered by the same 10-V supply we used for the mic bias, and it is built from parts available at RadioShack.

Jim Weir, WX6RST, philosophizes about using RadioShack parts in an aircraft project. *"I'll use any part that will do the job,"* he says. *"The source is not the question. FAR Part 21 allows the aircraft owner to manufacture any part so long as the installer/inspector is willing to go along with it. I've been recommending and writing about the use of easily obtainable electronic parts for more than 30 years. So far, nobody has called to tell me about a failure of a single part obtained from local sources."*

One last thing you need to know about headsets: The connectors on the headsets

are, again, right out of the Teddy Roosevelt era. The headphone plugs are standard 1/4-inch monophonic audio connectors, but the microphone plug is still a telephone device. The old PBX switchboards used hundreds of these strange plugs to operate, and the aviation folks decided that if it was good enough for Ma Bell, it was good enough for Orville and Wilbur's invention. The mic plug is sometimes mistakenly called a 3/16-inch connector, but the actual diameter is 0.206 inches. This is a specialty product that you will not find at RadioShack, but aviation parts houses sell them in both the male (cable) version and female (chassis mount) version. **Fig 3-23** shows which part of the plug does what.

Something else to note is that headphones normally used with ham transceivers are lower impedance than those designed for aviation use, so with aviation headsets the ham may have to increase the transceiver volume slightly. However, when using the low-impedance speaker output into any

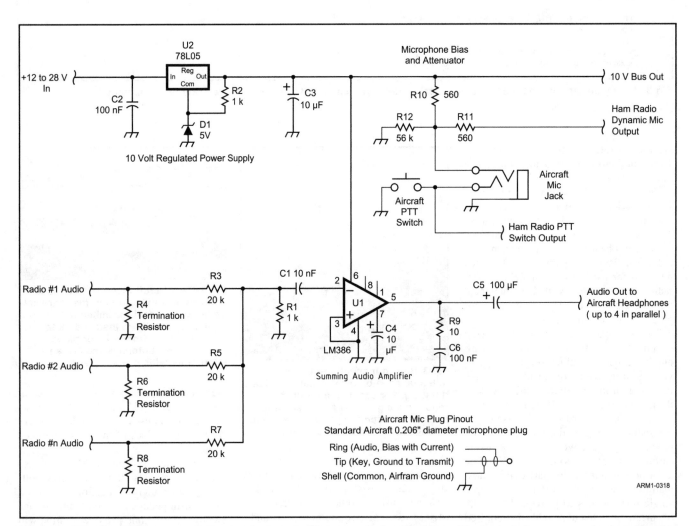

Fig 3-18—This circuit drives an aviation headset used with ham and aircraft radios. Also included are a summing amplifier for the output of two radios, plus bias and attenuation for the mic and a 10-V power supply.

Fig 3-19—A slightly modified aircraft audio distribution panel allows convenient integration of Amateur Radio transceivers in an aircraft.

Fig 3-20—Jim Weir's Cessna 182, N73CQ, is a testbed for his aviation electronic projects.

headset, turn the ham transceiver audio output down before connecting the headphone plug because the volume may otherwise be uncomfortably high.

Craig Cowles, WR7U, has another solution for the airborne ham's audio interface problem. His low-tech but highly effective solution is to wear a pair of inexpensive earbuds inside his comfortable, noise-attenuating aviation headset. With the volume controls set properly on the Amateur-Radio transceiver and the aircraft com radio, both inputs can be heard well. Earbuds are intended for listening to music or voice from a private radio or audio device, and they fit quite comfortably inside an aviation headset, provided you use the tiny round ones that fit entirely inside the outer ear. Some larger types extend slightly outside the ear and may be uncomfortable under an aviation headset. Good earbuds can be bought for about $10.

A variation on this tactic is to wear only one earbud for the ham radio connection. In any case, at least one adapter will be required for the 1/8-inch stereo earbud plug to make it compatible with 1/4- or 1/8-inch monaural ham transceiver output jacks. Depending on the ham transceiver headphone jack, a second adapter may be needed to route the radio monaural output into both sides of the stereo earbuds.

For the transmit side with this setup, the airborne ham could pick up and use the ham rig's dedicated microphone and PTT switch to transmit. Or the boom mic on the aviation headset and its associated PTT switch could modulate the ham rig as explained above. Many HF ham transceivers include an adjustable *monitor* feature allowing the ham to check on the outgoing audio. Pilots call this feedback system

sidetone, and most aviation transceivers include the feature.

Aircraft Audio Panels

If you participate much in aeronautical Amateur Radio, in time you will tire of plugging and unplugging your headset into a patch box to use the ham rig. After a few months of this, you will want an audio switching panel so that the flick of a switch goes from aircraft radio to ham radio (**Fig 3-19**). It's nice that almost all aircraft audio switching panels will accept the inputs and outputs of a ham radio almost as easily as they will accept aircraft radios. The headphone circuits will certainly provide the isolation that the little homebrew audio amplifier provided, and with a judicious amount of experimentation, the microphone circuits may also be made to work with the same techniques that were used in the patch box in Fig 3-18 (using a bias resistor and resistive divider).

Searching Jim Weir's web site, **www.rstengineering.com**, reveals an audio panel that is in kit form. Coupled with the technical knowledge that most hams acquire over the years, a simple modification of this type of panel will allow running a ham rig every bit as easily as an aircraft radio. An alternative is to use the LM386 summing amplifier in Fig 3-18 as a basis for homebrewing your own audio panel.

"But how do I get it approved?" you might ask. Jim Weir tells you to repeat the mantra: "*The owner of an aircraft may make parts for his own airplane. Say it again,*" he says. "*Believe it.*"

ANTENNAS

Forget the Yagis and the colinears, Jim

Fig 3-21—The removable fiberglass wingtip on this Cessna can conceal effective VHF and UHF antennas.

Weir says. "*Understand that at altitude, the only thing limiting your range is horizon, not gain. As a matter of fact, the amount of power needed to transmit 150 miles and be heard by a rather numb 1 μV receiver is something on the order of 10 mW. No, that's not a misprint. Ten milliwatts. The 'low' power setting on most transceivers is half a watt. That's a range of something around 600 miles. Forget gain. An omnidirectional antenna pattern is everything. Think groundplane. Think whip. Think simple.*" Weir tests radios, antennas and other equipment in N73CQ, his carefully maintained and named Cessna 182 (**Fig 3-20**).

He describes two aircraft antennas that let him work all call areas in the U.S. "*Too bad I didn't keep the QSL cards,*" he says. The first one is the (in)famous "inspection hole cover whip." This setup is based on a simple fact: All sheet metal aircraft have inspection holes about 4 inches in diameter cut into the skin so that we can get our noses inside the airplane and look for corrosion, broken parts and things that shouldn't be there, such as birds' nests. Inspection holes are literally all over the aircraft. Some are on the wings; some on the belly; and some on the tail. These holes are covered with a circular metal plate bolted to the skin.

The idea for the inspection plate antenna is to visit an aircraft scrap yard or to make a spare inspection plate to mount the antenna connector onto. Then if you ever want to restore the airplane to "new" condition (for display and judging, for example), all that is necessary is to pop the antenna plate off and reinstall the original plate.

Choose a location for your whip that is electrically optimum and aesthetically pleasing (there a few hams who think antennas are ugly). In general, this will mean an antenna on an inspection plate on the belly of the airplane. Please note that you need to take into account the attitude of the aircraft in both the takeoff and landing phases of flight (including normal compression of the landing gear in case somebody lands "firmly") to convince yourself that the antenna will not scrape the runway in any possible maneuver.

Look at Fig 3-16 again for the general assembly of the inspection plate-connector and the whip antenna itself. Note that the bulkhead-mount BNC connector on the plate allows the antenna to be connected/disconnected from one side of the plate and the coax to the radio from the other side of the plate.

The second antenna (the one that Jim Weir actually uses on his Cessna) takes advantage of two facts. One, most wingtips are made of fiberglass. Two, the last outboard rib of a metal airplane makes a completely acceptable ground plane.

Once again, the whip antenna is simply a 19-inch piece of brass brazing rod soldered to the center pin of a BNC connector and epoxied into place in the connector body. A mating BNC connector mounted in a hole drilled in the outboard rib is then wired, with the coax going to the radio. Easy on, easy off. See **Figs 3-21**. However, before drilling the rib, get the advice of your mechanic/inspector. The last thing you want is to be required have an aircraft shop replace or repair a rib rendered nonairworthy because of a misdrilled hole.

The Van's Aircraft RV series of homebuilts are sheet metal with fiberglass wingtips that can house antennas above the HF spectrum. Thousands of these aircraft are flying. See **Fig 3-22**.

A surprising number of modern aircraft are capable of supporting internal antennas for VHF and UHF ham operations

Fig 3-22—Thousands of sheet-metal Van's Aircraft RVs—like Craig Cowles' RV-6—provide homebuilder hams an opportunity for aeronautical mobile.

because their structures are transparent to RF. The majority of these aircraft are made of fiberglass or wood, and most are homebuilts. Well-known models include Burt Rutan's Long-EZ and VariEze designs that were built from plans in the 1970s and '80s. Lancairs and Glasairs are mostly fiberglass, as are the homebuilt aircraft from dozens of other kit makers. The number of all-wood aircraft is not large, but these airplanes (including the high-performance F.8L Falco and Periera GP4) also accommodate simple internal antennas. In addition to the homebuilts, a number of new factory-built airplanes are made of structural fiberglass. They include the Lancair Columbia series and the Cirrus SR line. These single-engine, four-seat, fixed landing gear airplanes are the fastest propeller-driven airplanes in their categories.

Two types of VHF and UHF ham antennas work well inside these aircraft. Quarter-wave verticals, which can be fed with RG-58 or similar small 52-Ω coax, may fit inside the fuselage of some of them. Line losses at VHF and UHF frequencies are considerable, but coax runs are short enough to eliminate this concern. These antennas require a metallic ground plane such as a thin aluminum plate or copper screen at the base of the antenna. The radius of the ground plane should be approximately one-quarter wave, but the exact diameter is not critical. These types of antennas may be installed after the aircraft is built. As always, builders must be absolutely certain that the coax feed line and the antenna itself cannot get loose and interfere with control cables, pushrods or other aviation-related systems, such as transponder antennas and navcom units.

The other appropriate antenna type is the half-wave dipole, particularly if it can be oriented vertically. It is often fed with 75-Ω coax and used with an impedance-matching device to accommodate the 50-Ω transceiver output. The quarter-wave ele-

ments can be made of wire or, for greater bandwidth, from copper tapes. The elements could be fastened temporarily inside fiberglass structures for tests of SWR and effectiveness. Once satisfactory performance is determined, the antenna can be installed permanently. Where access is possible, the successful antenna could be fiberglassed in place. An alternative procedure is to imbed several dipole antennas in a fiberglass aircraft during construction and experiment with the effectiveness of each. For example, a vertically polarized dipole could be imbedded in the tail's vertical fin.

An important point is to note that not all composites are RF-transparent. Specifically, carbon-fiber airplane parts absorb RF much like sheet metal does. If you own or are building an aircraft with areas of this exceptionally light and strong material, do not plan on having RF pass through it.

Corporate pilot Barry Brown, KØYLU, adds his advice on portable VHF/UHF aircraft antennas. *"For best results with VHF and UHF, use some form of mount to locate an antenna by a window or outside the aircraft. Remember that a flex antenna inside a metal fuselage will not radiate downward well at all. Small suction cup mount accessories are useful to locate a handheld flex antenna right at the window in the aircraft, especially for pressurized aircraft with no options to feed a coax cable to an outside antenna.*

For light, unpressurized, single-engine, fixed-gear aircraft, mounts can sometimes be fabricated to locate an antenna element outside the aircraft. The fixed-gear Cessna spring steel landing gear legs especially lend themselves to using a clamp-style mount placed on them with the coax fed into the cabin through the flexible rubber door seals.

I made such a mount easily out of two rectangular pieces of 1/4-inch insulating glass epoxy material with four bolt holes located at the corners. I used # 8-32 stainless screws with wing nuts for fast installation and removal. A 19-inch brass antenna element was mounted on and perpendicular to the half with the flathead anchor screw countersunk for insulating purposes. The coax center lead was attached on the element side, via a ring terminal between the element and glass epoxy plate. Another ring terminal was attached to the shield with its mounting screw threaded through the same glass-epoxy plate to contact the gear leg for the local ground contact. This setup goes on and comes off rental or club aircraft quickly and works well. Please don't try to add such mounts to retractable-gear aircraft!

Use simplex frequencies to avoid bringing up repeaters within line of sight, which is significant from only a few thousand feet above the ground. If you can't raise anyone on the popular simplex frequencies, consider a brief call on an area repeater and announce that you are switching to a simplex frequency."

ON THE ROAD VS IN THE SKY

Jim Weir's favorite two personal stories with aero Amateur Radio follow. *"I've probably logged 5000 aero ham contacts over the years, but there are two that spring immediately to mind as the most memorable. Some hams who haunt the amateur radio newsgroups know that I was one of 130 folks who ran for governor of California in 2003. Many had a QSO with me as I flew the length and breadth of the state in my Cessna 182 on what we jokingly referred to as the "prop-stop" campaign tour, reminiscent of the whistle-stop tours of the last century.*

What only a handful of hams (who have, for obvious reasons, requested to remain anonymous) have known is that we were always one step ahead of the eventual winner of that race (Governor Schwarzenneger) because we could track his campaign bus from the air. Arnold Schwarzenneger's top-secret campaign rally destination became evident from how the bus was proceeding. We had friends and fellow hams in each of the major towns up and down the state. We contacted them on simplex, and they came out to pick us up at the closest airport. We beat the governor-to-be to each of his rallies by at least half an hour, got our chance to talk to the local press first, and he could never figure out how we knew where he was going and how we got there ahead of him time after time!

It was the last weekend before the election, and I was down-state 500 miles from home and winging it up California's Great Central Valley at 0300 local time. I knew my XYL (Gail, KB9MII) would be worried about me, and I was still 300 miles away—well out of repeater range for home. I got on 146.52 simplex and called over Fresno, not really expecting anyone to answer at this outlandish hour. But there was a reply. He was a high school principal who was up early getting prepared for his day at work. He volunteered to call Gail and explain where I was and that I was coming home just as fast as the airplane would let me. Thank you, sir; that is the true spirit of ham radio.

We must have done something right. With zero dollars in my campaign war chest, a good airplane, and a good ham radio, I came in 30th out of the 130 who

ran. Some 4000 people (a lot of them hams, I'll bet) voted for me for governor. Bless you all."

Weir, WX6RST, welcomes discussion of airborne Amateur Radio and other topics on the Internet newsgroups **rec.aviation.piloting, rec.aviation.owning,** or **rec.radio.amateur.misc**.

A FEW MORE STORIES

Bob Hayos, K6CUK, shares some of his aero mobile experiences. *"Aeronautical mobile ham radio is perhaps one of the most interesting phases of the hobby. For me, such activity has involved some unusual operations.*

From an airship: Lighter than air (LTA), flights in non-rigid blimps have been possible. With the occasional availability of short demonstration flights, I was able to operate amateur radio while blimp mobile. The radio for these flights was a compact 10-meter AM BC-222 packset of late WW-II vintage. With a QRP output of about 1 W and a short telescoping whip antenna, numerous ground-wave contacts were made. The metallic-painted skin of the airship served as a reflector for the antenna on the radio. Unfortunately, few long-haul contacts were made due to poor propagation. Local reports were of strong signals varying in intensity, and the variation appeared to correlate well with the direction of flight. The airship's skin affected the antenna's radiation pattern. Additional tests using VHF-FM transmissions with a hand-held radio did not seem to be affected by the airship's skin. Operating from a site some 2000 feet above ground provided the usual strong local signal reports.

More recent operations from my Lake amphibian airplane provided many useful and interesting contacts. The radio of choice for this aircraft was a channelized VHF FM, 25-W unit permanently installed into the instrument panel. This radio, a standard 2-meter unit, was easily modified to extend its frequency coverage from 144 MHz up to over 160 MHz, which included common marine VHF frequencies and US Coast Guard channels. This modification allowed operations during USCG Auxiliary missions while contacting Coast Guard facilities and other surface vessels. The installation was wired directly into the plane's electrical system. Final approval for this installation was signed off by an FAA electronics inspector from a local FAA office despite initial concerns. It is interesting to note that other electronic equipment such as VHF direction finding gear and an amateur television

transmitter were also installed at different times in this aircraft."

Bob found an opportunity to demonstrate amateur television from his aircraft several years ago. For some time, US Coast Guard aviation personnel had considered the possibility of using aerial surveillance of maritime activities from small surface craft as well as from larger vessels. Bob felt this would be an excellent opportunity to put his ATV setup into practical use helping to evaluate the potential of various maritime activities.

Installation of a standard handheld color video camera coupled to a 70-cm video transmitter and a quarter-wave whip antenna was made easily in the aircraft. The receiving end of this equipment was a 70-cm down converter driving a standard television receiver with a similar simple whip antenna. The ground equipment was installed on a hilltop USCG station facility. Voice communication was established using a 2-meter link. Excellent high-resolution video and voice contact were completed from the aircraft to the ground station with only nominal blanking of video signals as the aircraft turned. Video recording of what the airborne camera saw was made available for later viewing. The aircraft was able to circle "targets of opportunity" and transmit real-time happenings. Although these tests were completed by hams using ham frequencies, they served to prove the value and feasibility of such surveillance to the government observers. Later tests were scheduled using similar techniques, but this time on military frequencies.

Sailplanes

Sailplane mobile is yet another aspect of aeronautical mobile ham radio that Bob has enjoyed. He notes that installation of any equipment in gliders is limited by extremely tight space and power requirements. However, a sailplane with an integral engine for self-launching allows tapping into engine electrical power. For all practical purposes in an unpowered glider, ham operation is limited to using a handheld radio with a self-contained antenna and power source.

Here is another aeronautical mobile radio adventure experienced by Bob Hayos, K6CUK:

"As a longtime ham operator and a member of the US Air Force for many years, I have had numerous opportunities to participate in aeronautical mobile amateur radio operations. Several years ago, while attending an old timer's reunion of sailplane pilots in Budapest, Hungary (an international yearly event held in a differ-

ent country each year), I jumped at the chance to combine my two favorite hobbies: flying and ham radio.

The event was held at the lovely country airport at Farkashegy, about 20 km west of Budapest. Several hundred pilots from many countries arrived bringing a large variety of beautiful sailplanes, many of a vintage dating back to the early '30s. Although most of my flying had been in powered aircraft, I had long before become qualified and licensed to fly sailplanes. The method of launching included aero tow where the sailplane is towed from the ground by an airplane. In this case the towplane was a Polish Wilga 35. The sailplane is then towed to about 3000 feet, where the sailplane pilot releases the towline and begins soaring in unpowered flight.

My opportunity to participate came after a brief checkout with one of the Hungarian pilots. Shortly after the checkout I was on my way in a Grobe medium-performance sailplane. Sitting in my lap was my Icom 2-meter handheld radio. At first I was very busy talking to Budapest air traffic controllers on the aircraft radio to advise them of my location and direction of flight. These most courteous people were ready to answer all aircraft radio transmissions in a variety of languages, including English.

My flight was to take me toward the Harmas Hatar Hegy area, and soon I was answering many calls on the 2-meter ham band. It seemed as if every Magyar ham wanted an aeronautical mobile radio contact in his logbook. The fact that I was signing 'K6CUK/AM/HA5 sailplane mobile' was, I suppose, the reason for the QRM. At any rate, it was a delightful experience, but one that had to end all too soon. The thermals were weak, making it necessary to turn back to Farkas Hegy field and prepare for a landing. As I entered the traffic pattern for the field I said my last 73 and promised to send QSLs.

The aero ham flight was the last of a number of soaring flights during my wonderful stay in Hungary. In addition to my air mobile radio contacts, I had several hundred QSOs on HF with a Yaesu FT-101 HF radio that I operated from a table outside our barracks at the airport. A longwire antenna placed high up in neighboring trees by my good friend Gyuri, HA5JI, served to keep 20 meters going strong.

My three weeks at Farkas Hegy Airport will always stay in my memory as one of the most fun-filled vacations of my life. In addition to meeting many pilots and amateur operators from many lands, I found how small this world really is and how alike we all are."

Fig 3-23—Pilots of powered hang gliders—known as *trikes*—often equip them with radios. Special attention must be given to noise reduction by using helmets with noise-canceling mics and sound-attenuating earphones.

Ultralights

Ham radio operation from an *ultralight* is much the same as that of gliders and sailplanes. Ultralights are single-seat, slow-flying, low-power aircraft weighing less than 254 pounds empty. Currently, there are two-seat ultralight trainers, but these are to be used only for flight instruction, so their legal usefulness as an Amateur Radio platform is questionable. See **Fig 3-23**.

The new *light-sport aircraft* (LSA) category will eliminate the waivered two-seat ultralight trainer as these aircraft are brought into the FAA-registered LSA category. In ultralights—because of weight, power and space limitations—few opportunities exist for successful hamming. VHF/UHF handheld radios on 2 meters or 70 cm are an obvious exception.

Model Aircraft

Although this chapter is primarily concerned with amateur radio communications while airborne, model aircraft operations offer a variant of this activity. For some time hams have successfully flown radio controlled model airplanes with an on-board television camera and transmitter relaying live video to the ground-based pilot on ham frequencies. If the television camera is positioned in the aircraft facing forward in the direction of flight, the transmitted video will appear to the control pilot on the ground as if he is in the aircraft looking ahead. When the images are real time, in color, with engine and wind sounds, the effects can be most dramatic and realistic. Whether this operation is actually aeronautical mobile radio is debatable, but few could deny that it is most certainly ham radio/TV airborne mobile and exciting to experience!

In addition to down-linking ATV, hams have long used designated portions of the 6-meter band for radio control of the air-craft itself. One advantage is that 6-meter control frequencies are considerably less crowded than the standard US 72-MHz RC frequencies. In large model aircraft clubs and at contests, there is often a considerable wait for an available 72-MHz control frequency, but there is seldom a wait for hams flying on 6 meters. Several of the popular RC system manufacturers offer systems for 6-meter operation.

OPERATING PROCEDURES

Jim Weir, WX6RST, suggests the following order of priority for airborne amateur radio use while you are flying the aircraft, and I agree:

a. Fly the airplane.
b. Communicate with Air Traffic Control.
c. Fly the airplane.
d. Communicate with other aircraft or aviation stations on the ground (control towers, Flight Service Stations, Unicoms) relative to flying.
e. Fly the airplane.
f. Navigate the airplane.
g. Fly the airplane.
h. Use the ham rig.

Despite the last-priority listing of aeronautical Amateur Radio, this mode of operating can be a source of considerable education and fun. Because an air-to-ground signal carries so much farther than usual line-of-sight communications, listen even more carefully than usual before transmitting. If you are on a repeater frequency, listen intently for the possibility that you are keying up more than one repeater. If that happens, it is usually best to clear that frequency immediately. As with flying an aircraft, operating an Amateur Radio station in an aircraft requires more than a bit of common sense. We would prefer not to hear of any amateur operating CW while flying an ILS approach in a single-seat airplane.

HOW TO GET STARTED

If the stories and technical details have whetted your appetite for some in-the-air, on-the-air radio time, there are a number way to proceed. Knowing a pilot with his or her own airplane is a good start. Most aircraft owners are happy to share their airborne experiences with others, especially when there is an incentive to do something interesting and unusual. In the US, a pilot with a private license cannot charge for taking someone flying, but direct expenses can be shared. That means that a passenger may pay for his or her share of the cost of gasoline, for example. That is often enough incentive for a pri-

Amateur Radio and Ballooning

By Brian Mileshosky, N5ZGT

Amateur Radio, as we all know, goes pretty much anywhere—on the trail; in the car; even in orbit. Another wonderful place to enjoy Amateur Radio is while afloat in a hot air or gas balloon! In addition to the fun created by marrying the two exciting hobbies, Amateur Radio also provides an invaluable tool to the balloon crew—chase communications.

Balloonists across the world take Amateur Radio to the skies to enjoy aeronautical mobile communications—what better place to make a few contacts while floating thousands of feet above ground on a cool, calm day? Altitude certainly makes for interesting simplex communications, and hams on the ground are always excited about aeronautical mobile QSOs. Amateur Radio also serves as a tool for a balloonist and the crew that chases him across the sky for later recovery. Many crews have traditionally depended on business-band radios to keep in touch while aloft. However, channel limitations have caused them to seek other methods of communication.

The EI2AIR QSL card leaves no doubt about what these Irish hams were doing! *(Photo courtesy EI5HW, from Feb 2005* QST.*)*

Amateur Radio provides the ability for crew and pilot to maintain voice communications throughout the flight via simplex operation or repeaters. Since the pilot has a bird's eye view of the ground below, for example, he can assist his chase team by describing certain roads that can be taken to pursue the balloon. The chase crew is also able to feed valuable information to the pilot, such as the location of power lines and other potential hazards that blend in with the surrounding terrain when viewed from above. In competitions, particularly those involving gas balloons, crews can provide weather information to the pilot to aid in decision-making. Equipment for keeping in touch via voice can consist simply of an HT aboard the balloon, and an HT or mobile in any of the chase vehicles.

In addition to standard voice communications, some pilots also take Automatic Position Reporting System (APRS) trackers along for the ride. Not only can the curious public track the progress of a floating balloon on their own APRS screens or via **www.findu.com**, but the balloon's chase crew can also utilize APRS to know exactly where it is at any time during the pursuit. Gas balloonists particularly turn to APRS during distance competitions for this very reason. All that's needed aboard the balloon is a GPS receiver, an HT and terminal node controller (TNC). APRS digipeaters and particularly Internet gateways help to relay the positioning information to the crew and event organizers.

Amateur Radio is aboard many hot air and gas balloons flown individually or at balloon rallies across the world. This is especially true during the Mecca of ballooning, the *Albuquerque International Balloon Fiesta*, where over 800 hot air and gas balloons simultaneously grace the New Mexico skies each

October. The Albuquerque Gas Ballooning Association operates two repeaters devoted to chase communications and as an on-air meeting point for hams, pilots and crew members. It also assists equipping APRS trackers on gas balloons participating in the Fiesta's annual cross-country distance race. This is for safety, and also so the curious public can see where the balloons are in near-real time.

Many ham clubs across the country join the fun by offering their services through providing communications support and tracking balloons for safety. For example, each year the Durango Amateur Radio Club partners up with the organizers of the *Snowdown* hot-air balloon rally to monitor the progress of all the balloons and make sure they land safely. Each club member has pictures of all participating balloons to aid in identification, and if a problem develops they provide assistance. Whether or not the pilots or crews have ham radio aboard with them, the club adds an invaluable level of safety to the entire event.

Other clubs across the country hold licensing classes for balloon pilots and crews, thus enabling them to take ham radio with them on future flights. If you're in a club that serves an area lucky enough to host balloon rallies, why not join the fun and offer your services?

Marrying ham radio with ballooning creates a whole new realm of fun, convenience and peace of mind. Enjoy making those distant contacts and having a reliable method of communication with your crew. Be safe, and may the winds welcome you with softness.

Links of interest:

Balloon Explorium: **www.explorium.org**.
Albuquerque Gas Ballooning Association: **www.gasballooning.org**.
Albuquerque International Balloon Fiesta: **www.balloonfiesta.org**.
APRS details: **www.arrl.org/FandES/ead/materials/videos.html**.

vate pilot airplane owner to schedule a flight.

Lacking a personal acquaintance with an aircraft owner, the prospective aero ham has some other opportunities. Businesses at airports that offer aircraft rental, flight instruction and other services are known as FBOs (fixed-base operators). Most small airports have one or more FBOs, and contacts for local flying clubs are often posted in the FBO office. Call the president of a flying club with an offer to share the cost of a brief flight while doing some handheld-radio hamming. This will probably result in an invitation or two.

A similar possibility is to attend a meeting of the local chapter of the Experimental Aircraft Association (EAA). While some EAA chapters are more into building aircraft than flying them, most have active pilots who own their own planes. Most welcome visitors and will help get you into the air with a qualified pilot. You can get the name and phone number of the president of the nearest chapter by calling EAA headquarters in Oshkosh, Wisconsin (920/426-4800). Ask for the chapters office.

A step beyond the ride-sharing method is to pay for a flight. Most FBOs still offer airplane rides. Hot-air balloon rides are another possibility. In the US, nearly every licensed hot-air balloon pilot has a commercial license, which means he or she can charge for a ride. Many balloon

Fig 3-24—Hams have made contacts from the air while riding in hot air-balloons. As in all forms of aeronautical Amateur Radio, hams need permission from the pilot in command.

owners subsidize their flying hobby by charging. Some balloon operators advertise in the telephone yellow pages. The disadvantage of hamming from a hot-air balloon is that they seldom fly very high or long because of the limited supply of propane. See **Fig 3-24**.

Yet through an active 2-meter or 70-cm repeater, a balloon-borne ham is likely to

have enough contacts to make the effort worthwhile. Besides, based on a French tradition, passengers are frequently offered champagne at the end of a flight! (The French were the first people to fly, and they used hot-air balloons.) Producing a special QSL card with a color photo of the balloon flown is simple and practical with the advent of digital cameras and color printers.

Whatever method of getting in the air is used, it is important to talk with the pilot and get permission to operate before arriving for the ride. You can emphasize that aeronautical Amateur Radio is not particularly unusual. It is also safe and perfectly legal with the "captain's permission." Added incentive for commercial rides—you could point out—is that you will mention with whom you are flying when you do it. Under the current interpretation of FCC rules, that kind of promotion is legal because there is no financial benefit to the ham.

WRAPPING UP

From these stories and technical details—both current and historic—you have a picture of an Amateur-Radio activity that relatively few hams have experienced. And it's worth noting again that there is no need to get airborne to have the fun of aeronautical mobile hamming. More than half of the participants are hams on the surface, and they are having almost as much fun as their up-in-the-sky contacts.

CHAPTER **4**

Motorcycle Mobile

Al Brogdon, W1AB, was licensed as Novice WN4UWA, and then General class W4UWA in 1952, and as Amateur Extra class in 1956. His other call signs have included K3KMO, N3AL, DL4WA, DJØHZ, LX3TA, and M1M. Almost all of Al's operating has been on HF CW—ragchewing, DXing, contesting, mobile CW and CW traffic nets. He has operated portable from TI2, KH6, KL7, VY1, PJ7, EI, GM, BY and on the 1987 MADRAS Tangier Island DXpedition (KT4A). An ARRL member since 1953, Al has been a Life Member since 1971. He operated mobile CW from his Honda Gold Wing from 1992 to 2001, and now rides a Honda Helix touring motor scooter.

Al has authored many articles in the various Amateur Radio magazines since his first article was published in *QST* in September 1958, and now writes the "75, 50 and 25 Years Ago" column. Al worked as an electrical engineer for 11 years and as a technical writer and editor for 29 years. He worked for two defense contractors until his retirement, and as an editor on the *QST* staff for four years. He moved to New Hampshire in 1998. Al's other interests include aviation (a private pilot since 1970), music (cornet, trombone, and tuba player in Dixieland and Bavarian oompah bands) and travel.

Fig 4-1—Riding along—enjoying the fresh air, sunshine and Amateur Radio.

INTRODUCTION

Riding a motorcycle is very pleasurable. The rider can enjoy the perfumes of nature—such as the scent of fresh-cut grass or the fresh aroma following a rainfall. He or she can feel the cooler air when riding into a wooded area, where the trees' transpiration lowers the air temperature. He can see more of his surroundings than he would see driving the same route in an enclosed vehicle. It's easy to get bitten by the motorcycle bug!

While riding, the driver's mind is free to roam. For a radio amateur, the mind sometimes thinks, "*Wouldn't it be nice to have a rig on the bike to pass the time on these road trips?*" See **Fig 4-1**.

In this chapter, most of the discussion will deal with generalities rather than specifics, and I will suggest ideas that you can consider for your two-wheeler. The main reason for not dealing with specifics is the great range of variation from one motorcycle to another, even within the same manufacturer's product line, and most certainly from one manufacturer to another. Most notably, the space available for mounting rigs and antennas will vary considerably from one cycle to another.

A second reason is the variation from rig to rig in ham radio gear. There is considerable variation in the mounting requirements and accessibility requirements for various amateur transceivers and accessory equipment.

The intent of the text and the photographs in this chapter is to provide ideas for the reader, which he or she can then implement with his or her cycle and ham rig. This puts the ball in *your* court to figure out exactly how and where you're going to mount the rig, the push-to-talk switch, the antenna, the power and antenna leads, etc. The hams I have known who are also interested in motorcycling are usually pretty smart guys, so I have confidence that you, the reader, will be able to get it together—on two wheels.

I will cite examples of devices and gadgetry that are commercially available. These are not to be taken necessarily as recommendations, but rather as examples of what is available. When you start configuring your own motorcycle-mobile system, you should perform thorough searches—both on the Web and with your local cycle dealers and ham radio dealers. Find the various possibilities that are available to you.

To add spice—as well as practical information—and to stimulate your thinking, the chapter includes descriptions of the setups used by several motorcycle-mobile hams and tales from the trips they have taken and the operating they have done. Read them carefully and you can pick up more tips and ideas of how to implement your own motorcycle mobile setup.

In this chapter, I'll use the notation "/MC" (pronounced "Slash MotorCycle") to indicate "motorcycle mobile."

SAFETY FIRST

The first and last consideration of riding a motorcycle must always be *SAFETY*. The cyclist is much more vulnerable to harm than the driver of a large, enclosed vehicle. A cyclist is also less visible to other drivers, not only because of his smaller size but also because many drivers look *only* for car- and truck-sized vehicles when they perform a visual scan for other traffic.

In the USA in particular, there are far too many drivers who give driving and traffic only minimal attention and who don't always observe the traffic laws or the rules of the road. You can never fully anticipate what some inattentive or road-raging driver might do that would endanger your cycle and your life.

America's roadways are, therefore, fairly dangerous places, especially for cyclists. So beware, and be wary! Always keep in mind the First Rule of Motorcycling, as quoted in the sidebar "The First Rule of Motorcycling."

Adding an amateur rig to the motorcycle mix makes it necessary for the cyclist to become even more cautious and to carefully observe all safety considerations. He must always make the operation of the motorcycle his *first* priority. If the rider has any excess brainpower left after that task is accomplished, that extra brainpower can be used to operate ham equipment.

When an unexpected traffic situation arises—and they always do—the ham rider must quickly and automatically switch ham radio out of his thinking processes. Experience will make this brain switching automatic. Operate the cycle and watch the roadway and other traffic *first*; operate the ham gear *second*. To operate /MC, an ama-

---

> ## The First Rule of Motorcycling
>
> Be aware of every other driver and vehicle within your range of vision, whether his vehicle is in motion or not, assume that he or she might do something that would put your life at risk, and take steps to stay out the danger zone he might create.

teur should have considerable experience in four areas:

(1) As a cyclist
(2) With the specific cycle he or she is riding
(3) As a ham operator
(4) With the specific equipment he will use on the cycle.

Experience will translate into the smallest possible amount of conscious thought required to operate the cycle and the radio equipment. Furthermore, and for the same reason, the ham equipment should be either simple to operate or, if it is of the newer and more complex variety, operated in a simplistic manner. The rider cannot take his attention from the roadway and traffic to scroll through menus and make selections or changes while riding.

It's best not to try to operate ham equipment when riding in even moderately heavy traffic or during inclement weather. At those times, the rider needs to be fully alert to the highway and other traffic.

GETTING STARTED WITH AN HT

The easiest way to get a taste of motorcycle mobile operation is to operate using an FM handie-talkie (HT) on 2 meters or other VHF/UHF bands. As noted in the Safety First discussion, the HT should be set up for operation in a simplistic manner—the only controls the operator should routinely adjust while moving are the volume control and the push-to-talk (PTT). If other adjustments such as frequency changes are desired, the new /MC operator should stop his cycle and put down his kickstand before making them.

In the following sections, we'll look at the various aspects of operating /MC that you should consider when you start making plans to operate from your two-wheeler. See **Fig 4-2**.

Earphones

Let's first consider the question of whether to use one earphone or two earphones. Motor vehicle laws in most states

---

Fig 4-2—It's easy to mount an HT on a motorcycle.

prohibit the use of earphones in *both* ears when operating a motor vehicle. However, as a practical matter, cyclists almost universally use earphones on both ears for their CB radios, entertainment systems, and intercoms. I have yet to hear of a police officer ticketing a cyclist for having earphones over both ears, so getting ticketed must be fairly rare.

I leave it to the reader to decide whether he or she wants to be a law-abiding citizen or to listen with both ears. As a practical matter, the discussions in this chapter will always speak of a pair of earphones.

There are a number of earphone types that can be used underneath a helmet when you're getting started with /MC. Wait a minute ... you say you don't wear a helmet? *Please* close this book right now and never *ever* read this chapter again. A helmetless rider should not operate motorcycle mobile. There are already enough greasy spots on the pavement of our roads and highways.

Anyway, back to earphones... One good possibility is to use the small earphones of the "earbud" variety. The earbud earphones fit snugly in the user's ears and will not interfere with putting the helmet on. Another type of earphone that works well is the type with a loop that goes over the ear to hold the earpiece in place.

If you want to spend a little more money, you can buy earbuds that fit better than the usual type, which will have either a "universal" ear insert or several from which you can choose for the best fit to your ears. For example, the Shure E3c Sound Isolating Earphones come with ear inserts of different size, so you can pick the pair that best fits your ear.

If you want to go First Class, you can buy earbuds with custom-molded earpieces, such as the UE5c made by Ultimate Ears (on the Web, see **www.ultimateears.com/UE-5c.htm**). The UE5c is a more modestly priced model of the custom-molded earpieces developed for professional musicians to use as on-stage monitors.

If you live near a large music store, you could check with them to see what they might have in the way of custom-fit earbuds.

Otherwise, you can find and order molded earbuds on the Internet.

Typically, custom-fit earbuds include a molding kit that you use at home to make an impression of the inside channel of your ears. You then mail that impression to the manufacturer, who produces the custom-fit earpiece at no further cost.

However, in some cases, the manufacturer wants you to have the impression made by an audiologist, so check that out before you buy. The total cost could get pretty high if you need the services of an audiologist.

You can get into /MC by using something as simple and available as the typical headband earphones that come with every piece of portable stereo equipment. Most households have a few dozen of that type lying around. The downside of using those earphones is that that are likely to slip askew on your ears in use.

After you've tried /MC operation and you want to improve hearing the receiver audio, you should consider installing earphones in your helmet. Some helmets are available with earphones already mounted in them. Various helmet-mounted earphones are available from cycle dealers and cycle shops.

One excellent source of helmet-mounted headsets (microphone and earphones) is the J&M Corporation. Go to **www.jmcorp.com/index.html** to see what J&M has to offer.

Microphones

The motorcycle mobile ham should not use a handheld mike with a push-to-talk switch on it—the driver needs both hands to operate the cycle. Possible choices of a mike include

(1) An earphone/mike headset that can be worn under the helmet
(2) A boom mike attached to the helmet
(3) Earphones that pick up the voice via bone conduction and also act as the mike. Mikes for helmet mounting are available at motorcycle dealers and via the Internet, or you can fabricate your own.

A foam-rubber cover on the mike is quite useful for reducing wind noise into the mike. It's even better to have a noise-canceling mike, to reduce both the wind noise and cycle noise. Again, noise-canceling mikes can be found at ham radio dealers or on the Internet.

Most cyclists agree that you shouldn't try to use VOX for /MC operation. There's just too much noise on and around the cycle for reliable triggering of the transmitter by your voice, yet *not* having it triggered by other sounds.

The Push-to-Talk Switch

An external push-to-talk switch for the HT can be mounted on the handlebar and positioned so it can be pressed by the thumb or forefinger without removing your hand from the handgrip. It's better to have the PTT on the left handgrip, so it won't interfere with throttle operation. A rocker or toggle PTT switch with a center-off position makes a good PTT switch. You can use the momentary contact in one direction for short transmissions and the positive (locking) contact for longer ragchew type of transmissions.

If your cycle is one of the larger touring cycles that has an on-board audio entertainment system and CB radio, there usually is a control box for the CB and audio located near the left handgrip, so the left forefinger can pull a trigger switch downward to transmit. It's easy to mount an A-B toggle switch on the front panel of that box to switch the PTT circuit between the ham transceiver and the CB. If your cycle doesn't have an existing CB PTT circuit you can tap into, you'll need to make a bracket for a PTT switch.

First decide whether you want to use your

(A)

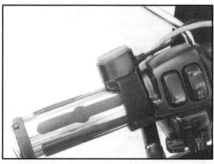

(B)

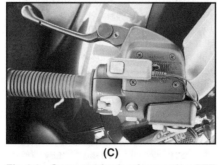

(C)

Fig 4-3—Some methods of mounting PTT switches on a motorcycle.

thumb or your forefinger to activate the switch. Then sit on the saddle and visualize where you can mount the PTT switch and what kind of bracket you'll need to mount it to the cycle.

For PTT use, enclosed and waterproof switches are ideal. But you can usually put shrink tubing on the exposed external parts of the switch to provide a reasonable level of waterproofing. See **Fig 4-3** for some ideas on mounting a PTT switch using Velcro.

Here's an idea related to /MC operation that you might find useful. For many years I used an inexpensive throttle lock on my Gold Wing. It's a simple friction device with adjustable clamping force and with a lever to lock or unlock the throttle lock. I set the friction of the lock just barely high enough to hold the throttle setting when I take my hand off the throttle. I can then twist the throttle easily against the lock's light friction to adjust the throttle setting. Various types of these simple throttle locks are available from motorcycle dealers and repair shops.

The advantage of using a throttle lock is that you can, for periods of a few seconds, take your right hand off the throttle for any reason, including operating the ham rig.

The Antenna

There are several possibilities for a simple antenna setup that would give you a first taste of VHF/UHF FM voice operation from your motorcycle. First, it isn't a great idea to use a rubber duckie for /MC operation. Many hams who operate VHF/UHF using HTs and rubber duckies are not aware of how inefficient rubber duckies are. And the smaller the rubber duckie, the less efficient it is.

It's much better to have at least a quarter-wave whip, a half-wave antenna (such as those made by Larsen), or a coaxial dipole on your motorcycle. It's even better to have an antenna with some gain, such as a $^5/_8$-wave whip, a collinear array, or a properly built J-pole.

Note the phrase "properly built J-pole." Most of the magazine articles about J-pole antennas give exact dimensions for building your own, using a fixed tuning stub at the antenna's feed point. However, when you try to replicate the antenna described in the article, small variations in measurement or materials from those used in the original (even a different brand of twin lead, which will often have a different velocity factor) can cause variations in antenna feed-point impedance and a resulting mismatch between the antenna and your transceiver.

Therefore, if you build a J-pole antenna, you should build it with an *adjustable* shorting bar on the tuning stub. You can then adjust the stub for the best match between the antenna and the coaxial cable. Of course, you should tune the antenna for a proper

Fig 4-4—Some members of the Motorcycling Amateur Radio club and their rolling antenna farms.

match with the antenna in place on the motorcycle.

The Motorcycling Amateur Radio Club (MARC) of southern California worked with engineers at Comet Antennas to develop a VHF/UHF antenna for /MC use that would be mechanically rugged enough to survive long-term use on a motorcycle. The result is the Comet HP-32 mobile antenna for 146 and 440 MHz. Roger D. Rines, W1RDR, reports that MARC members have had virtually no failures while using the HP-32, and that they view it as "indestructible." See **Fig 4-4**.

If you fabricate your own antenna and/or antenna mount, here's a way to check it for vibration problems: Have a buddy ride alongside you and take a look at your antenna at various highway speeds, and on various highway surfaces (the good, the bad, and the ugly). If he sees serious vibration, you might want to swap cycles with him and take a look for yourself.

If there is excessive vibration, you can take steps to stabilize the antenna and mount, so the vibration won't precipitate a premature failure. If you use a tall antenna for your /MC setup (such as a collinear), you can use a monofilament fishing line to keep the antenna in a near-vertical position.

It's always a good idea to have a low-impedance RF ground strap from the base of your antenna mount to a nearby place on the motorcycle frame. The best thing to use for a low-impedance ground strap is braided strap at least a half inch wide. See **Fig 4-5**.

If you can't find braided wire to use as a ground strap, here's how to make some from a length of large-diameter coaxial cable (such as RG-8). Carefully remove the black jacket from a length of coax, taking care not to cut any of the small wires that make up the braided shield. Then compress the length of the braided shield (which will make its diam-

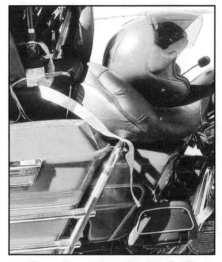

Fig 4-5—Note the ground strap that runs from the antenna base to the antenna mounting clamp and then on to the motorcycle frame.

eter larger) and slip the braid off the inside of the cable. Stretch the braid out to full length again and flatten it, and you will have an excellent low-impedance braided ground strap.

Here's an interesting grounding experience that I had when operating mobile from my Toyota van, but which is applicable to /MC antennas. In 1994 I built up a 4½-foot-long mag-mounted antenna mount to place on the van's roof, where I had mounted five single-band whips. I ran five runs of coax, one from each antenna, to an antenna selector switch beside the rig. If you would like to see a photo of that big mag-mount, look at the cover photo of the September 1994 issue of *QST*.

When I tuned each antenna on frequency and tried the antennas on the air, they worked quite well. *But...* I had ignition interference in the receiver that wasn't there when using

my previous single-band antenna mounted on the left rear side of the van.

After puzzling over this problem for a bit, I added a low-impedance ground strap from the rear of the mag-mount down and inside the rear door of the van, connecting it to a good vehicle ground point. The ignition noise went away. And the tuning of the antennas did not change.

The grounding provided by the shields of the five runs of coax was good enough for the antennas to work well for transmitting and receiving, but adding the separate DC ground eliminated the ignition noise.

Therefore, I suggest that a low-impedance ground from your /MC antenna base to your cycle frame would be a good idea—just in case it might be needed.

Antenna Mounting

Because of the vibration present in motorcycle handlebars, you should avoid mounting antennas on them. It's much better to mount the antenna to the cycle's body at a very stable location, on the trailer hitch (if you have one) or to the cycle frame itself.

If you plan to use an antenna larger than a simple whip, look for a place on the motorcycle frame or trailer hitch to attach a sturdy bracket to which the antenna can be mounted. Although a trailer hitch is a dandy place to mount an antenna base, it will put the antenna base (which does most of the radiating) quite low. One possibility when mounting an antenna on the trailer hitch is to fabricate a mast (perhaps four or five feet long) to elevate the feed point of the antenna, then mount the antenna at the top of the mast.

If you use a mast to elevate the antenna, two simple but effective antenna possibilities are

(1) A ¼-wave vertical with a counterpoise consisting of three downward-sloping radials

(2) A coaxial dipole (see *The ARRL Antenna Book* for construction information on either of these antennas).

If you use radials as a counterpoise on your motorcycle antenna, be sure they are located and arranged in a manner to minimize the possibility of poking yourself or someone else in the eye!

Many VHF/UHF hams who operate /MC clamp an antenna mount on the chrome-plated luggage rack that's typically on top of the center-mounted rear trunk. See the photos in **Fig 4-6** for some examples of antenna mounting systems used by MARC members.

MARC found a manufacturer who custom-made brackets for them for mounting antennas on Gold Wing cycles, and another bracket for mounting PTT switches on Gold Wings. Point your Web browser to **http://marc-hq.org/40.technotes/40.htm** to see the description of those two brackets and for ordering information.

Hams who operate VHF/UHF are accustomed to the ease of use of a mag-mounted antenna base on their cars and trucks. Oops! There isn't much steel in a motorcycle body for a mag mount to hold onto. If you want to try a mag-mounted antenna on your cycle, there's a simple way to do it if you have a center-mounted trunk (as does the Honda Gold Wing), but no luggage rack on top of the trunk. This will work for smaller antennas, such as a ¼- or ⅝-wave whip.

Get a strong magnet and place it on the underside of the point where you want to mount the antenna. You can use thin double-sided tape or a low-tack adhesive to hold the magnet in place while you position the antenna's mag-mount base directly above it. The two magnets will have enough mutual attraction to hold the antenna firmly in place.

There are many other possibilities for mounting a VHF/UHF antenna on your cycle. Take a look at your cycle and let your imagination roam free. For further ideas, you can consult *The ARRL Antenna Book* and *ARRL's VHF/UHF Antenna Classics,* as well as other antenna books and articles in *QST* and other Amateur Radio magazines.

Many hams who contemplate /MC operation on 2-meter FM are concerned about the adequacy of the motorcycle as an antenna counterpoise. I found in my /MC operation on the HF bands that the antenna performance was excellent even on 80 and 40 meters. If the motorcycle is an adequate counterpoise on the lower HF bands, its performance at 2 meters will also be excellent.

Securing the HT when Riding

The HT can be either carried on the rider's person, or mounted on a bracket on the cycle. Let's first consider carrying the HT on the rider's body, since we're trying to make it simple for you to start operating /MC.

A simple approach of where to stash your HT when cycling would be to stick it in your jacket or shirt pocket. That would work, but you must ensure that the pocket will not allow the HT to fall out. Also, it might be difficult to find the volume control if you need to adjust it.

Fortunately, there is an excellent solution to the problem of how to carry the HT on your body. There is a product available from various manufacturers called a "chest harness," "shoulder harness," or "tactical radio patrol harness." The harnesses were originally developed for the Ski Patrol, and have since become widely used by police, fire and EMT personnel. It appears as if the Amateur Radio community hasn't yet caught on to how useful they are.

These harnesses are usually made of sturdy nylon, which makes them quite durable. The harnesses hold the HT on the wearer's chest (via a system of chest straps and/or shoulder straps), in a position that makes it easy to reach the radio's controls. The harnesses have pockets for one HT, and some have pockets for two HTs. Others have additional pockets for pens and other supplies.

I did a search using Froogle, searching on the words "chest radio harness." and found quite a few Internet suppliers of radio harnesses, with various configurations and features. Most of the Web pages you will find will have descriptions and dimensions of the harnesses, so you can determine which harness is best suited to use with your own HT. The good news is that these radio harnesses are quite reasonably priced.

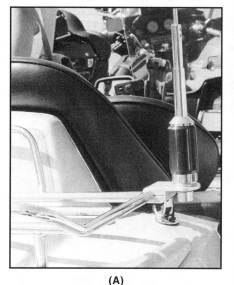

(A)

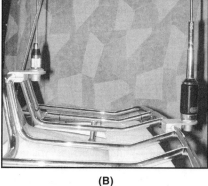

(B)

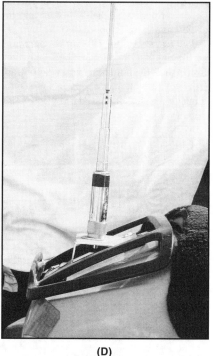

(D)

(C)

Fig 4-6—Some schemes that can be used to mount a VHF/UHF antenna on a motorcycle.

Now let's consider how the novice /MC operator could mount his HT on the cycle itself, which provides some advantages. There are a number of commercially available systems available for mounting various types of gadgetry on motorcycles.

One versatile series of mounting devices is the R-A-M ("Round-a-Mount") system, which modestly claims that you can "mount nearly anything, anywhere." The R-A-M system is a two-piece system with various types and configurations of base pieces and various types of "device cradles." You can choose the base and the cradle that best suit your cycle and HT using the information on R-A-M's Web page, **www.ram-mount.com/.** The page **www.ram-mount.com/hamradio.htm** specifically addresses the mounts that are useful for ham gear.

Other manufacturers' mounting systems are available for use with motorcycle accessories. Check with your local cycle dealer and ham radio dealer to see what is currently available.

You could also fabricate a mounting system for your HT with hardware from your local hardware store. The result might not be quite as attractive as a commercial mounting system, but it could work well.

The very simplest (and least expensive) mounting system to use for mounting the HT to the cycle would be a bracket you could make from sheet aluminum and mount on your cycle, either on the handlebars, on top of the gas tank or around the console or dashboard area. (Remember to avoid obstructing the driver's view of the dashboard and the cycle's indicator lamps).

The bracket for the HT can be something as simple as an open-face shelf fabricated from a sheet of aluminum, with Velcro straps (such as the type commonly used for bundling cables) holding the HT in place on the shelf. That type of Velcro strap can be extended by connecting them end-to-end until you get a long enough strap for your application. See **Fig 4-7**.

If your cycle is not in the touring cycle class, and there's no console area in front of the driver, another possibility is to buy a tank

Fig 4-7—Simple Velcro straps can be used to secure your HT to your motorcycle.

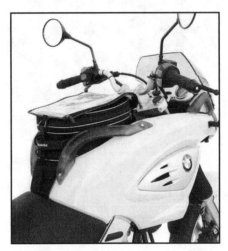

Fig 4-8—A tank bag can be used to hold your HT on your motorcycle.

bag that mounts on the gas tank, and place the rig in it. See **Fig 4-8**.

Powering the /MC HT

The simplest and most obvious way to power an HT on your cycle for /MC operation is to use the rig's internal batteries. If you do that, it's advisable to carry two spare batteries. (The First Rule of Batteries is that "A battery's charge always runs out at the worst possible moment.")

Another method of powering your HT is to use the HT's external 12-V power supply connected to the cycle battery. If you do that, you can run the external antenna feed line and the external power line together, so there's less chance that your body or motorcycle parts get tangled up in multiple cables.

Still a third way to power the HT is to have a physically small 12-V battery—preferably one with greater amp-hour capacity than the rig's internal batteries—in a convenient location together with the HT's 12-V power supply. These two items could be secured in one of the cycle's smaller compartments, its saddlebags or its trunk. Another motorcycle battery would be a good choice for this external battery.

Taking Your /MC Show on the Road

So there you have it! That gives you some ideas on how you can use your HT with a fairly simplistic antenna to try VHF/UHF FM voice operation while you ride on two wheels. **CAUTION:** You are likely to get hooked on /MC operation. If that happens, read on for ideas about more extensive installations and more varied operation.

HF ON A MOTORCYCLE

After you've tried /MC operation using a simplistic HT setup for a while, you may want, as chef Emiril Lagasse says, to "*kick it*

Fig 4-9—A Kenwood TM-741A mounted on a motorcycle.

Fig 4-10—An Alinco DR-610 set up for /MC operation.

up a notch." Most of the following remarks can be applied both to VHF/UHF FM and to HF SSB or CW operation with transceivers that run 25 to 100 W.

"*CW Operation from a motorcycle?*", you may ask. Yes, it has been done. Descriptions of HF CW operation from cycles will come later in this chapter. For now, let's look at the second level of /MC setups, beyond the simple HT stage. See **Fig 4-9**.

At this point, we are looking at ham transceivers that are physically larger than HTs, that required higher DC primary current and which might have detachable faceplates. All these factors bring forth new requirements that the /MC operator must deal with. But they bring benefits in both operating ease and fewer dropped contacts.

Mounting a Bigger Transceiver

When mounting Amateur Radio equipment on a motorcycle, you must be sure that the installed equipment and its cables do not interfere with the motorcycle's existing equipment, controls, indicators or readouts.

The motor-vehicle laws of most states require that installed equipment must not obstruct the view of the cycle's dashboard, *when the driver is seated*. It doesn't count if the driver has to lean to one side or raise himself off the seat to see the speedometer, high-beam indicator, turn-signal indicator and other dash indicators and readouts. The driver must be able to see them all when he is seated in an upright position. And some police officers have been known to stop cyclists and check out this point of motor vehicle law.

The handlebars of a motorcycle are usually subject to more vibration than is the area around the gas tank, console and dashboard. This is true even with the Honda Gold Wing, a very smooth cycle. For that reason, it's best to avoid mounting a transceiver on the handlebars. A rig or the rig's removable control head can be mounted on the handlebars, but the area around the console would be better place to mount it.

Compared to a VHF/UHF HT, the larger size of a transceiver that also covers HF in the 25- to 100-W class, requires a strong, secure, and dependable mounting system. Some, but not all, of the larger VHF/UHF and HF transceivers are still within the size that can be accommodated by a mounting system such as the R-A-M system discussed earlier.

If you have access to a machine shop, or if you are willing to spend the money to have a commercial machine shop fabricate a mount for you, you can have a custom mount made to fit your transceiver and your cycle. Or you can fabricate one yourself using commonly available hardware from your neighborhood hardware store. See **Fig 4-10**.

You must pay careful attention to the possibility of vibration problems. The vibration

Fig 4-11—A rain cover was made for this transceiver from a piece of clear plastic tubing.

from even a smooth-running cycle can cause a mounting system to develop problems. For example, I used common hardware to fabricate a mounting system for my HF CW /MC setup on my Honda Gold Wing (discussed later in this chapter). Even though I used lock washers and stop nuts to bolt the hardware pieces together, there were a couple of pesky places where the vibration was obviously focused, and where the nylon stop nuts kept oozing loose. I had to replace one nut-and-bolt combination on one bracket five or six times before I found a method that would hold securely despite the vibration.

That is the kind of problem you discover when you start riding with ham gear on board, and you can try various possibilities until a solution is finally reached. Something else you should consider about transceiver mounting is how to protect the transceiver (or removable control head) from rain and from theft or vandalism. A quick-disconnect mounting system for both the mechanical and the electrical connections, either commercially made or self-fabricated, will allow you to quickly unhook the transceiver or faceplate and stow it in a trunk or saddlebag, when needed.

Or you can fabricate a rain cover for the transceiver or removable control head. One simplistic approach is to use one of the various high-quality refrigerator dishes or storage boxes made of heavy-gauge flexible plastic as a rain cover. You can place the cover of the box under the transceiver mount and route the wiring through the cover. You should carefully seal the holes that the power, antenna, mike and control cables pass through, so water won't be driven into the box from underneath. Then you can quickly pop the plastic container onto its cover to

protect the rig from unexpected rain. If you use this approach, you should have a way to secure the cover for the rig when it is in place, even with something as simple as a small bungee cord.

If you want a more elegant solution, there are various other approaches, such as a hand-made, waterproof plastic box with a waterproof cover—the same idea as the refrigerator dish, but more polished in appearance. See **Fig 4-11**.

A transceiver with a detachable control head and remote cable simplifies the installation. The control head requires less space and there is less mass to it, as compared with the entire transceiver. Because of its smaller size and mass, it's easier to find a place to mount a bracket to hold it. The radio itself can be stowed in a side compartment of the cycle. When installing the control head remember that the cycle's controls and dashboard indicators must not be obstructed by the radio components.

When placing the main body of the radio in an enclosed saddlebag or trunk, remember that it can generate a fair amount of heat. Check the ambient temperature inside the trunk/saddlebag after a lengthy period of operating, to be sure the heat isn't building up to a level that could damage the radio. If it does build up too much, you will need to provide some means of air circulation (either ram air when riding or with a small circulating fan) to keep the radio cool.

It's also a good idea to provide mechanical isolation between the inside of the trunk/saddlebag and the radio, to minimize the vibration reaching the radio. This can be done simply by using a block of foam perhaps 2 inches thick to place the radio on, then securing the radio to the foam pad and one sidewall.

When mounting a control head, always take special care to route the cable between the head and the transceiver body so that it will not be pinched, and so it will not be pulled too tight. On some radios, pinching the cable between the control head and the body of the radio can cause a short-circuit that will blow a surface-mount fuse in the body of the radio. Because most hams are not set up to work with surface-mounted components, blowing that fuse will require a trip to the manufacturer's service depot. Remember that many manufacturers now charge a minimum service charge of about $125 to replace that $10 fuse. You certainly don't want that expense, so be very careful with the placement of the remote cable!

Powering a Higher-Power /MC Rig

While it's possible to power HTs from either their internal batteries or external batteries, using a transceiver that runs 25 to 100 W output requires an external power source. The most obvious possibility is to connect the transceiver to the motorcycle's battery.

Don't try to power the medium-powered ham rig from the cycle's accessory 12-V jack. Those accessory jacks are not usually wired with large enough power wiring to support the power requirements of medium-power rigs on an ongoing basis.

Here's a tip for a source of heavy-gauge 12-V power cable for any higher-power transceiver—whether home station, automobile mobile, or motorcycle mobile. It's quite likely that you have had to buy large-gauge wire at some time to use for the primary power cable supplying your 100-W transceiver, whether at home or in a mobile setup. If you've done this, you've learned that large-gauge wire is fairly expensive.

A good and inexpensive source of two-conductor large-gauge cable with rugged insulation is available at your neighborhood hardware or automotive store, or at your local Wal-Mart. Buy one of the less expensive sets of automotive jumper cables (be sure that the conductors are copper, and not aluminum). Cut off the battery-post clamps on each end of the jumper cable and there you have it—a long piece of heavy-gauge, well-insulated two-conductor cable.

In addition to having large-gauge conductors, battery jumper cables have heavy-duty insulation on them. Furthermore, there is usually some sort of coding molded into the rubber jacket so you can see the difference between the positive and negative leads, a nice bonus.

When running the battery cable from your motorcycle battery to the rig (as well as the coax from the rig to the antenna), mount the cable along the frame under the bodywork. Usually you can run it alongside the cycle's wiring harness.

Always put a fuse in both the positive and the negative lead of the cable, near the motorcycle battery. A second set of fuses where the cable connects to the rig is also a good idea.

In my own HF motorcycle mobile, I connected the TS-140 on my Gold Wing directly to the cycle battery. I could run the rig's full 100 W output on CW when the engine was running. Trying to run the full 100 W with the engine off would drag the battery voltage so low that the rig would become unhappy, but it was possible to run about 30 W with the engine turned off.

An alternative way to power a higher-power rig on larger motorcycles is to find enough space to mount a second battery with about the same amp-hour rating as the cycle battery. The saddlebags or trunk provide plenty of space for an extra battery. The external battery can either be separate from the vehicle battery and electrical system,

with a switch to connect it to the vehicle's battery for charging, or it can be connected directly in parallel with the cycle's battery, to keep its charge topped off.

A second battery can also be connected to the cycle battery through a battery isolator. The advantage of a battery isolator is that you can operate the ham rig without worrying about running down the voltage on the cycle battery. Battery isolators are made up of a pair of power diodes that allow DC power to flow from the alternator to both the cycle battery and the accessory battery, from the accessory battery to the rig, but *not* from the cycle battery to the rig.

The alternator charges both batteries when the engine is running (see the circuit diagram for a battery isolator). It would be easy for you to make your own isolator using two diodes rated at about 25 A. If you make your own isolator, mount the diodes in a rugged,

---

Motorcycle Antenna Modeling

Dean Straw, N6BV, the Editor of this book, is an experienced antenna modeler. As he and I were discussing via e-mail the writing of this chapter on motorcycle mobile operation, his curiosity caused him to offer to model some antenna patterns. Dean said:

"I did some quick calculations using a very simple model for a motorcycle/ bicycle and for a standard compact car—both with center-loaded mobile whips that are a maximum height of 9 feet above ground. I did this for 7.05 and 21.2 MHz. Very interesting—both the car's and the motorcycle's patterns were maximum in the forward direction across the front of each vehicle. Most intriguingly, the motorcycle signal was pretty close to the car, particularly on 15 meters. The patterns and comparisons are shown below.

Pretty cool, eh?"

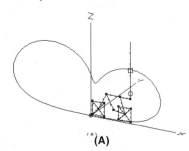

(A)

Fig A—Computer-predicted radiation pattern for a 15-meter antenna on a motorcycle.

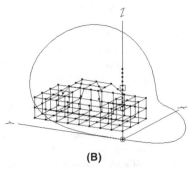

(B)

Fig B—Computer-predicted radiation pattern for a 40-meter antenna on a car.

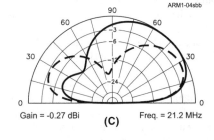

ARM1-04sbb

Gain = -0.27 dBi   (C)   Freq. = 21.2 MHz

Fig C—Comparison of the predicted radiation patterns of 15-meter antennas mounted on an automobile (solid line) and motorcycle (dashed line).

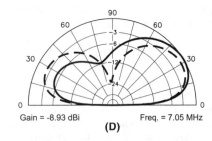

Gain = -8.93 dBi   (D)   Freq. = 7.05 MHz

Fig D—Comparison of the predicted radiation patterns of 40-meter antennas mounted on an automobile (solid line) and motorcycle (dashed line).

weatherproof enclosure. See **Fig 4-12**.

If you use a battery isolator, there will be a voltage drop of 0.5 to 1.0 V across each of the two diodes. Therefore, the voltage appearing at the rig's power connector will be only about 12.5 to 13.0 V, rather than the usual 13.8 V.

Commercially made isolators are available, usually in small packages. Sure Power Products makes a 25-A isolator (model 122) in a package that's about $1 \times 1 \times 0.75$ inch, small enough to allow it to be tucked away near the accessory battery. Sure Power makes larger isolators, if you need a higher current rating. Sure Power isolators are available at the larger NAPA and Pep Boys stores.

Once again, for the power lead between the cycle's battery and the added battery, you can use conductors made from an inexpensive battery jumper cable. And you should always place fuses in both the positive and negative leads of the cables, with a pair of such fuses near each battery.

Another gadget that is often helpful when operating a medium-power rig on a motorcycle is a device called a "power minder." Power minders, available from automotive and cycle shops, warn the rider when the battery voltage starts to drop too low. Some power minders have an adjustable warning level, so you can set the voltage at which you will receive a warning. The price of power minders is usually in the $20 to $30 range.

The K3KMO Motorcycle HF Antenna

On my Gold Wing motorcycle, my /MC antenna was a whip made up of Hustler components, mounted on the trailer hitch. The whip consisted of (starting at the base insulator) an MO-4 (22-inch) mast, a quick-disconnect fitting, then a standard MO-3 (54-inch) mast, an RSS-2 resonator spring and the resonator for the desired band.

I didn't use a spring at the base of the antenna, to keep the antenna from being blown out of the vertical plane at highway speeds. If the antenna hit a low tree branch or other obstruction, the spring at the bottom of the resonator would allow the top of the antenna to flex with the blow.

To further stabilize the mast in a vertical position, I made two braces from narrow strips of polystyrene that secured the MO-4 mast to the license-plate bracket. With that arrangement, if I wanted to reduce the height of the antenna to fit under some low overhead (such as a parking garage), I could use the quick-disconnect fitting to remove all but the MO-4 mast from the cycle.

You might wonder why I added an MO-4 short mast to the standard MO-3 mast. Here's why.

Most of the radiation from a loaded mobile antenna is done from the base section of the antenna below the loading coil, so I figured that adding another 40% to the bottom length of the antenna would help the antenna efficiency. I never did make comparative measurements, but the longer antenna worked quite well on all bands, 80 through 10 meters. If it works, leave well enough alone.

While on my summer 1992 trip to Alaska, I used a single resonator at any given time, and I would stop riding to change resonators when changing bands. After I was back home again, I added a multiple-resonator plate to the top of the mast, so I could switch across three bands without having to stop. I placed the lowest-frequency resonator in the vertical line above the mast and the two other resonators in two of the three holes in the adapter plate, with the resonators in a trailing-V configuration.

Getting all three resonators tuned to the desired points in their respective bands is time-consuming and something of a pain in the saddle, but it can be done. The best and fastest procedure is to tune the lowest-frequency resonator first, and proceed through the other two resonators in frequency order (for example, first tune 40 meters, then 20 meters, then 15 meters). After you have done that, when you go back to the two lower bands you will find that the center frequencies of the two lower-frequency resonators have moved, because of the mutual coupling among the three resonator coils and the change of the inductance of each resonator coil as you adjusted it.

At that point, make a note of how far and in which direction (higher or lower) the two lower-frequency resonators have moved in frequency as you have adjusted the highest-frequency resonator. This will help you anticipate how much and in which direction you should retune the lower-frequency resonators the second time, so they will tend to fall nearer the desired center frequencies.

After about four passes, on average, you will get all three resonators tuned to the desired frequencies. Be sure and lock the lock nuts securely, so the top stub won't slip and require you to start tuning all over again.

Another consideration about Hustler resonators is that their high-power Super Resonators have lower Qs and, therefore, wider usable bandwidths than the standard resonators. For that reason, I used the Super Resonators for 80 and 40 meters to get slightly wider usable bandwidth. Hustler's specifications state the bandwidths of their resonators versus Super Resonators for 80 meters as 30 versus 60 kHz; for 40 meters, 50 versus 80 kHz.

On my 1992 trip to Alaska, I carried a section of large-diameter PVC pipe with screw-off end caps, bungeed to the top of the tent trailer. I used that to stow replacements for every component in the antenna, in case of antenna damage. I also stowed the resonators in that carrier when they weren't in use. You guessed it: I didn't have any antenna damage on the trip, and I never had to use any of the spare antenna parts. See **Fig 4-13**.

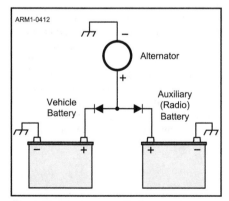

Fig 4-12—Two power diodes can be used to make a battery isolator for your cycle. Both batteries will be charged by the alternator, and the ham rig will not drain power from the vehicle battery.

Fig 4-13—Note the large-diameter PVC tube that held spare parts for the K3KMO/MC antenna setup.

Earphones

Now that you're getting more serious about /MC operation, you should consider buying helmet-mounted earphones. Many of the available motorcycling helmets can be purchased with headphones built into them. Check with your local cycle dealers or search on the Web to determine what's available.

Another possibility, and one that I used for my helmet-mounted earphones, is to install a pair of the larger stereo earphones in your helmet. There are usually cutouts in the foam-rubber padding of the helmet where earphone elements can be installed. I used a pair of stereo headphones that I had on hand that had a broken headband but were electrically intact (like many other hams, I never throw away anything that might later be of use—if you don't believe me, ask my wife!).

I peeled back the fabric lining of the helmet around the ear locations, glued the headphones into the helmet lining's cutouts (which the earphones fit perfectly) and replaced the fabric. It was a simple job, and the earphones work well.

ELECTRICAL NOISE FROM THE CYCLE

If your cycle either has a radio (entertainment or CB) already installed, or if it was built to accept a radio as an accessory to be added later, it's likely that the ignition system will have noise-suppression features built into it. If not, you may install a ham rig and find that you have ignition or other electrical noise in the ham receiver. If that happens, use the noise-suppression techniques that are also used on four-wheeled vehicles, and which are described earlier in this book in Chapter 1.

MOTORCYCLE MOBILE LOGGING

Call me old-fashioned, but I still keep a log of all my HF ham contacts (and the most noteworthy of my 2-meter FM contacts). At the minimum, for each contact I log the date, time, frequency band, the operator's name and location and signal reports sent and received.

In communications support of public events or following a natural or manmade disaster, you will often need a way of writing down messages for delivery or making notes. Therefore, consider how you can make notes

using (permanent, waterproof) pen and paper.

When operating /MC on my Gold Wing, I used a small clipboard attached to the top of the gas tank with self-adhesive Velcro strips to hold a pad of log sheets. When I would look down to make a log entry, I would always keep mental track of the amount of time my eyes were off the road and make sure I keep it to a minimum.

When you are traveling on an Interstate highway at 70 mph, you move 103 feet every second—the length of a football field in 3 seconds. I always try to limit having my eyes down and off the roadway to no more than 2 seconds—and then only if there were no other vehicles within 500 feet of me.

Another way to log or to make notes is to use a small recorder that records on cassettes or mini-cassettes, or in a digital memory. You can even use the same mike you're using for the ham rig, switching it from the rig to the recorder for logging.

INTEGRATED AUDIO SYSTEMS

Many upscale motorcycles, especially the touring cycles, have on-board audio systems with mikes and earphones for the cycle's intercom and entertainment radios, cassette players, CD players and/or satellite radios. When the rider puts on his helmet with built-in earphones and a boom mike, he can listen to the entertainment stereo and can push the talk button to override the stereo and speak with his rear-seat passenger. The rear-seater can also push her talk button to speak with the driver (for example, to suggest that he stop at the next service station and ask for directions, since they appear to be lost).

It would be convenient to integrate the ham audio—both receiver audio and mike input—with the existing on-board audio system. Many hams have performed this integration.

You will need to get information on the cycle's wiring, either from the cycle's shop manual or from a helpful cycle shop. Also, you must check to be sure that the impedance of the cycle mike mounted on the helmet is a close match to the mike-input impedance of your ham rig. You can find helpful information on integrating the ham audio with the existing audio system on MARC's Web site (**http://marc-hq.org/00.htm**).

Before trying to integrate the cycle's au-

dio system with the ham rig's audio (mike and earphones), you should consider the trade-offs:

(1) How much time and work would be involved in the integration, versus

(2) How much the integration would enhance your /MC operation.

Hams who have integrated the ham audio with the cycle's audio system usually use a switch to switch the audio (mike and earphones) between the two systems, and to switch the PTT line between the cycle's CB and the ham rig.

POTENTIAL VIBRATION PROBLEMS

Even the smoothest-running motorcycle generates a significant amount of vibration. You should anticipate that there may be problems caused by that vibration, and you must take steps to minimize the possibility that it will affect your /MC operation or damage your ham equipment.

The first consideration is that of vibration causing mechanical problems in the mounting system for the ham gear. If you see (or feel) any mechanical vibration of the ham equipment as you ride, you should look into ways to change the mounting system and eliminate the movement. When checking for this type of problem, remember not to take your eyes off the road for more than 2 seconds at a time.

When I was operating /MC on HF CW, I asked one of my CW friends (who I knew had "good ears") to listen to the frequency stability of my transmitted signal while I was at highway speeds. It turned out that there was a constant but very slight upward creep of the transmitted frequency. Apparently the creep was caused by a slight but steady mechanical movement of the main tuning dial caused by vibration. To eliminate that frequency-creep problem, I pressed the Frequency Lock button after I selected the operating frequency.

Then I asked for another check with the Frequency Lock turned on. The other operator listened critically to the transmitted signal and said there was an almost undetectable wobbling of the transmitted frequency (less than 10 Hz) caused by the vibration from the cycle at highway speeds. That amount of wobbulation was so small that no one would notice it, unless asked to listen for it.

Resources and Adventures

MARC—THE MOTORCYCLING AMATEUR RADIO CLUB

MARC, organized in June 1992 in Irvine, California, has an excellent Web page at **marc-hq.org/** (The gif file of the little guy riding his Gold Wing across the top of the home page is fun to watch!). I am indebted to Roger D. Rines, W1RDR, for information he provided on MARC that appears in the following discussion. See **Fig 4-14**.

MARC states that the club is *"Dedicated to providing service to the community while combining the two hobbies of motorcycling and ham radio."* With that aim, they provide communication support of many charity and public events in their area, such as bicycle rallies. Motorcycles give the hams the mobility they need without crowding event participants or onlookers, and their use makes it easier to talk with participants, as compared to riding alongside in a car. Roger suggests that *"If you install a radio on your motorcycle, try charity support. It's a lot of fun and you'll be helping a lot of people."* And, of course, that kind of public service gives Amateur Radio high visibility and positive recognition.

When supporting charity events, MARC members use VHF and UHF FM voice. Their Web page gives lots of practical information on the transceivers and antennas they have found to work best for their purposes. They also offer mounting brackets for antennas and PTT switches that you can buy from MARC. You can find descriptions and photos of those brackets on their Web site.

Roger points out that just about any manufacturer's transceivers will work for /MC, but features that should be considered include the size of the controls and switches (it's more difficult to push small buttons with gloved hands), the readability of the displays in strong sunlight and the availability of remote control heads on the transceivers. The transceivers most widely used by MARC members include the Kenwood TM-47, TM-V7, and TM-D700A radios.

Progressive MARC members are now installing APRS systems on their cycles. They have found that APRS is a valuable asset to manage their communication support for various events.

Installing APRS is very simple today; some rigs have the software and hardware built into them to accept a GPS input and provide APRS operation. For example, Kenwood's TM-D700A is a dual-band rig that has APRS built in. After a GPS receiver is connected to the TM-D700A, only a few simple button clicks are required to make the

Fig 4-14—MARC members on roving motorcycles and at a base station provide communications for the ADA Tour de Cure.

APRS system operational. See **Fig 4-15**.

Kenwood's TH-D7A(g) HT is another APRS-ready rig. Roger uses this rig on his Gold Wing to provide APRS. When he leaves the cycle, he can quickly disconnect the HT and carry it with him to remain in radio contact (but without APRS). Roger points out that it's best to have a separate antenna on the cycle for APRS, but the APRS can share a dual-band antenna with the transceiver.

For general information on APRS, see the ARRL publication, *APRS—Moving Hams on Radio and the Internet.*

K7ZUM/MC—MAN OF MANY MILES

Ken Knopp of Gresham, Oregon, started operating /MC on HF SSB in 1993 as KA7ZUM. Later, he dropped the A from his call sign, via the vanity call sign program, to become K7ZUM.

Ken had ridden street bikes for about

20 years, but he kept thinking it would be nice to have a touring cycle and take long trips around the States, with one of his two young sons on the rear seat. And so Ken bought his Gold Wing and soon installed a TS-50 on it. Ken still uses the Kenwood TS-50 and Gold Wing cycle that he started with—now 11 years and 102,000 miles later (at the time of this writing).

After buying the Gold Wing, Ken started thinking about adding a ham rig to the cycle. However, at that time there weren't any small rigs commercially available. One afternoon, Ken got a phone call from his friend Lee Strahan, K7TJR, telling him to check the current issue of *QST* to see Kenwood's new—and very small—TS-50. Ken looked at the ad, and then "beat a path to his nearby HRO store" to buy one.

Ken operates a machine shop, so he fabricated a mount for the TS-50, using the Kenwood MB-13 quick-release mount bolted to a custom-made crosspiece attached to the handlebars. He made the crosspiece

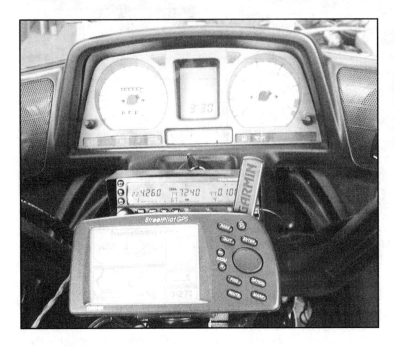

Fig 4-15—Some examples of MARC members' APRS-equipped cycles.

Fig 4-16—K7ZUM's motorcycle mount for his TS-50.

from scrap pieces of 6061 T6 aluminum that he had on hand. See **Fig 4-16**.

Ken first cut out cardboard templates for the various pieces of the mount, trimmed them to the correct sizes and shapes, and then used them as guides for cutting the aluminum pieces. He then used a belt sander for final trimming and shaping. He made the clamps that connect the mount to the handlebars by putting aluminum blocks in his lathe and drilling the hole for the handlebars, then cutting each clamp into two pieces that could be clamped securely onto the handlebars.

Ken can remove the TS-50 from the cycle quickly by uncoupling the several electrical connections and releasing the MB-13's two clamps (one on each side of the TS-50). Ken reports that he can take the TS-50 out of the saddlebag, clamp it on the mobile mount, make the electrical connections and be on the air in less than one minute.

Unfortunately, Kenwood no longer makes the very utilitarian MB-13 quick-disconnect mount that Ken uses, but there may still be some units in stock at Kenwood dealers. The current square-U-shaped mobile bracket (Kenwood J29-0422-13) will also work, but the user would have to remove four wing bolts to disconnect the TS-50 from the bracket (and then keep up with the four wing bolts until time to remount the rig), rather than just pushing two quick-release levers.

Ken showed up at ARRL Headquarters in Newington one day in the summer of 1996 to turn in some DX cards for DXCC credit, riding his Gold Wing. At that point, he was halfway through a 7300-mile trip that took him through 22 states. Ken's son Dustin, then 10 years old, rode on the rear seat during that trip.

Ken fell into the habit of taking a long cycle trip each summer, with one of his two sons on the rear seat —one son one year and the other son the next. See **Fig 4-17**. Since that time, Ken's two sons Dustin, now KD7BSW, and Jordan, now KC7TWZ, have grown into young men and are riding their own cycles. The Knopps have taken a number of trips around the country, seeing many of the National Parks and the sights of the United States and Canada. Ken has ridden over 70,000 miles on these jaunts around the country with Dustin and Jordan. They have toured from Oregon to the East Coast several times, and up and down the Rockies from Canada to Mexico several times.

Dustin is, at the time of this writing, on active duty with the US Navy, stationed on board the carrier USS *Enterprise*. On one recent trip, Ken and Jordan rode their two cycles from their home in Gresham, Oregon, to Norfolk, Virginia, where Dustin was stationed. Dustin hopped on his cycle and joined Jordan and Ken for some three-Knopp motorcycling.

I admire Ken for staying close to his sons as they grew from youngsters into early manhood, sharing good times as he showed them the pleasures of cycle touring. Obviously, I'm not the only one who feels that way. Ken says that when people he met along the way learned that he was on a long road trip with one of his sons, the reactions of the people were always along the same lines, and were usually expressed with these few words, "*If only I had done something like that when my kids were young.*" See **Fig 4-18**.

Ken says that he almost always creates quite a stir on the air when he signs "motorcycle mobile," and he has no trouble making contacts on the road. Ken operates HF SSB while in motion and operates both SSB and CW from the Gold Wing when he isn't in motion.

Ken very often impresses people with how

Fig 4-17—Jordan, KC7TWZ, with his cycle (r) and Dad K7ZUM's motorcycle mobile Gold Wing (l).

Fig 4-18—The cycling Knoops (l-r): Jordan, KC7TWZ; Dustin, KD7BSW; and Dad Ken, K7ZUM.

Fig 4-19—K7ZUM's "secret weapon"—three ¼-wave radials that can be attached to a stainless steel bolt near the base of his mobile whip for fixed operation.

well and how far he can make radio contacts with his TS-50. On one of his road trips with Jordan, Ken had stopped at a motel in New Castle, Wyoming. The two had just returned to the motel from dinner. Ken pulled his TS-50 out of the saddlebag, hooked it up and fired up on his usual hang-out frequency of 7203 kHz. Several of his friends had been monitoring and they called in to say hello and chat a bit.

Also staying at the motel were quite a few cyclists on their way to the big rally at Sturgis. One of the young women from that group walked by and asked Ken if he was "*talking to someone down the street with [his] CB.*" Ken replied that it was a ham radio station, not CB, and that he was talking with friends up and down the West Coast. The women looked at Ken's Oregon license tag and asked "*You mean back in Oregon?*" Ken said, "Yep," but the woman didn't quite believe it.

On his next transmission, Ken told his buddies that he was demonstrating the rig to another cyclist and asked each of them to introduce himself, giving their name and location. One by one, they said hello to Ken's visitor and told her that they were in Oregon… and Washington… and California.

The woman was amazed. As Ken continued the roundtable with his buddies, the woman brought a big bunch of other cyclists to hear Ken's conversation with people 1000 to 1500 miles away—from a motorcycle. Ken made some good points for ham radio!

Ken reports that incidents like that have taken place many times, leaving in his /MC trail a lot of people who were impressed with what Amateur Radio can do. Who knows; maybe some of them will get ham licenses and put rigs on *their* cycles.

When Ken operates while not in motion,

he has a simple but very effective "secret weapon" that increases his antenna performance. He made up three ¼-wave radials for 40 meters using insulated #22 wire, terminating one end of each radial with an alligator clip. To use his portable three-wire counterpoise, he connects the alligator clips to a stainless screw a couple of inches in front of the base of his mobile antenna and rolls the radials out in three directions. Ken says that he consistently gets better signal reports when using the counterpoise. See **Fig 4-19**.

Things don't always go 100% okay when riding a cycle. On one trip with Jordan (on his own cycle), Ken's rear tire went flat on a secondary road. He had to take the rear of the Gold Wing apart to drop the wheel out of

the bottom. A young man on a Harley stopped and told Ken there was a cycle shop in the next town that could fix the tire. Ken called the shop on his cell phone and asked them to come pick up the tire, repair it and bring it back. With that help, Ken and Jordan were soon on the road again.

Ken notes that, true to legendary southern hospitality, several motorists stopped to see if Ken needed help, while he was waiting for the tire repair. And, sure enough, more than one of them asked… See **Fig 4-20**.

Ken is a firm believer in motorcycle safety. Here, in Ken's own words, is why: *"I have been riding street bikes for 37 years. On July 3, 1977, I was involved in a very serious and very freaky motorcycle accident. I ended up with a broken back,*

Fig 4-20—"Say, why do you have that fishing pole on the back of your motorcycle?"

cracked neck, collapsed lung, partially paralyzed leg and numerous other things.

I spent almost four months in the hospital. When I got out, I was in a body cast for another six months. It was well over a month after I was out of the hospital before I relearned how to walk. Continuing pain reminds me of that freak accident on a daily basis, so I ride motorcycles very safely.

There are a few things that I will not do when operating motorcycle mobile. The first is that I will not operate the radio in any type of traffic, or on a city street. As far as I am concerned, there are way too many things to keep track of while riding, and riding in traffic safely, needs the driver's undivided attention... period!"

That's just one example of why motorcyclists need to always think SAFETY.

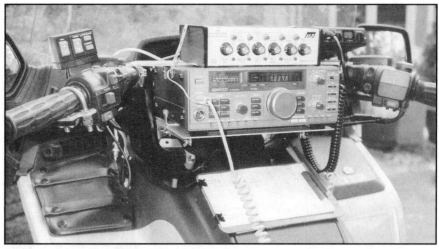

Fig 4-21—K3KMO/MC—A Kenwood TS-140 and an MFJ-484 memory keyer. Note the left-handed paddle mounted to the left handgrip with a large U-bolt.

HF CW FROM K3KMO/MC

Yes, Virginia, it *is* possible to operate CW from a motorcycle in motion. I'm one of the people who has done it, starting back in 1992, when my call sign was K3KMO.

In the summer of 1991, I learned that there would be a motorcycle rally along the length of the Alaska Highway in 1992, to celebrate the Highway's 50th anniversary. That sounded interesting, but I made no plans to make the ride from my home in Maryland to Alaska and back. At the time, I owned a 250-cc Honda Helix, a large and upscale "touring" class of motor scooter.

That fall, while I was having some scheduled maintenance done on my Helix at a nearby Honda dealership, I wandered into the showroom to look around. I noticed a new Gold Wing being offered at an unusually low sale price. When I asked why the price was so low, the salesman told me that they wanted to get it out of the showroom fast, since the end of the summer riding season was here. I told him I would buy it if he would put a trailer hitch on it for that price. He said, "Okay." The deal was done.

After becoming the owner of the Gold Wing, during the winter of 1991-92, I occasionally thought about the motorcycle rally up the Alaska Highway. Still I didn't make any plans to ride in it.

Spring came. The Gold Wing went on the road again. In the exuberance of spring riding, I did decide to ride the Gold Wing from my home in Maryland to Alaska and back that summer. Furthermore, I would put an HF CW rig on the Gold Wing, combining two of my long-time interests. I was *really* exuberant.

At that time, I had already been working mobile CW from my Toyota pickup truck and Toyota van for 12 years. I had operated in radio contests when I was on the road. I had worked over 150 countries from my

Toyota van with the 25 W from my little Ten-Tec Century 22. It all added up to enough mobile CW experience to make me think that a transition from operating mobile in a van to operating mobile on a motorcycle should be fairly easy. And it was.

By the way, one requirement for operating mobile CW while driving, regardless of the type of vehicle, is that you must be able to copy CW comfortably in your head at a speed of at least 25 wpm. You must be "fluent" in Morse code, so that you "speak" it and understand it with almost no conscious thought, just as you speak and hear your native language. Don't try to operate mobile CW until your CW skills are up to the task.

In the spring of 1992, I mounted my Kenwood TS-140S (with its 500-Hz CW filter) onto the Gold Wing. I bolted the TS-140 to a heavy-gauge aluminum plate and made a mounting system that held the aluminum plate onto the handlebars using common hardware items (steel angle brackets, compression type hose clamps, etc).

I secured an MFJ-484 Grandmaster memory keyer to the top of the TS-140 with a metal strap that ran across the top of the keyer case to hold it in place. The mounting system wasn't very pretty, but it was solid and it worked well.

I could remove or replace the mount and rig in about two minutes by loosening two clamps and disconnecting three electrical cables, so I could stow the rig in the Gold Wing's trunk for security, if need be. See **Fig 4-21**.

My wife Maggie made a waterproof cover for the rig and keyer from plastic-coated fabric, with Velcro straps to attach it over the rig. I soon learned that I could operate in moderate rain by folding the front of the cover back so I could see and operate the

front-panel controls.

To make a keying paddle, I mounted two microswitches back to back on a small aluminum plate, and attached the plate to the left handlebar grip (I'm left-handed) using a large U-bolt. When my hand was wrapped around the left handgrip in riding position, my forefinger and thumb fell onto the microswitches. Keying was easy—and I could key the rig while keeping both hands on the handlebars. See **Fig 4-22**. (Remember: Ride and operate *safely!*)

K3KMO/MC—Maryland to Alaska and Back

My mobile installation on the Gold Wing was completed a few weeks before my scheduled departure for Alaska, so I was able to operate /MC for about a hundred miles before leaving for Alaska. Those few miles of riding turned up a couple of minor requirements for changes in the mounting system. I had to make further minor changes to the mount on the way to Alaska. But, finally, everything held together okay.

And so I pulled away from my driveway, bound for Alaska on the Gold Wing, towing a pop-up tent (and storage) trailer, with my son Pat (then 15 years old) on the back seat. The August 1992 issue of *QST* had published a "Stray" announcing my plan to cycle to Alaska and back, listing the spot frequencies I would monitor—7037, 14037, 21037, and 28037 MHz. The strategy of spending much of my operating time on spot frequencies paid off. Quite a few hams who read that "Stray" called me for a contact. And there were a lot of casual contacts with hams who were quite surprised to hear that the CW they were copying was coming from a motorcycle in motion.

The ham activity during the ride across

the continent was fun, as well as helping keep me mentally alert as I drove along on the many miles of monotonous Interstates. Pat and I were fortunate in that we had excellent weather most of the 32 days of our trip. Most days were clear and sunny, with only occasional short bursts of inclement weather. With one exception, the rain showers and thunderstorms came and went quickly, and we almost always found a place under cover (highway overpasses, friendly trees or roadside shelters) to wait out a passing rain shower or storm.

That one exception to our good luck with weather was two solid days of steady but moderate rain in the Yukon, just before we reached Alaska on our northbound ride. Riding in the rain for two days isn't a lot of fun, but that made the days of sunshine that followed all the more welcome. See **Fig 4-23**.

The ride to Alaska was a unique adventure for me. The scenery was beautiful. I met and chatted with a lot of nice people while on the road—residents of Canada and Alaska as well as fellow travelers along the Alaska Highway. Pat and I saw wildlife now and then, but we never saw a bear alongside or in the road. When you're riding a motorcycle, that's just as well.

Hamming was a great way to pass time during the long hours of the ride to Alaska and back. Several of my Maryland friends—Aaron, W3GIS; John, K3NNI; Walt, KX3U; Tom, W3BYM; Ray, N4GYN; and especially Jack, W3TMZ—made contact with me often during the trip. If I asked, they would call my wife Maggie with reports of my progress. Jack, W3TMZ, was the final contact at the end of the Alaska trip—we signed off just as I was reaching my driveway, 32 days and 10,500 miles after the trip began.

While Pat and I were riding to Alaska, Maggie flew to Hawaii with a friend for a week's vacation. When she was in Honolulu I had a nice chat with Peter, KH6CTQ, also in Honolulu. Peter delivered a message from me to Maggie at her hotel. Sometimes ham radio comes through!

I made a very interesting contact when I was on the way home but still in Alaska, with John, KL7GNP, in Anchorage. Mount Spurr, a volcano 78 miles from Anchorage, had erupted the day before and John told me how he was having to stuff towels under the doors of his home and seal the windows with tape to minimize the possibility of the very fine ash from the eruption getting into his home and his electronics equipment. (If you want to read a good account of that eruption and see some photos of it, go to **www.gly.bris.ac.uk/ www/teach/virtrips/volcs/Spurr.html** .)

For more photos and information about my Alaska /MC adventure, see the article about it in the July 1993 issue of *QST*. See also **Fig 4-24**.

You will recall that I used a memory keyer in my /MC setup. This made it easy to "troll" for contacts. (This same technique can be used for other mobile operation, and even home-station operation.) I would record a short CQ in one of the keyer's memories— "CQ CQ CQ de K3KMO K3KMO/MC K" and set the keyer to repeat the call after a pause of 10 seconds. When the CQ had sent its final K, I would have 10 seconds to listen for a reply, using the RIT control to tune slightly to either side of my frequency as I searched for a caller. (If you troll for contacts like this, please use your RIT to tune

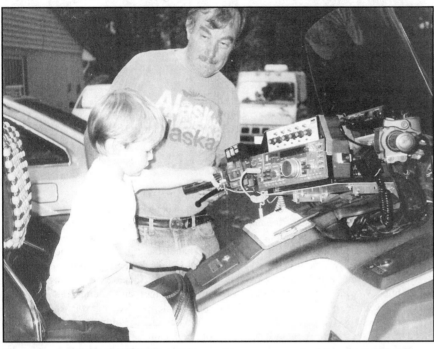

Fig 4-22—Jack Colson, W3TMZ, watches as his young son Andrew plays with the K3KMO/MC keying paddle and listens to Morse on the monitor (photo from 1992).

Fig 4-23—Looking back to the west on I-70, toward the storm clouds the cycle had just been through.

Fig 4-24—Al, K3KMO/MC (now W1AB), taking a break during his ride around the interior of Alaska via Routes 2, 3 and 1.

around your frequency for possible callers. Not every ham will get his transceiver exactly on your frequency.)

If someone replied to my CW, I would push the Reset switch on the memory keyer to stop it from calling again, and would begin the chat with the caller. If no one called then the keyer would automatically start another CQ. I would let the CW repeat about 10 times. If no one had called by then, I would stop CQing and tune around to find someone else calling CQ.

DXing from K3KMO/MC

My results with /MC DXing on CW were quite good. I found that I could break through pileups quite well with 100 W and a mobile antenna. I worked quite a bit of long-haul DX on 40 and 20 meters, where the competition can be pretty fierce.

My most memorable DX contact from the Gold Wing took place at 1018 UTC, August 3, 1993, as I was riding my Gold Wing to work in Maryland. I was going to work much earlier than usual to do some pressing work and the hams I usually chatted with on 40 meters during my morning commute were still asleep.

When I tuned to the low end of the band looking for a contact I heard Yuri, 4K1F, entertaining a well-managed pileup. Yuri was the radio operator on an expedition to King George Island in Antarctica. I tuned around to find the frequency where Yuri was listening and switched to split-frequency operation. When he finished the contact in progress, I called him. Yuri came right back to me on my first call and gave me a 599 report. We had a short but solid chat, and he went on to work others.

I had assumed that Yuri might be following the not-unusual practice of handing out 599 reports to everyone, but then I heard him give someone else a 579. *Whoa!* I must have had a good signal in Antarctica.

Later, I realized that I had worked Yuri via a seriously good gray-line path, which explains why my /MC signal was so strong. Nevertheless, the contact shows what can be done with a mobile signal, even from a motorcycle. Also, this gives a firm indication that a motorcycle makes a quite adequate counterpoise, even on frequencies as low as 7 MHz.

Reactions of Other Hams

Most of the time, when I would tell a new

contact that I was operating mobile CW from a motorcycle, their replies fell into three categories:

(1)   Quite a few of them would be ask, "*Did you say you are on a motorcycle?*"
(2)   Others would obviously be slightly worried as they asked, "*OK on the motorcycle. You aren't in motion, are you?*"
(3)   But quite a few of the operators would accept the fact that I was operating CW while riding my Gold Wing without blinking an eye. Hams in that group would immediately start asking questions about where the rig was mounted, where the keyer paddle was mounted, what kind of antenna I was using, etc. That was especially true of the hams who were also motorcycle riders.

But then there was the ham I contacted who was whiling away a nice Sunday afternoon, probably doing other things in his shack as he made contacts on 40 CW... I called CQ, and that ham replied. We had our first exchange—the usual name, location and RST. He copied my name, because he called me Al, but he didn't react to my saying that I was operating from my motorcycle. I thought perhaps he hadn't copied my statement to that effect.

During the second exchange, I made the point more clearly, telling him that I was in Kansas, "riding my motorcycle eastbound on I-70." He still showed no reaction to my /MC comments. During the third exchange, I said that I was "cruising along on two wheels on my motorcycle at 75 mph." Still no reaction.

Finally I decided he wasn't listening to what I was saying, but was just talking to me each time I stood by. At the end of his third transmission, he declared that he was QRU and would say 73. After the two of us signed off, I listened on frequency a bit. I heard him call CQ and start another contact, which he also didn't listen to—he never responded to or commented on anything that ham said, either.

KA7W/MC—ANOTHER HF CW OPERATOR

On August 5, 1993, while I was riding my Gold Wing to Alaska, Tom Wilson, KA7W, of Aztec, New Mexico, called me on 20 meters from his home station and we had a nice chat.

Three hours later, as I was still cruising down the road, Tom called in again—but this time, much to my surprise, he was operating motorcycle mobile from *his* Gold Wing! I believe that QSO between KA7W/MC and K3KMO/MC may have been the first ever (and perhaps *only*) motorcycle-to-motorcycle CW contact with both motorcycles in motion. (If you, the reader, can cite other such contacts, I would like to hear about them; please send an e-mail report to **W1AB@aol.com**.)

In those days, Tom was operating /MC in a fairly simplistic way, but one that worked well. He mounted a homebrew antenna on the tongue of his motorcycle trailer. He used a keying paddle fabricated from two microswitches mounted back to back, which he mounted on his right handlebar grip, so he could rest his hand on the throttle while he was keying the rig.

In Tom's first /MC setup he put his Kenwood TS-130 in the trunk. He could then make contacts with stations that were in his receiver bandpass—a simple fixed-channel operation that worked well. Later, Tom operated /MC with an Icom IC-735 held onto the rear seat with bungee cords. He could reach around behind his back and operate the controls by feel. Tom's most recent /MC rig is an Alinco DX-70.

Tom (as do other cyclists), enjoys how easy it is to find a contact when he announces that he's operating motorcycle mobile. He reports that he entertains pileups of other hams wanting to talk with someone on a two-wheeler.

Tom tells of generating a pileup one time on 20-meter SSB, during a good band opening between Oregon and Australia. He worked one Australian ham after another after another for a long time until he finally had to shut down, leaving a pileup of Australian hams still calling. That's the kind of excitement that a station operating /MC can generate.

KC3VO—*EXTREME MOTORCYCLING*

It isn't unusual for hams to get interested in a particular area of Amateur Radio and push their station to the limit. Bob Curry, KC3VO, of Adelphi, Maryland, is the best example I know of in terms of putting together a massive, high-power, multi-mode, multi-band and excellent /MC setup.

Bob has been riding motorcycles since 1965 and has operated /MC since 1971. He was first licensed in 1969. For the past 10 years, he has operated /MC from his 1994 Honda Gold Wing SE and he pulls a small trailer behind the cycle. He has various transceivers and amplifiers mounted on the cycle and in the trailer, with several batteries and a 120-V gas-powered ac generator (installed in the trailer) to power the ham gear and some other accessory equipment!

KC3VO is the chief engineer

for WHUT-TV in Washington. His technical expertise and his knowledge of available products, both amateur and commercial, have helped him make some excellent and very interesting choices of equipment for his /MC ham station. The following description will give the reader a broad outline of what Bob has done to set up his /MC station.

The 12-V Electrical System

KC3VO replaced the Gold Wing's standard alternator with a 1200-W truck alternator. The new alternator is an after-market item modified for use on Gold Wings by Compu-Fire. Compu-Fire uses a standard Delco truck alternator and replaces its front casing and pulley with a new casing and coupler to mate with the Gold Wing's alternator drive shaft. The new alternator fits into the same space occupied by the original alternator and it will power virtually all of the ham gear. Bob makes the good point that the replacement Delco alternator can be bought at most auto parts stores. If the alternator failed, it would be not only easier but also less expensive to replace than the original-equipment alternator would be. See **Fig 4-25**.

Bob beefed up his Gold Wing's battery capacity by adding three more 12-V batteries—a 30 A-h battery in the trailer, and a pair of 17 A-h batteries (one in each of the Wing's saddlebags). See **Fig 4-26**.

When he needs even more electricity for the ham equipment, he can use a 700-W Kawasaki generator mounted in the trailer. The generator provides both 120 V ac and 13.8 V dc. The generator has an electrical starting system with its own internal battery, or it can be pull-started with the usual rope and pulley arrangement. See **Fig 4-27**.

In the small compartment on the Wing just to the left of the dashboard, Bob mounted a control panel for the generator. The panel has a push button for electric starting and a pilot lamp to show when the generator is

running. He can start the generator while cruising along the road, if necessary.

Although it hasn't ever happened, KC3VO points out that if he runs the voltage of the vehicle and rig batteries so low that he doesn't have enough battery power to start the cycle, he can fire up the generator and quickly charge all the batteries. See **Fig 4-28**. You can see the large-gauge, two-conductor cable that can carry up to 100 A (at 13.8 V) between the cycle and the trailer (**Fig 4-29**).

Bob carries a pair of folding solar panel arrays made by Global Solar, each of which provides 30 W at 12 V dc on a sunny day. The solar panels are mounted on a cloth backing that becomes the carrying case when folded, and the array folds neatly into a package about 10 × 12 × 1 inch in size. He can run the arrays either singly or in parallel for 12 V output or run them in series for 24 V. He seldom uses the panels, but they can be unfolded and put into service within a couple of minutes, if needed. See **Fig 4-30**.

The Transceivers, Transverters and Amplifiers

Bob's current main rig (as of January 2005) is a Yaesu FT-100D, with the rig in the right saddlebag and the control head mounted on the center cover plate just behind the windshield. He use the FT-100D as the exciter for 1.8 through 432 MHz, and as a driver for transverters for 1296 and 2304 MHz and 10 GHz.

The Gold Wing has the Honda CB radio installation, with a remote-control head inboard of the left handlebar grip. The push-to-talk switch is a large trigger switch that can be reached without taking your left hand off the handgrip. Bob mounted an A-B toggle switch on the front of the CB control head to switch the PTT between the CB and the FT-100D. See **Fig 4-31**.

When he gets caught in the rain while in motion, the FT-100D head is protected from the rain by the windshield. If he is not in motion, he places a simple Ziploc freezer bag over the control head to keep it dry.

Although Bob usually runs the FT-100D barefoot (100 W output) on the HF bands, he can switch in a Tokyo Hy-Power HL-700B (available only in Japan) that he has modified to deliver 1000 W output on the HF bands. When he wants to run the full legal limit of 1500 W output, he uses a modified Yaesu Quadra VL-1000 amplifier for 160 through 6 meters.

On 144 MHz, KC3VO can run either the FT-100D barefoot at 50 W or switch on a TE Systems amplifier to run 400 W output on

Fig 4-25—KC3VO/MC's aftermarket 1200-W alternator on his Gold Wing.

Fig 4-26—KC3VO/MC has mounted an extra battery in each saddlebag, and the main body of his Yaesu FT-100D in the right saddlebag.

Fig 4-27—The KC3VO/MC trailer carries a Tokyo Hy-Power HL-700B linear amplifier (r) and a 700-W Kawasaki generator (l).

Fig 4-28—The KC3VO/MC auxiliary generator control panel (in the small compartment to the left of the console). The larger red push button is the starter switch for the generator, and the smaller green pilot lamp directly below the starter switch illuminates when the generator is running.

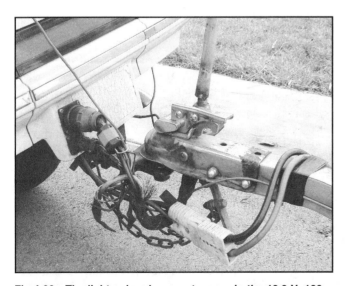

Fig 4-29—The light-colored connectors are in the 13.8-V, 100-A circuit between KC3VO/MC's gold Wing and its trailer.

Fig 4-30—KC3VO/MC's folding solar panel can produce 30 W at 12 V.

SSB or 50 W output on FM. He modified the amplifier for better performance and higher output.

We come next to 432 MHz, where KC3VO can run the FT-100D barefoot with 20 W output on FM, or he can switch in another modified TE Systems amplifier to run 225 W on SSB. Moving right along, on 1296 MHz he uses his FT-100D into an SSB/DB6NT transverter module to drive a solid-state SSB Electronics linear amplifier at 100 W output on all modes. On 2304 MHz he uses another German-made transverter to run 20 W on all modes. And finally on 10 GHz, Bobs feeds the HF transceiver into a DB6NT transverter followed by a 2-W amplifier from SSB Electronics.

KC3VO has so much ham equipment that can be mounted in the Gold Wing and its trailer that it won't all fit at the same time. Therefore, he mounts only the equipment he expects to use at any given time. He has installed two cooling fans in the rear wall of the trailer, to limit the heat build-up inside the trailer when the lid is closed. See **Fig 4-32**. Bob uses a helmet-mounted mike for the various transceivers, with switch selection of the rig to which the mike and PTT are connected.

The KC3VO/MC Mobile Antenna Farm

KC3VO has the standard Gold Wing antennas—a CB antenna on the left side and an AM/FM broadcast antenna on the right side, both positioned just behind the rear seat. He also has a Little Tarheel II whip on the rear of the Gold Wing. That antenna, rated at 200 W, can be tuned to any frequency from 3.5 to 54 MHz.

When he runs the HL-700B amplifier, he switches to a Tarheel 200 mounted on the rear of the trailer. The Tarheel 200 is rated at 1500 W and also covers 3.5 to 54 MHz. See **Fig 4-33**. On the top of the Gold Wing trunk, in the center, he uses either a Diamond 770 antenna for 144 and 432 MHz, or a Diamond NR2000NA antenna for 144, 432 and 1296 MHz.

When Bob is operating while not in motion, he augments his omnidirectional mobile antennas with other antennas that provide gain and directionality. He places an "antenna plate" on the top deck of the trailer, and mounts a mast on the plate. His fixed-motorcycle antennas include the following:

- For 50 MHz, a Saturn-6 Halo (from the 1960s).
- For 144 MHz, either a WIMO Big Wheel antenna (available from SSB Electronics) or a 4-element Yagi.
- For 432 MHz, a WIMO Little Wheel antenna.
- For 1296 MHz, either a homebrew antenna modeled after the WIMO antennas

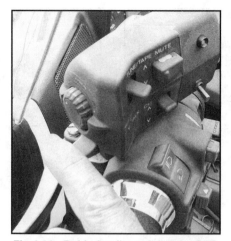

Fig 4-31—Bob's forefinger is on the PTT trigger switch. The small toggle switch on the lower right of the radio control box switches the PTT between the CB and the FT-100D.

(he calls this his Micro Wheel) or a 25-element loop Yagi (6-foot boom) from Directive Systems.
- For 2304 MHz, a 34-element Yagi (5-foot boom) from SSB Electronics. For 10 GHz, there is either a 17-dBi horn or an 18-inch dish.

KC3VO can't mount all these antennas on the antenna-plate mast at the same time. Rather, he puts out only the antennas he needs for the operation at hand. He also carries a video camera and a small handheld TV receiver (HT sized) on the cycle, so he can operate amateur television when not in motion. With ATV, he runs either 10 or 200 W on 420 MHz (AM), 1.5 or 50 W on 1250 MHz (FM), and 1 W on 2400 MHz (FM).

Other On-Board Equipment

KC3VO has an audio system on his Gold Wing that is similar in size to his suite of ham equipment. For entertainment radios, there is the AM/FM broadcast receiver that came on the Gold Wing SE and an XM satellite receiver. Both receivers can be switched to drive stereo amplifiers that will deliver 1200 W peak of stereo audio output to several stereo speaker pairs and a subwoofer.

Bob can use his audio system as a remote public address system when he is helping with emergency communications. He carries a small FRS HT in the Gold Wing at all times. He can tune the FT-100D to the HT's frequency and feed the FT-100D receiver's audio to the entertainment system's audio amplifier. He can then use that remote-mike PA system to make announcements to people within a radius of 100 feet or more, even in a noisy environment. See **Fig 4-34**.

KC3VO/MC on the Road; On the Air

Of course, Bob uses the /MC setup for his

Fig 4-32—Two fans, one of each side of the license tag, provide positive circulation through the cycle trailer, to keep the equipment in the trailer cool.

routine hamming, both in motion and while standing. (You will recall that he has enough battery power to operate for long periods of time with both the cycle's engine and the trailer-mounted Kawasaki generator shut down.)

He often attends hamfests to show off his Gold Wing setup (I first met Bob when he had the cycle at the Gaithersburg Hamfest in 1993). He has ridden his Gold Wing to the Dayton Hamvention several times and to the Richmond (Virginia) Frostfest one time—the Frostfest is held in January each year and that's *cold* riding!

As a member of the Potomac Valley Radio Club, one of the country's major contest clubs, KC3VO is a serious contester. He often uses the /MC setup as a Rover in VHF contests, and he has enough power and good enough antennas to run up excellent scores. Bob has made some excellent long-haul contacts on 10 GHz. Since he carries a GPS receiver in his cycle, he can determine his position with accuracy when roving.

Speaking of 10 GHz, Bob reports an interesting situation on that band. He has found that, as a practical matter, he can't operate on 10 GHz when in motion. The Doppler shift of signals on 10 GHz is 300 Hz per 10 mph (or 1800 Hz at 60 mph). Because the Doppler shift varies with the Gold Wing speed, tracking the Doppler-shifted signal with the receiver tuning (at *both* ends of the contact) would be extremely difficult.

KC3VO likes to have the /MC at the ready for emergency communications support of any local emergency where hams are called on to help. The /MC can operate on any frequency that might be used for emergency communication—with high-power transmitters and good antennas. It has a high-power public-address system and it can provide TV images to a command center.

Keep in mind that all of this capability is built into a motorcycle and its small trailer. That's why I call KC3VO's setup "*Extreme /MC*"!

KC3VO's Power Trailer

Although Bob's power trailer doesn't have anything to do with /MC, it is worth

An Interesting Antenna-Pattern Observation

Several times while operating /MC, W1AB noticed something interesting that confirmed the theoretical radiation pattern of a vertical antenna. We've all read the antenna theory and studied the plot of a vertical antenna's radiation pattern. Close to the antenna, the ground-wave signal is strong. Then the ground-wave signal falls off and there is a large, doughnut-shaped dead zone around the vertical, in which you hear little or no signal from the transmitting antenna.

Beyond that dead zone, you start to receive the skywave signal via ionospheric refraction. Several times, as W1AB left work in the afternoon on his Gold Wing, he contacted a ham who lived near his workplace. They would have good signals between them via the ground wave.

Then as he rode farther away from the other ham, the signals would drop off and they would lose contact. That signal drop-off occurred fairly quickly, within a mile or two of travel. Multiple contacts on the same frequency band showed that the signal drop-out always occurred at about the same place during his ride.

If the two stayed on frequency, by the time he reached his home town, W1AB would start to hear the other ham's signal with a weak signal, but one that was starting to build in amplitude. Once again, observation confirms theory.

Fig 4-33—The two 3.5 to 50 MHz antennas used by KC3VO/MC. The antenna on the cycle is the Little Tarheel II, rated at 200 W; the antenna on the trailer is the Tarheel 200, rated at 1500 W.

Fig 4-34—Bob carries various pieces of equipment in his Gold Wing's trunk. Note the forward-firing subwoofer for his audio system, and the fire extinguisher.

describing to readers who are interested in emergency communication. The trailer, towed by his SUV, is outfitted as a power source for emergency use. The PVRC has used his power trailer for Field Day several times with as many as 51 transmitters, and those FD uses have served as excellent tests of the power trailer's capabilities.

As you might expect, Bob's power trailer is as extravagant as is his Gold Wing! It has eight 100 A-h batteries connected in series-parallel to provide 400 A-h at 24 V. It has a 6-kW Honda water-cooled, electric-start generator on board and an inverter powered by the 24-V battery bank that can supply 4 kW continuous output at 120 V ac (8 kW peak for up to 30 seconds). The combination of the generator and the inverter can be used for a total continuous load of 10 kW.

He has step-up and step-down transformers that can be used for power distribution at either 240 or 480 V ac, in addition to the usual 120 V ac. Using the higher voltage lowers the current on the distribution lines, thereby lowering the I^2R power loss in the lines.

The power distribution system can be set up to run from the generator, with the batteries and inverter acting as an uninterruptible power supply. The capacity of the battery and inverter is large enough to run the multi-transmitter Field Day stations for up to a half hour. Therefore, it isn't necessary to carefully watch for the time the generator must be refueled—You only have to be aware that when the generator runs out of gas and stops, you have 30 minutes to gas it up and get it on line again.

Then, as the crowning touch, the power trailer has a large array of solar panels that can provide over 1 kW of power at 24 V to charge the battery bank. As usual, KC3VO thinks *big* and executes *well*, following good engineering and amateur practices.

And Once Again, SAFETY

As I mentioned at the beginning of this chapter, the first *and last* consideration of riding a motorcycle must always be SAFETY. I hope you will consider taking ideas from this chapter and installing a /MC setup on your cycle. When starting out with /MC, it's best to observe the KISS principle—"Keep it simple, stupid!" Then, as your experience level progresses, you can get into more complex ham setups and enjoy the two-wheel hamming even more.

But, no matter what hamming you do from your motorcycle, please remember these points:

- Always ride safely.
- Don't let the operation of your ham equipment divert your attention from your primary task—that of riding safely.
- And last but not least: *Laissez les bons temps rouler… sur une moto!*

ACKNOWLEDGMENTS

Sincere thanks to the experienced and knowledgeable hams who helped me with information, ideas, and photographs for this chapter on motorcycle mobile operation:

- Roger D. Rines, W1RDR, long-time member of the Motorcycling Amateur Radio Club, for furnishing the photographs of MARC hams and activities that appear in this chapter, and also for technical and mechanical information about some of the common problems MARC riders have encountered and resolved. See **Fig 4-35**.
- Bob Curry, KC3VO, who rode his Gold Wing partway around the Washington Beltway on a cold winter's day to meet me for conversation and photographs of his mega motorcycle mobile setup. By the way, did you figure out—through learning how intense Bob is about Amateur Radio and how well he treats himself by assembling so much excellent and expensive ham and cycling equipment—that Bob is a single guy?
- Ken Knopp, K7ZUM, whom I met at

Fig 4-35—Roger D. Rines, W1RDR, and his wife Sharon enjoy the view at Pescadero Beach, California.

ARRL HQ in 1996, when he rode his /MC Gold Wing across the country to deliver some QSLs for DXCC credit. Ken supplied me with the photos of himself and his two sons, along with some very interesting commentary about his adventures with them.

- Tom Wilson, KA7W, for the information on his /MC CW operating, and for

the history-making first QSO between two motorcycle mobiles in motion. (Now if I could just figure out how to get him to QSL that contact…)

- Although she isn't a ham, my wife Maggie has always been very supportive of my Amateur Radio activities. An example: In 1971, I single-operated W3USA in the CW Sweepstakes contest to help the club score of the Potomac Valley Radio Club. When Maggie came to the station to pick me up, I was dragging because I had been awake for most of the contest period. Maggie asked how many sections I had worked. I replied that I had worked all sections except the Yukon, but I was ready to quit, even though it was an hour before the contest period ended. She urged me to keep looking for the Yukon until the contest ended. Danged if I didn't find a Yukon station 20 minutes later—and work him for a Clean Sweep! For that, and for all her support of my ham radio and other activities over the past 34 years, I offer my sincere thanks to Maggie. She even made a certificate for me to commemorate the accomplishment, awarded by WAR—Wives of Amateur Radio—a certificate that still hangs on the wall of my shack. See **Fig 4-36**.

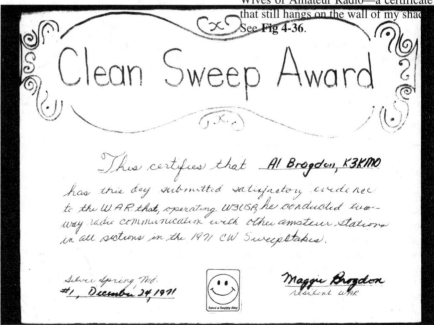

Fig 4-36—WAR award (Wives of Amateur Radio).

Motorcycles in *QST*

A search for "motorcycle" in the *QST* index resulted in some interesting hits. Few of them were about /MC operation with the cycle in motion, but it was interesting to see what *QST* had to say about motorcycles over the years:

November 1934

The first mention of motorcycles found by the search was in "Operating News" in this issue. The item "Amateur Radio Aids Motorcycle Run" tells how Amateur Radio was used to relay the results of an American Motorcycle Association race for the first time, on March 11 of that year.

The half-day "trial run" was from New York City to Peekskill and back. W2CLB was set up at the NYC club headquarters, manned by W2CLB and W2CJI. W2BNJ (portable) was set up in a delivery truck at the Peekskill end of the run, manned by W2BNJ and W2EMQ. The hams successfully relayed the results of the first leg of the round-trip, so the results could be posted promptly at AMA HQ. But, true to form for March weather in New York, the QST item concludes with the comment that "A snowstorm the preceding day made the run extremely difficult."

April 1938

Another report in "Operating News" tells about the Oakland (California) Radio Club furnishing communication for the Oakland Motorcycle Club for a motorcycle endurance run held on January 29 and 30. Three stations—W6EGM, W6ELW and W6OT—operated continuously for a period of 12 hours 32 minutes to pass along progress reports during the 145-mile, figure-8 race. Other operators who participated included W6LVS, W6MMK, W6OMC and W6ZM. An interesting operational note is that both 1.75 MHz and 56 MHz were used for the phone communication.

A footnote to this *QST* item: W6ZM was Herbert Hoover, Jr, son of the US President who was involved with Amateur Radio regulatory efforts during the time he was the US Secretary of Commerce. Later W6ZM was President of the ARRL. His son, Herbert Hoover, III, obtained the call sign W6ZM after his father passed away.

One evening in the early 1990s, I had a nice CW chat with the current W6ZM as I was mobiling along in my Toyota van. During the conversation, he told me of going to Germany to visit his son, and of the two of them renting BMW motorcycles and whizzing along the German autobahns at speeds up to 100 mph. The W6ZM-motorcycle connection continues…

July 1952

In "Correspondence from Members," Lee Lamascus, W6JQP, points out that, because of the current emphasis on Civil Defense communication, the importance of motorcycles with radios on board should be addressed in the pages of *QST*. Lee makes the point that motorcycles can negotiate terrain that would stop four-wheeled vehicles, especially after a natural or manmade disaster and that they can do it while using less gasoline.

October 1962

A photo in "With the AREC" shows WA6SXG operating WA6HBU/6 on 6 meters on Ranger Peak, as part of the communications support provided by the AREC of Southern California to the Road Riders of Southern California for their motorcycle run earlier in the year.

July 1988

An item in "Strays" reports that K1YYP and KA1RED will operate a QRP motorcycle DXpedition in June and July in Inuvik, in the far north of the Northwest Territories. Their setup will include a Heathkit HW-9, Curtis keyer and a collapsible vertical for 40, 30, 20 and 15 meters.

August 1992

An item in "Strays" alerts readers to listen on for K3KMO/MC on 7037, 14037, 21037, and 28037 kHz as he rides his motorcycle from Maryland to Alaska and back, from July 26 through August 26. (The announcement produced the desired results, with quite a few hams looking for and finding K3KMO/MC while on the road.)

July 1993

K3KMO tells the story of his 32-day, 10,500-mile ride from Maryland to Fairbanks, Alaska, and back on his Honda Gold Wing. Operating a Kenwood TS-140 and memory keyer mounted on the handlebars, during the ride K3KMO made frequent contacts with friends in Maryland, had long ragchews with hams throughout the country and worked plenty of DX. While Al was driving, navigating, and hamming, his son Pat rode on the back seat, listening to music on his Walkman and reading Stephen King paperbacks.

September 1993

KØKS comments on K3KMO's /MC installation, as described in his article two months earlier. Ken makes the good point that the motor vehicle laws of most states require that the driver must be able to see the speedometer, turn-signal indicator and high-beam indicator when seated in the riding position.

August 1995

AA1EP describes "A Simple 2-Meter Bicycle/Motorcycle Mobile Antenna," made by building a twinlead J-pole antenna and securing it to the fiberglass pole of a bicycle safety flag. His construction and mounting schemes are easy to replicate by the bicyclist or motorcyclist ham.

September 1996

"Up Front in *QST*" reports that KA7ZUM (now K7ZUM) showed up at ARRL HQ one bright and sunny summer's day to have some QSL cards checked for DXCC credit. Ken, with his young son Dustin on the rear seat, had ridden his Honda Gold Wing from Oregon to make the delivery.

December 1996

An item and photo in "Strays" reports that the Motorcycle ARC, then based in southern California but now with chapters throughout the USA, celebrated its fourth anniversary in June. [For the latest information on MARC, and interesting photos of the members' /MC installations and the club's activities, point your Web browser to **www.ba-marc.org/site-map.htm** .]

February 2002

The cover shows Charles, KD5KA, on his motorcycle, with the article "A Virginia Ham Goes West" telling of his QRP HF operation from various locations during his cross-country motorcycle trip in the summer of 2001. Charles and his wife (KG4LRG) were joined on the trip by two other cycles—one carrying Charles' son Jeremy (KG4IEK) and his wife, and the third with Gary (KG4OFU) and his wife. When riding along the road, the cyclists communicated via 2-meter FM and also via CW messages tapped out on the cycles' horns.

For a follow-on story of a later motorcycle trip Charles took, go to **www.arrl.org/news/features/2002/11/26/1/**.

CHAPTER 5

HF Unplugged

As predictably as the swallows return to Capistrano, Amateur Radio operators take to the hills on the fourth full weekend of June. Tons of aluminum get hoisted into the air; miles of wires and rope get strung in the trees; countless batteries, tuners, keys, tents, backpacks, coolers, and yes, even radios are hauled to temporary locations. Of course, it's ARRL Field Day, the most popular operating event of the year!

To a large and growing segment of the ham population, however, such operation is no longer limited to just one weekend. For us, *any* day can be Field Day.

We call ourselves HF Packers, HF Trekkers and other similar names. For the most part, we have one thing in common though. We enjoy operating from non-traditional locations with temporary antennas, no commercial power and with gear that we've carried to the site. These are by no means hard and fast rules. For some, sitting at a backyard picnic table with an extension cord for power and sitting in the shadow of that "BandBlaster" monster beam atop a 200-foot tower is "roughing it."

For others, nothing but hiking to the summit of a remote mountain for a week of hamming will do. Most of us fall somewhere in-between. In this chapter we'll focus on operating with a complete station that you can carry with you, set up and take down in a reasonably short time, and that requires no commercial power. Once you try *HF Unplugged*, you'll be hooked. As a bonus, the experience and expertise you gain will make you a real asset to your club or group on Field Day weekend. And of course, what could be better preparation for disaster and emergency communications setup?

If you're bedeviled by antenna restrictions, living in a small apartment, looking for new Amateur Radio frontiers or you just love being outside whenever you can, read on!

John Bee, N1GNV, wrote this chapter. He was first licensed in 1989 and holds an Amateur Extra ticket. John earned a Bachelor of Science degree from Fordham University, Bronx, NY. After several years as ARRL Advertising Manager, he founded Quicksilver Radio Products, a distributor of ham radio parts and accessories. He enjoys designing "gadgets, gizmos, and doodads" to meet the real-world challenges that hams encounter on a regular basis.

A FEW WORDS ABOUT SAFETY

When erecting antennas in the field (as at home), make doubly sure that they cannot *ever* come in contact with power lines. Don't let, *"It'll be okay for a few hours"* be your last words.

Remember also that rechargeable batteries can deliver a tremendous amount of current in a short time—easily enough to start a fire if they're shorted. Always fuse power leads as close as possible to the battery. Never pack a battery away with any possible chance that the terminals can short. A "Car-B-Cue" is a guaranteed way to ruin your day!

Radial wires, feed lines, power cords and the like should never be deployed where folks will be walking. If you leave a wire where someone can trip over it, rest assured that someone *will* trip over it.

We hams use many methods to get our antennas up as high as practical. Slingshots, bow-and-arrow, and rope-tied-around-a-rock (as well as other systems) are all commonly used. Whatever your choice, make sure that no one is anywhere within range of your projectile.

And familiarize yourself with the RF Exposure guidelines. You'll often be operating with antennas much closer to you than they'd be in a typical permanent installation. Fortunately, you'll also—for the most part—be operating at low power.

And finally, please make sure that you have the necessary permission to operate from your chosen location. If it's private property, don't even think about setting up without the blessing of the property owner. If it's a public park, a few minutes spent with the ranger or other authority may save you a lot of grief and aggravation later. Let them know what you'll be doing and when. Invite them to visit your setup. Stress that you won't be erecting any permanent structures and that when you leave there'll be no trace that you were ever there. And then make sure that happens.

Remember also that you may be called upon to explain to the public what you're doing. This is a wonderful opportunity for some great Amateur Radio public relations. When I'm out operating Unplugged, I try to make sure that I have a few copies of some basic information that I hand to anyone interested. I include contact and meeting information for local clubs, licensing classes, VE sessions, and of course the ARRL's address, telephone number and Web Site—**www.arrl.org**. You never know when those seeds will flower.

BALANCING THE BUDGETS

We're all familiar with budgets. Operating Unplugged means balancing several, often conflicting, demands. Most often, you'll find yourself juggling power, weight and bulk, time and money. While there may be no perfect solution, some judicious choices and realistic expectations will lead to more fun and less frustration. *"You can work the world with 5 W and a good antenna,"* say the QRP enthusiasts. *"Flea power and a compromise antenna are exercises in futility,"* counter the skeptics. The truth generally lies somewhere in-between.

If you already own an HF radio, you can almost certainly try getting Unplugged with very little additional expense. As we'll see, suitable batteries are often available for next to nothing. Effective antennas can be constructed with little more than a length of wire and perhaps a few parts from the local hardware store. Once you have some experience, you'll be able to make better decisions about acquiring new equipment. Avoid the temptation of just throwing money at the problem.

POWER TO THE PEOPLE

Your power budget consists of two related parts—RF power out and dc power to produce it. In most cases, your dc power source will be rechargeable batteries. Obviously, the more RF power you generate, the faster you'll drain the batteries.

Rechargeable batteries come in a wide variety of types or *chemistries*. Some of the more common ones are Lead-Acid, Nickel-Cadmium, and Nickel-Metal Hydride, as well as newer technologies like Lithium-Ion and Lithium-Polymer. They're rated in Amp-Hours (Ah) or Milliamp-Hours (mAh).

To oversimplify, a 7 Ah battery should be able to operate a 7-A load for one hour, a 3.5-A load for two hours or a 14-A load for 30 minutes, until the battery is dead. Most often, "dead" is defined as 10 V. In practice, a new 7-Ah battery will last longer than 2 hours with a 3.5-A load, but less than 30 minutes with a 14-A load. As the battery ages, its capacity diminishes until it will no longer supply its rated voltage for any time at all. At that point, it's time to replace the battery.

Lithium-Ion, Lithium-Polymer, and other exotic chemistries offer the highest power density—that is, the most power per pound and/or power per cubic inch—of any generally available rechargeable batteries. Unfortunately, they are also very expensive and require special charging techniques. Be careful! Lithium cells have been known to overheat or even *explode* when charged improperly. For those reasons, very few HF Packers have adopted them. Look for these cells to decrease in price as they're more widely used, and for chargers to become more widely available.

Nickel-Cadmium (NiCad) is an older technology and has been largely supplanted by Nickel-Metal Hydride (NiMH). NiMH cells have very good power density and are available in many configurations. For example, most battery packs for handie-talkies use NiMH cells. They also come in standard sizes: AAA, AA, C, D, and 9-volt. All of them, except 9-volt, are 1.2 V per cell, so you'll need to have at least 10 cells in series to make a nominal 12-V pack. Of course, if your rig requires less than 12 V, you can use fewer cells. Still, using these cells can get expensive in comparison to the Lead-Acid type. As this is being written in early 2005, AA NiMH cells cost about $2 each for 2000-mAh. On the other hand, D cells with 4 to 5 times the capacity of the AA cells go for about $5 each. Chargers for them are widely available and are not terribly expensive.

Lead-Acid batteries come in two basic types. *Wet* cells have a liquid electrolyte that requires periodic replenishment with water. Automobile and marine batteries are often made this way. The electrolyte is a highly corrosive acid solution, which can spill out if the battery is not kept right side up. Wet cells also produce hydrogen gas when charging—a serious explosion hazard without proper ventilation. For those reasons, they're not the best choice for portable operation.

The other type of Lead-Acid battery is a *sealed lead-acid* cell. These use a gelled electrolyte with no provision for adding water. They are, therefore, non-spillable. These batteries are often called *gel cells*. While there are some minor chemistry variations among them, we'll treat them as one class. Gel cells have the lowest power density of any of the battery types under consideration. However, they do offer several advantages that make them the most popular option for HF Unplugged operation. They're cheaper than the others (sometimes free!). They're widely available in a variety of sizes; they're easy to charge; they're fairly resistant to physical abuse; and they're reasonably tolerant of over- and undercharging.

We'll confine our discussion to 12-V gel cells, since that's what most rigs require. They're available in sizes ranging from small 1-Ah units, 50+ pound 80-Ah behemoths, and there are even much larger ones, designed to run industrial equipment like forklifts and pallet jacks. However, not many of us are going to carry a 600-pound battery very far!

So how much battery capacity do you need? The answer depends on several factors. How much current does your rig draw? What is your operating style? How long do you want to operate before recharging your battery? Okay, I'm not a big math fan either, but a little time with a pocket calculator will help you to make a decision. Let's work through a few examples.

Assume that in a typical session, you'll transmit 15% of the time and listen 85% of the time. Adjust these figures to match your own style if necessary. My ICOM IC-706 MKIIG draws 2 A on receive, and 20 A on transmit at full power (100 W). My average current draw is therefore (2 A × 85%) plus (20 A × 15%) = 4.7 A. I'll need about 5 Ah of battery capacity for each hour I want to operate.

If I crank the power back to minimum (5 W) it still draws about 6 A on transmit. My average then becomes (2 A × 85%) plus (6 A × 15%) = 2.6 A. By contrast, my Yaesu FT-817, a 5-W maximum rig, draws about 2 A on transmit and 400 mA on receive, for an average of 640 mA. At 5 W output, I'll get about four times as much time out of the same battery by using the '817 versus the '706 set on low power. This is by no means a slap at the '706. It's a great rig, and I've had a lot of fun with it. It does, however, point up the importance of considering your power requirements when operating Unplugged. We'll discuss rigs a bit later in this chapter. Meanwhile, back to batteries.

The advantages to gel cells that were outlined earlier make them the overwhelming choice for backup batteries in alarm systems, emergency lighting and similar applications. They do have a finite life, however. To comply with fire codes and insurance regulations, alarm and security companies need to replace these batteries on a regular basis. Many companies simply replace them on a schedule and don't test them. Not surprisingly, many of them still have plenty of usable life.

In many cases, the alarm company needs to pay a recycling company to take the old batteries away. Over the past few years I've obtained several hundred batteries from a local company. Oddly enough, when I show up at their office with a box of doughnuts or a pizza (pepperoni is optional), they're more than happy to give me as many old batteries as I want to take away. You'll often see batteries at hamfests, too, but you can expect to pay between $5 and $10 each for used but good 7 Ah cells. Reputable sellers load test them and stand behind them—and responsibly recycle the dead ones.

By a wide margin, the most common size of battery used in alarm systems is 7 Ah. This also happens to be a great size for HF Unplugged operation. They measure about 6 × 4 × 2½ inches and weigh about 6 lbs each. They fit into belt packs sold at most major department and discount stores. **Fig 5-1** shows a pack I built for portable use. With its 7-Ah battery I can easily get 5 or 6 hours of operating time with my FT-817. (On a side note, it's also quite handy for use at public-service events where I need to use a 2-meter or dual-band HT for an extended period. Battery packs for my HT are about $50 each for a 1600 mAh pack. I'd need four of them to get the operating time that this pack gives me and that I built for a whole lot less than $200!)

Note that the cable attached to the battery in Fig 5-1 is simply a short, fused cable, terminated in an Anderson Powerpole connector. In the middle is a "Y" with both a cigarette lighter socket and a Powerpole. This allows me to power two devices at the same time, or charge the battery while using one device. The lighter socket comes in especially handy for powering cell phones. There's room in the pack for a charger and a cable for my rig, plus a few other odds and ends. Without the charger, I can even fit a small HT and keep up with the local gang on 2 meters while I'm up on the mountain operating HF Unplugged.

Most smaller gel cells—up to about 10 or 12 Ah—come with "Faston" tab connectors. They're most often $^{3}/_{16}$ inches wide, although some have $^{1}/_{4}$-inch tabs. Mating female connectors—$^{1}/_{4}$-inch will work on both sizes—are available from most electronic suppliers, including RadioShack.

The color of the insulation on the connectors denotes the wire size they accept. Yellow is for #12 and #10; Blue for #14 and #16; Red for #18 and smaller. I always wire my batteries with #12 or #14 wire, so I use the yellow ones (although, in a pinch, I've used the blue ones with #12 and had no problem.) The connectors are often called *solderless terminals*, requiring only a crimp.

I prefer the belt-and-suspenders approach, and both crimp and solder them to the wire. Since I use Anderson Powerpoles for all of my 12-V dc connections, I make a short, fused pigtail with female Fastons on one end and Powerpoles on the other. Then I solder the Faston connectors to the battery terminals. Soldering is not necessary, but I think it makes for a more secure connection.

When the battery finally dies, it's simple to unsolder the terminals and move the pigtail to a new one. Larger batteries use a nut and bolt through a lug fitting for their terminals. For those I use ring terminals—again, yellow for #12—crimped and soldered to the wire. However, since the nut and bolt make a good electrical connection I see no need to solder the ring terminal to the battery terminal.

At this point you may be thinking, "*Great, 7 Ah is a convenient size and the price sure is right. I'll wire up two in*

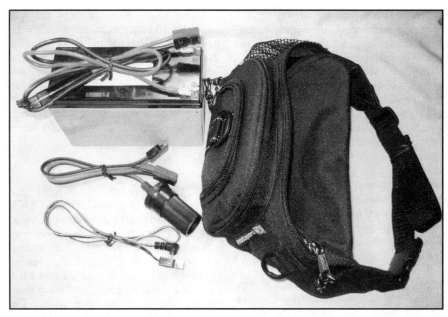

Fig 5-1—Gel battery pack with convenient "fanny pack" nylon carrying case and Powerpole connectors built by N1GNV for use in the field.

parallel and double my operating time." A great idea in theory, but it doesn't work out that way in practice. One battery will always be at least little weaker than the other. The stronger battery will then expend some of its energy trying to charge the weaker one. You could build an isolation circuit with a few diodes to prevent this, true. But it's much simpler to just use them one at a time.

A word to the wise here. I'm often asked at hamfests for a used battery for someone's home security system, computer backup system, etc. In my opinion, this is the epitome of penny-wise and pound-foolish. The alarm company replaced the batteries I offer because they had reached the age where they could not be considered reliable. Spend the money on a new battery for your security system, replace it every 3 years and use the old one for playing radio. In fact, approximately 30% of the used batteries I get are dead and will no longer hold a charge. That really makes a good case for regular replacement, doesn't it? In a similar vein, I'd be hesitant to employ used cells in any mission-critical situation. Fortunately, gel cells rarely fail instantaneously. Instead, you'll notice a gradual reduction in the time before you need to recharge them.

To test the batteries, you need to charge them (more about charging later) and then place them under a heavy load. I use a resistor pack designed to draw about three times the battery's Amp-Hour capacity. For a 7-Ah battery, I use three 2-Ω, 25-W resistors in parallel, giving 0.66 Ω rated for 75 W. A fully charged battery should be a bit over 13 V. By Ohm's Law, I=E/R.

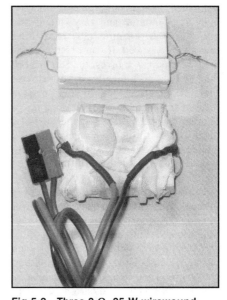

Fig 5-2—Three 2-Ω, 25-W wirewound power resistors wrapped in a rag and secured with cable ties are ready to be immersed in cooling water. This makes a convenient test fixture to ensure that surplus gel-type batteries are still up to snuff.

13/0.66 = 19.6 A, which is close enough to three times the 7-A rating.

Sharp-eyed readers will note that 13 V × 19.6A is over 250 W, far above the resistors' rating. In **Fig 5-2** you can see that I wrap the resistors with a rag and secure it with zip-ties. I then place the resistor pack in a bowl with a small amount of water to serve as a poor man's heat sink. For the brief period that I test the batteries, this is acceptable. Even at that, the resistors do get quite warm. For the record,

I would never use such an arrangement in anything but the most intermittent service.

After charging, let the batteries stand for at least an hour to settle. Attach a voltmeter. It should read somewhere close to 13 V. Now attach your resistor load. The voltage will drop. If it stays above 10.5 V after 15 seconds, your battery is probably good. If the voltage drops much below that, recycle that battery and try another one. Why not just use your 100-W radio that draws about 20 A to test them? I believe in some cases undervoltage can be just as harmful as overvoltage to a rig. I sure don't want to risk an expensive radio to test a battery! Resistors cost a whole lot less than radios. Get a few like-minded hams together to share the cost and it's even less.

So how do you charge these things, anyway? Well, there's the *official* answer, and then there's the *"what most of us do"* answer. Chargers designed specifically for gel cells are available from several sources. There are also many circuits published, should you choose to build your own. They monitor and vary voltage and current, often keeping tabs on battery temperature as well. If you're buying new cells, it may pay to invest the $50 to $100 that the chargers typically cost, or to invest the time and effort to build your own. On the other hand, if you've obtained a number of used batteries for the cost of a pizza or less, you may wish to take a simpler approach. Purists will no doubt shudder at my system but it has worked well for me for many years.

Wall warts, those black cubes that plug into electrical outlets to charge and power our ever-growing collection of gadgets are ubiquitous. I don't know about you, but I can't bring myself to discard them even when their associated devices attain Silent Key status. They're nearly always labeled with output voltage and current. Chances are good you've got one or more rated at 12 Vdc, with current ranging from 300 to 700 mA. Fortunately for us these are not regulated power supplies. With no load, you'll probably measure them at 14 to 15 V. That's fine for charging those gel cells. *Important: Make sure to check the maximum voltage that your equipment is rated for. If your wall wart is above or close to that value, don't take a chance. Charge your batteries off-line, and then connect them to the equipment.*

What about the current? As it turns out, gel cells are best charged at no more than a C/10 rate —that is, 1/10th of their rated Amp-Hour capacity, or 700 mA for a 7-Ah battery. Charging them at much more than that can shorten their life. Charging them at much less than that requires longer

charging times. Keep in mind that gel cells, like all batteries, do have some self-discharge. Allowed to sit they will eventually go dead. With a very large battery and a very small charger, it's possible that the charger won't keep up with the battery's discharge rate. This will be a rare occurrence though. The worst things you can do to your gel cells are:

- To overcharge them
- Discharge them too deeply
- And let them sit dead.

At a C/10 rate or less, you can leave them on charge more or less indefinitely. While many rigs will operate at voltages as low as 7 or 8 V, I don't like to draw my batteries down below 10 V. From 10 V it takes about a 15-hour recharge at C/10 to bring a battery back to a fully charged state. I make it a practice to charge mine at least every three months. During the warmer months I'm using them a lot more often and so I recharge them more frequently.

If your various budgets permit, look into the emergency jump starters available from auto parts, hardware and discount stores. Mine has an 18-Ah battery, jumper cables for starting the car, a cigarette lighter socket (very handy for running a low power rig), a built-in light and even an air compressor. It also came with a charger for 110 V ac and a charger that plugs into a cigarette lighter socket. At about 20 pounds, it's a bit much to carry for any distance, but great for those times when I want to operate for an extended period without commercial power.

If you expect to be away from commercial power for an extended period, you might consider solar power to recharge your batteries. However, good solar panels are not cheap and they will almost certainly require a charge controller of some type. Of course, they'll also add to the weight and size of your portable station. An extended discussion of the various options is beyond the scope of this chapter, but a little digging on the Internet will provide a wealth of information. A good starting place is at Sunlight Energy Systems. Mike Bryce, WB8VGE, is the owner. Several of Mike's articles on solar charge controllers have been published in *QST*. See **www.seslogic.com**.

POWERPOLES TO THE PEOPLE

Anderson Powerpoles have justifiably gained a wide following in the Amateur Radio community for dc power connections. They have quite a few advantages over the binding posts, banana plugs, molex-type or Jones-type connectors that many of us have used in the past. Perhaps most important is that they're genderless. There's no

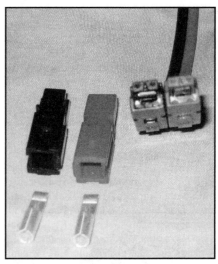

Fig 5-3—Powerpole connectors. "Tongue-Top-Right-Red" is a mnemonic coined by N1GNV to help him remember the correct color coding.

male or female—every one is exactly the same whether it's used on a source (like a battery) or a device (like a radio). They can handle high current (up to 45 A) with no problem. They slide together on a dovetail joint, making it virtually impossible to connect them backwards, even in the dark.

Fig 5-3 shows the Powerpole components and the standard configuration. Looking into the connector, with the metal contact tongue on top, the red housing is on the right. "Tongue-Top-Right-Red" is how I remember it. If you build yours this way you'll be compatible with just about any other ham using Powerpoles. But always double-check to make sure that the other guy built his correctly.

Powerpoles easily accommodate a range of wire sizes. Anything from #24 to #8 is no problem. Anderson actually makes three different sized terminals—15, 30 and 45 A. All three fit into the same housings and each will mate with the others just fine. I don't bother with the 15 A terminals. When I build with smaller gauge wire, I just fold it over once or twice to get a good amount of copper into the barrel. #14 and #12 wire fit easily into the 30-A terminal. #10 is a tight fit but can be done. Don't worry if one or two strands don't go in. If that would really make a difference, then you need to be using heavier wire anyway.

You'll definitely need the 45-A terminals if you're using #8 wire. With just a little practice, they're easy to install and when properly assembled are extremely reliable. Unlike other connectors, they're designed for repeated plug/unplug cycles, a very handy feature when you're assembling and disassembling a station frequently.

There's some debate about whether Powerpoles should be crimped, soldered or both. Folks have told me that they only crimp, or only solder, and they don't have problems. I can't say that they're wrong. On the other hand, I've built over 3,000 cables and have crimped and soldered every one. I've never had a failure so I'll keep doing both. If you do solder them, make sure not to get any solder on the outside of the terminals. That can make it difficult or impossible to get the terminal into the housing. I use a Klein crimping tool that typically sells for about $30 to $35. I've been told that the Klein tool has the same dimensions as a Thomas and Betts crimper, but I have no direct knowledge of that. West Mountain Radio, **www.westmountainradio.com**, makes a crimping tool specifically designed for the Powerpoles. It's a little more expensive, but it's as close to foolproof as you can get. If your club or group is changing over to Powerpoles it's a worthwhile investment.

Powerpoles make it a snap to move equipment from place to place. You can move a rig from the shack to the car to the park, and back, just as easily as you'd move a lamp from room to room. Almost everything I own that runs on 12 V has a Powerpole on it. I keep a selection of Powerpole cables in my "go-kit", including some with common dc connectors, two with alligator clips and a few just wired with nothing on the other end. I'm reasonably confident that I can get/supply power from/ to just about anything that has/uses 12 V.

Fused distribution panels (think outlet strips) are available from West Mountain Radio and from MFJ, **www.mfjenterprises.com**, among others. A wide variety of assembled cables is available from Quicksilver Radio Products, **www.qsradio.com**. For Unplugged operation, the distribution panels are probably overkill. It's easy enough to build or buy a "Y" cable to supply two devices from one battery. In fact, you can make "Z" or "W" cables using the same technique.

Be aware that #16 is the largest wire that you can double up in a Powerpole housing. Of course, if you make splices outside the housing you have no such limitation. **Fig 5-4** shows the schematic diagram for a "Y" with battery-voltage monitoring LEDs. Using the zener diodes specified, both LEDs will light when your battery is fully charged. The green LED will dim and then go out as the battery drops to about 11.75 V. The yellow LED will dim and then go out at about 10 V. With a little experience this will give you a good idea of the state of your battery. Use different values for the zeners and different colors for the LEDs to suit your own preferences.

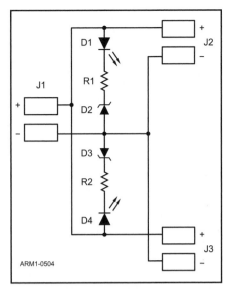

Fig 5-4—Schematic for voltage monitor and "Y" splitter.
J1, J2, J3—Anderson Power Pole
R1, R2—470 Ω, 1/4 W
D1—green LED
D2—11-V zener diode
D3—9.1-V zener diode
D4—yellow LED

Fig 5-5 shows my own "BattMon" built into a commercially available Lexan case.

RIG ME UP, SCOTTY

Choosing a rig for Unplugged operation again involves balancing the budgets—power, weight and bulk, time and money. If you already own an HF radio, chances are very good that it operates on (nominal) 12 V dc. For your first few forays into the wild, you may wish to use your existing equipment. It may not be ideal, but it will give you a good taste of the experience. Depending on your source of power and how often and long you want to be Unplugged, this may be all the rig you'll need.

If you decide that going Unplugged is something you'll enjoy on a regular basis and your finances permit, you're in luck. Most of the major manufacturers offer multi-mode rigs that cover all of the HF bands and are just begging to be used in the great outdoors. Further, there are many smaller manufacturers making single-band kits or assembled rigs. These are primarily CW only and most run only a few hundred milliwatts output but there are exceptions. MFJ, for example, makes a line of single-band SSB-only transceivers, as well as CW-only rigs. While most Unplugged operation is either CW or SSB, with a laptop computer and a sound card interface there's no reason why you can't enjoy your favorite digital mode just like you do at home. Just remember to factor in the laptop's requirements in your power budget.

You may have guessed that most Unplugged operation is done with low power. Many refer to it as QRP, although strictly speaking QRP is no more than 5 W output. Some of the popular rigs used outdoors will put out 10 or 20 W, still much less than a standard 100-W home rig. And yes, you can make plenty of contacts with low power. Change in power is most often expressed in decibels (abbreviated dB). A change of 3 dB is a doubling of your power. S-meters on our rigs are also calibrated in dB. One S-unit is 6 dB. A 6 dB change (3 dB plus 3 dB) means a

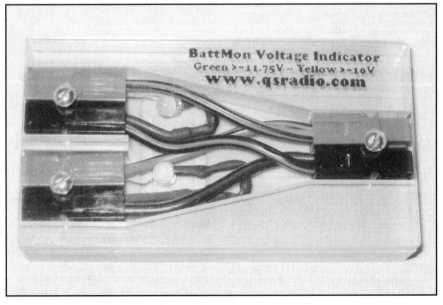

Fig 5-5—"BattMon" battery monitor built by N1GNV. Visit www.qsradio.com.

quadrupling of power. The difference between 5 W and 100 W is 13 dB—just a little more than two S-units. All things being equal, if you're S9 with 100 W, you should be S7 with 5 W.

In all fairness, this doesn't always work out so neatly. You can be down in the mud with 5 W and armchair copy with 100 W. But remember that putting out that 100 W will put a big dent in your power budget in a hurry if you're carrying batteries to your location!

In no particular order, here's a sampling of the current crop of multi-mode, multi-band rigs that are especially suited for Unplugged operation. Remember that manufacturers frequently update, replace and retire their models. Keep a sharp eye out for closeouts and don't overlook used equipment. Bargains abound if you're in the right place at the right time. Don't feel that you need to invest in a new rig to join in the fun. One friend has even been known to take his 1970s vintage Kenwood TS-520 (tube finals and all!) out for an afternoon in the park. Balance those budgets, use what you have, and get out there and make contacts. But if a new rig is in the cards:

- The introduction of Yaesu's FT-817 in 2000 was a milestone event in the evolution of Unplugged operation. This all-mode, all-band radio includes 2 meters and 70 centimeters and is a marvel in a package just a little bigger than the HTs that many of us dangled from our belts not very long ago. It's a true QRP rig, pumping out a maximum of 5 W. See **www.yaesu.com**.
- ICOM's IC-703+ is a larger unit. It looks and feels like their popular IC-706 and covers HF and 6 meters, although it does not have the '706's 2-meter and 70-centimeter coverage. It does, however, include a capable internal automatic antenna tuner. Output power is 10 W maximum at 13.8 V and 5 W at 9.6 V. See **www.icomamerica.com**.
- SGC offers the SG-2020. It's similar in size to the IC-703 and covers all of the HF bands. Its 20-W signal can sometimes make the difference in snagging that contact and not. See **www.sgcworld.com**.
- If you'd rather build your own kit rig, take a look at Elecraft's highly regarded K2. The basic kit covers 80-10 Meters, with up to 15 W on CW. They offer a wide range of optional add-in modules including 160 meters, SSB, an internal antenna tuner and 100 W output. The K2 is similar in size to the IC-703. Elecraft also offers their K1 and KX-1 CW-only kits if you'd prefer a smaller package. See **www.elecraft.com**.

- Ten-Tec's legendary Argonaut is a bit larger than the other rigs and not nearly as battery-friendly as the radios listed above. However, it does include quite few bells and whistles and is easily upgradeable with a software download. The Argonaut covers all of the HF bands with up to 20 W. See **www.ten-tec.com**.

All of the above radios have been reviewed in *QST*. ARRL members can view those reviews online at the League's Web site. The respective manufacturers' Web sites provide detailed information and you'll find user groups for all of them on the Internet too. Looking through the postings will give you a good feel for the experience of real-world users and should help you make your purchasing decision. This is by no means an all-inclusive list, but does provide some choices of gear particularly well suited to going Unplugged. Depending on your own combination of power, weight and bulk, time and money budgets, there's almost certainly a rig out there that will fill your needs.

MAKING WAVES

OK, we have a rig, we have power for it, it's connected and we're just about all set. Now we need an antenna. Easy, right? Actually, yes. Or as difficult and as complicated as you want to make it. Get three average hams together and you'll get six opinions about the best kind of antenna. Get three Unplugged hams together and you'll get at least twice as many opinions! If you keep two basic rules in mind, you'll be way ahead of the game.

- Rule #1: You cannot break, bend, tweak, spindle, fold or mutilate those Pesky Laws of Physics.
- Rule #2: When you discover, design, or develop an antenna that defies those Pesky Laws of Physics, please refer to Rule #1.

Antennas are not black magic. They always obey the Laws of Physics. A specific antenna, erected in exactly the same way, in exactly the same environment, will always radiate in exactly the same way. Conversely, any change in the way an antenna is erected, or in its environment, will almost certainly result in at least some change in the way it radiates. Notice the word "radiate." That doesn't mean "perform." Performance is also governed by propagation. That's a subject for another chapter in another book, but we can do a lot to take advantage of what the gods of propagation deign to give us.

If you already have an HF antenna up at home, you probably know the drill. Build one from plans in *The ARRL Antenna Book*

or plans that were published in *QST* or in another magazine or on the Internet. Perhaps you purchased a commercial antenna. The procedure is the same—hang it in the air, take some measurements, bring it down for tweaking and pruning, hang it back up, measure some more, tweak some more, measure and tweak, measure and tweak, until you're either happy, satisfied or just plain out of patience. It can be a time-consuming process, but most times that antenna will be up for several years without needing more than a bit of periodic maintenance.

When putting up an antenna for Unplugged operation, it's a completely different story. In most cases, every time you set up an antenna, the electrical environment will be different. Antenna height, ground conditions and any nearby objects will all interact to change the way your antenna behaves. For example, its impedance, and therefore the SWR presented to your transmitter, will change. Changing bands will, of course, also affect the impedance. A fundamental difference between an antenna you put up at home and one you use for Unplugged operation is that in the field you typically won't have time for all of that measuring and tweaking. Bottom line—You can spend a whole lot of time getting an antenna "just right," and then only a little time on the air. Or you can spend a little time getting an antenna "close enough," and a lot of time on the air. It's your choice.

Here are a few basic antenna truths to keep in mind. If you don't understand (or believe) any of them, a little time with *The ARRL Antenna Book* is definitely in order. You do own a copy, don't you?

- A low SWR does not mean you have a good antenna.
- A high SWR does not mean that you have a bad antenna.
- A low SWR does not mean that your antenna is resonant, although it could be.
- A resonant antenna does not necessarily have a 50-Ω impedance.
- Antenna tuners do not "tune" antennas, whatever that means, nor do they make bad antennas good.

In most cases, your transceiver wants to see something close to a 50-Ω resistive impedance, which is a 1:1 SWR for a 50-Ω coax, to put out its maximum power with minimum distortion. As you move away from 50 Ω, the rig's automatic protection circuitry will begin to reduce power to prevent damage. It's time for those purists to shudder again. If I can get an antenna tuned to an SWR somewhere between 1.5:1 and 2:1, my rig is probably happy enough to put out full power and I'm happy. I quit fiddling and start making contacts. Depending on the circumstances, if I can't get it to tune below 2:1, I might try making some adjustments to the antenna and see if I can get a better match. If not, I'll change bands and try again.

There are a few basic ways to make those adjustments to get closer to a 50-Ω impedance. You can make changes to the antenna itself. Lengthening it, shortening it, raising it, lowering it, and reshaping it will all change the impedance. Again, this can be quite a long process. Or you can change the length and/or characteristic impedance of the feed line, in essence creating a matching section. For an excellent explanation of this, see "My Feed Line Tunes My Antenna" at **www.arrl.org/members-only/tis/info/ pdf/9111033.pdf**, a true classic from Byron Goodman, W1DX (SK). But hauling around a lot of different length cables and swapping them in and out doesn't sound very practical either, does it?

Fortunately, there are alternatives. We can place variable components—either coils or capacitors—in the antenna. For example, a coil in series with an antenna element will make it electrically longer. If we vary the inductance by tapping the coil at different points, we're varying the electrical length of the element and thus the impedance presented by the antenna to the feed line. If we can keep the coil at a height that we can reach, our tune-up procedure can often be quite easy and brief. This is generally done with an alligator clip on a "wander lead."

We could also insert a matching network at the transmitter end of the feed line. Since our impedance can vary greatly with each setup and each band change, we'll want the components of our matching network to be variable, with at least one inductor and one capacitor of significant range. And guess what? That's exactly what an antenna tuner is! It takes whatever impedance is presented at the transmitter end of the feed line and transforms it into the 50 Ω that the rig wants to see. That's all it can do. It can't do anything about whatever mismatch exists between the antenna and the feed line (the degree of that mismatch is the definition of SWR). It can't do anything about an antenna system that has a lot of ground loss or feed-line loss. All it can do is present a 50-Ω impedance to our rig. It's up to us to design an antenna system that minimizes ground and feed-line loss resistance and maximizes the antenna's radiation resistance. After all, it's the radiation resistance that squirts our signal out into the ether for others to hear.

There are several small, lightweight antenna tuners available. MFJ offers a few different models that won't add too much

weight or bulk to your Unplugged station. LDG at **www.ldgelectronics.com** makes a compact automatic tuner called a Z-100. Although it does require power to tune, its use of latching relays means that it draws nearly zero current once tuning is complete. This is a very popular unit with Unpluggers. Many newer transceivers include automatic antenna tuners, too. They're typically limited in tuning range and often unable to handle much more than a 3:1 SWR. However, if you can get your SWR reasonably close to that, autotuners built into your radio can make things quite easy and require one less piece of equipment for you to carry.

My very first Unplugged antenna was about as simple as you can get. I took a piece of stranded wire about 120 feet long, and attached a banana plug to it. Why 120 feet? Because that's how long the wire was when I hauled it out of the junk box! I attached a banana plug to it, and inserted that into the SO-239 socket on my antenna tuner. I tied a rope around a rock, snaked it through some trees and pulled the wire up. I found a frequency that the antenna tuner could match and started calling CQ. I did manage a few contacts, but it took a whole lot of calling for very few answers. Improvement was clearly needed.

Another trip to the junk box yielded a length of 4-conductor telephone wire. I twisted one end of the wires together and soldered an alligator clip on it. I cut the cable to 66 feet. Then I took about an inch out of one wire at 33 feet from the alligator clip; an inch out of another wire at 16$\frac{1}{2}$ feet; and an inch out of the fourth wire at 8$\frac{1}{4}$ feet. I attached the alligator clip to a ground screw on the rig and stretched the cable out more or less underneath my wire in the trees. I found that I could now tune the antenna on many more bands. And even better, I was making a lot more contacts. The telephone wire acts as a *counterpoise*, and the lengths of the conductors are roughly $\frac{1}{4}$-wavelength on 10, 20, 40, and 75 meters. It was not a great antenna, but it was about as cheap and easy as you can get, and best of all, it did get out. More counterpoise wires, fanned out as best you can, will improve performance.

If you have the space, the time, the supports and the means to get it

up, it's hard to beat a $\frac{1}{2}$-λ dipole at a height of 30 feet or more. The problem with a dipole is that it's a single-band antenna, although you can often press it into service on other bands. A better alternative is an 88-foot dipole (for 80 through 10 meters) or a 44-foot dipole (for 40 through 10 meters) fed with "window" open-wire line. Although neither antenna length is $\frac{1}{2}$λ on any amateur band, both present a reasonable SWR that's typically within range of small tuners. Both antennas also have fairly consistent radiation patterns over different bands, allowing you to erect them to favor your desired direction. L.B. Cebik, W4RNL, has quite a bit of information on this antenna at **www.cebik.com**.

My favorite method of getting antennas in trees is a combination of a slingshot and a fishing reel. In **Fig 5-6** you can see my setup mounted on a piece of $\frac{1}{2}$-inch PVC pipe with hose clamps. If you paint the weight with some fluorescent pink spray paint, it's usually quite easy to spot amidst the foliage. Ten-pound test fishing line is generally adequate for pulling light ropes up into the trees. I use $\frac{3}{32}$-inch Dacron-polyester rope, although for a temporary antenna that's even more than necessary. With small gauge wire, even mason's twine will work just fine for temporary antennas. Wire gauge has almost no effect on performance and neither does insulation on the wire unless the elements will come into contact with leaves or branches. Larger sizes will generally be stronger, but will add weight and bulk, a real consideration if you've already got a lot to carry. Use what you have and keep an eye out at hamfests for great deals on surplus

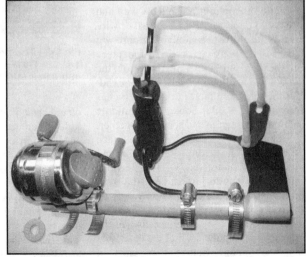

Fig 5-6—Slingshot and fishing reel system mounted on $\frac{1}{2}$-inch PVC pipe with hose clamps. This is used to launch wires into trees. Note metal washer used as weight. This has been painted a bright pink color to facilitate spotting it in dense foliage!

wire. I'm still kicking myself for passing on that large spool of #26 Teflon insulated wire at a very attractive price.

Textbook illustrations of horizontal wire antennas nearly always show them suspended in nice straight lines. If you can hang yours this way that's great. But it's certainly not necessary, or often even possible, to have such a perfect antenna. Try to get the middle portion of your antenna as high and in the clear as possible. That's generally where most of the signal radiates from. Don't worry if you need to bend, zigzag or droop the ends. If there's only one tall support available, try configuring the antenna as a sloper or even as a vertical dipole.

At the low antenna heights that our Unplugged antennas are often limited to, vertical antennas can provide better signals at low radiation angles. This typically means better DX performance. One very popular antenna is the ¼-λ ground plane with a loading coil in the middle to tune 20 through 10 meters, and a telescoping whip as the top section. It's easily constructed, and can be made in pieces small enough to fit in a standard briefcase or backpack. **Fig 5-7** shows the basic design. All of the parts are made from schedule-40 PVC plumbing or electrical pipe. No piece is longer than 17 inches, so it's easily transported. **Fig 5-8** shows the complete antenna, along with a Yaesu FT-817 all-band all-mode rig, a 7-Ah gel cell battery and a small antenna tuner packed inside of a briefcase and ready to take out in the field. **Fig 5-9** is a drawing of my tunable vertical.

For the ground counterpoise, I try to use at least four radial wires. Ideally, I'd have four ¼-λ wires for each band that I plan to work. That can mean a lot of wires! Here's a solution that works reasonably well and is easy to deploy. I use #24 speaker wire for this but any two-conductor zip cord you have will be fine. Start with a 25-foot piece of speaker wire. Cut just one of the conductors at 8 feet 4 inches from one end. Cut the other conductor at 8 feet 4 inches from the other end. Separate the middle section, leaving the end sections together. You should have two identical cables, each 16 feet 8 inches long with a second attached conductor that's 8 feet 4 inches long. These lengths just happen to be 1/4 λ on 20 and 10 meters respectively.

To attach them, I just slip the ends under an automotive hose clamp and tighten the assembly around the shell of the PL-259 at the feed point. Rather than carry a screwdriver, I simply use a coin to tighten the clamp. You could shorten the longer sections of the radials by rolling them up in a small coil, so that they'd be ¼ λ on the

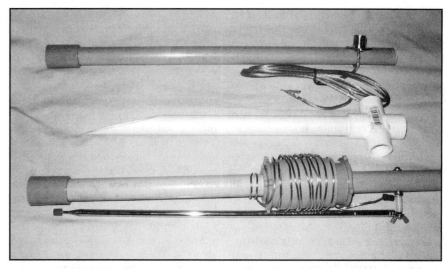

Fig 5-7—N1GNV's tunable portable vertical made from schedule-40 PVC plumbing.

Fig 5-8—The portable vertical ready to travel, complete with coax, ground radials, Yaesu FT-817 transceiver, antenna tuner and gel-battery pack. A complete station in a small suitcase!

other bands. I've tried that and noticed very little difference. By adjusting the coil tap and the whip length, I can usually get a good match on 10, 12, 15, 17, and 20 meters.

If you can elevate the radials at least a few inches above the ground, the antenna will perform better. In fact, if you can elevate the whole antenna so that the radials slope down toward the ground, you may find it much easier to get a good match to 50-Ω coax. However, this may mean that the coil and its tap will be out of reach, requiring you to lower it down for adjustment. Just remember to keep all of your wires and feed lines away from

anyplace where they could present a possible hazard.

The loading coil is made from a 1¼-inch PVC coupler, with appropriate reducing bushings to attach the ½-inch tubing I use for the rest of the structure. Loading coils are much more efficient when their diameter is approximately equal to their length, which is why I built my coil on the larger-sized coupling. I used 12 turns of #14 wire, although the wire gauge is not at all critical. It's ugly but it works just fine and cost just a few dollars. Commercial coils can be hard to find and can also be quite expensive.

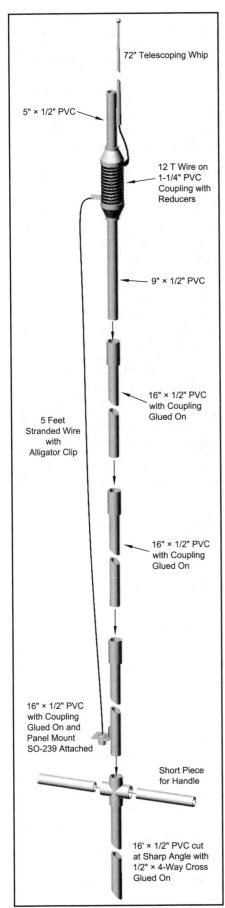

72" Telescoping Whip

5" × 1/2" PVC

12 T Wire on
1-1/4" PVC
Coupling with
Reducers

9" × 1/2" PVC

16" × 1/2" PVC
with Coupling
Glued On

5 Feet
Stranded Wire
with
Alligator Clip

16" × 1/2" PVC
with Coupling
Glued On

16" × 1/2" PVC
with Coupling
Glued On and
Panel Mount
SO-239 Attached

Short Piece
for Handle

16' × 1/2" PVC cut
at Sharp Angle with
1/2" × 4-Way Cross
Glued On

Fig 5-9—Drawing of portable vertical antenna.

The telescoping whip is 72 inches long, and is mounted on a #6 screw to allow it to swivel down for storage. A Velcro strap holds it more or less vertical when the antenna is set up. The top of the loading coil is also fastened to the screw using a ring terminal. If you can't find a 72-inch whip, use the longest one you can get. More whip and less loading coil makes for a more efficient antenna.

You may also be able to extend the whip's length by adding a piece of wire with an alligator clip and tying it off to a convenient support. Ideally the extension wire will be vertical, but if it runs off at an angle it's not going to make too much difference. If you add enough wire you may find that you can also tune the antenna on 30 and 40 meters. I haven't had much success with it on 80 meters, but you won't know unless you try, will you?

The feed point is made from an SO-239 panel mount connector attached to 1/2-inch PVC with a #6 screw. Again, this may not be the most elegant way to do it but it works just fine. I soldered a 5-foot length of stranded wire to the center pin. An alligator clip on the other end attaches to the coil, allowing you to tap at various positions to find the best match. This 5-foot length of wire, plus the 6 feet of telescoping whip, is about 1/4λ on 15 meters. Shortening the whip will get you to 1/4λ on 10 and 12 meters. Thus, you'll only need to tap the loading coil for 17 and 20 meters. Simply attach the alligator clip at the top of the coil to take it out of the circuit.

The antenna is supported by a piece of PVC cut at a sharp angle and driven into the ground. A PVC 4-way cross fitting allows handles to be attached. These are generally sufficient to drive the stake into the ground by hand. There are any number of other schemes you can use to accomplish the same purpose, as well. I haven't found it necessary to guy the antenna when built as presented here. However, I have made serviceable ground anchors for other projects using the same type of sharpened PVC stake. Just to be clear, this is *definitely not* a technique for permanent installations or where significant stress on the guy ropes is possible. Put safety first, as always. The rest of the mast is simply a few more pieces of PVC with couplings to hold them together.

Schedule-40 PVC plumbing pipe is a wonderful portable antenna-building material. It's cheap and widely available; it's very easy to work with using standard hand or power tools; and it comes in a wide array of sizes and quite a few different fittings. When you're experimenting with a new design, fit the pieces together

without glue first, to make sure it will work mechanically as you expected. Be aware that PVC cement sets extremely quickly. Within about 5 or 10 seconds of cementing, it's usually impossible to get the pieces apart. For the antenna described above, I attach one fitting—a coupling, tee, cross, etc—to each piece. This makes assembly and disassembly much simpler and keeps the number of loose parts to a minimum. A trip to the local hardware or home improvement store should give you a wealth of ideas for what's possible. Take a look through *The ARRL Antenna Book* and you'll find many other designs that can be adapted to Unplugged operation with just a little ingenuity.

You may already be driving around with a great Unplugged antenna. Most mobile HF antennas can be readily used in the field. Most, but by no means all, mobile antennas are electrically 1/4 λ long and use the body of the car as the counterpoise or ground plane. They can be single band, with a built-in loading coil, or they can have a variable loading coil enabling them to be tuned over a number of bands.

Monoband, helically loaded whips are available for each of the HF bands with the possible exception of 160 meters. I've never seen one for Top Band, although others have reported that they have indeed used them. They're often referred to generically as "Hamsticks," although Hamstick is a trademark of The Lakeview Company, located in South Carolina, at **www.hamstick.com**. Hamsticks and similar antennas consist of a 4-foot fiberglass section with a loading coil around it. A 4-foot stainless steel whip fits on top of it. This can be adjusted in length to provide the best match at your frequency of operation. All of these antennas have a 3/8-24 threaded base. This is a standard mounting thread, used on many ham, commercial and CB antenna mounts. That means that you have a wide variety of mounts to choose from. In addition to the usual ham suppliers, don't overlook RadioShack, CB stores and truck stops as sources for these items. You'd be surprised at the number of different mounts that are manufactured. Many of them can be easily adapted to Unplugged operation. Remember that you'll need to provide a counterpoise. You can use the same techniques as described above, or even better, devise your own.

Since Hamsticks are electrically 1/4 λ long, you can use two of them to make a dipole, too. Commercial products are available to do this but it's an easy thing to build yourself. **Fig 5-10** shows one way to do it. A panel-mount SO-239 connector is attached to a piece of polycarbonate sheet

(Lexan is one brand name of poly-carbonate sheet). Do not use acrylic sheet (Plexiglas is one brand name of acrylic sheet). One wire, with a $^3/_8$-inch ring terminal on one end, is soldered to the center pin. Another wire, with one $^3/_8$-inch ring and one #6 ring, is secured by one of the #6 mounting screws to the flange of the SO-239. Each $^3/_8$-inch ring is attached to the $^3/_8$-24 threaded coupling nut and its bolt. The Hamsticks screw into the coupling nuts (a coupling nut is simply a long nut). The U-bolt allows you to mount the assembly to a mast. Many variations on this theme are possible. To gain greater height, for example, you could suspend it from a tree or other tall support with a rope.

Lexan is another one of my favorite materials for building portable antennas. Note that in the section above, I cautioned against using Plexiglas. Plexiglas is much more brittle than Lexan, and will often crack or shatter when being drilled or cut. I've never had that happen with Lexan. I've also had Plexiglas dipole mounts similar to the one in the photo crack in half but never a Lexan mount. Not surprisingly, Lexan is quite a bit more expensive than Plexiglas.

However, there's good news. Like the batteries from the alarm company, you may be able to get all you need for free. A visit to the local window glass company, armed with your trusty box of doughnuts or a large pizza, may yield a good number of scraps that would otherwise be headed for the landfill. The most common sizes of Lexan are $^1/_8$, $^3/_{16}$, and $^1/_4$ inch although thinner and thicker sizes are sometimes available. I use the $^1/_4$-inch stuff for most of my projects. By the way, those sizes are nominal, just like lumber. A piece of $^1/_4$-inch Lexan is most likely 0.220 inches thick, but in most cases that makes no difference. You can cut the material with a hacksaw or with a plywood blade in a power saw. It can be drilled with any standard drill bit. Although I've never had reason to use a router on it, I suspect that it would not be a problem. *Always* wear eye protection when working with it, as with anything else.

Multi-band mobile antennas most often employ some type of tapped-coil scheme for matching on various frequencies. Some require you to physically move a tap on the coil. Examples of this include the Outbacker series of antennas and the Texas Bugcatchers. Motorized versions are also available. They're often referred to as *screwdriver* antennas, since early models used a cordless screwdriver motor to move a sleeve over a fixed coil, effectively moving the tap point. Modern versions, like those from High Sierra Antennas, at **www.cq73.com**, use in-

Fig 5-10—A panel-mount SO-239 connector attached to a Lexan sheet with adaptors to allow back-to-back Hamstick whips to form a dipole.

dustrial-quality motors but the principal remains the same. These are also $^1/_4$-λ verticals, and so require a counterpoise as outlined above. Because of the weight of the motors and the need for battery power for tuning, they're not particularly well suited for situations where you're carrying your gear for any distance. However, they can be good performers when mounted on a portable tripod with an adequate ground plane and they have the advantage of covering many bands without the need for a tuner.

Several companies produce antennas specifically designed for Unplugged operation. Budd at W3FF Antennas, **www.buddipole.com**, markets the *Buddipole* portable dipole, as well as the Buddistick portable vertical. Vern at Super Antennas, **www.superantennas.com**, also offers horizontal and vertical models as well as a portable Yagi.

I GOT A LINE ON YOU, BABE

OK, now we have a rig with dc power, and an antenna. Hmmm, what are we missing here? Oh yeah, we gotta connect them! And that's why we have feed line. As with antennas, myths and misinformation about feed lines abound. And as usual, your choice comes down to balancing conflicting factors. In this case, it's more or less a choice of weight and bulk versus line loss and, to a lesser extent, convenience.

Although there are many types of feed line, we'll limit our discussion to 50-Ω coaxial cables and 450-Ω windowed line, often referred to as "ladder line." The four most common types of coax you'll run into, listed in ascending order of loss (but descending order of size/weight and power handling capability), are RG-213, RG-8X, RG-58 and RG-174/RG-316. Assuming that you'll be running no more than 100 W out, any of the four types are more than capable of handling the power.

Loss in feed lines is generally specified

as so many dB per 100 feet. Loss increases with frequency, so you need to look at the figures for the band(s) you'll be using. It's a linear function of the length of the line. For example, a 50-foot cable has half of the loss, in dB, of a 100-foot cable of the same type at the same frequency. Quoted specifications almost always assume a matched line; that is, the feed line is terminated in a load equal to its characteristic impedance. Actually (for those shuddering purists) the *definition* of a cable's characteristic impedance is the resistive impedance that results in a 1:1 SWR at any point along the line. In other words, our 50-Ω coax needs to be feeding a 50-Ω antenna for the loss figures to be accurate. As we move away from a matched condition—as our SWR rises—loss will also rise. The feed-line chapter in *The ARRL Antenna Book* is an excellent resource for understanding this topic, and contains several graphs, tables and equations to help you estimate the loss in your particular situation.

Although RG-213 has the least loss of any of the cables listed above, it's also the largest and heaviest. At HF frequencies, the loss difference between RG-213 and the smaller cables is probably not significant enough to worry about. *The ARRL Antenna Book* says that RG-213 has 1.25 dB of loss per 100 feet at 30 MHz (less, of course, at lower frequencies). In comparison, RG-58 has 2.5 dB of loss at the same frequency. A typical HF Unplugged station might use a 33-foot run of coax, so the difference in loss is just about $^1/_3$ of the difference in the 100-foot lengths, or about 0.4 dB. That's essentially no difference at all. However, RG-58 is a tad less than half of the diameter of RG-213 and thus weighs only about one quarter as much, since the weight of a round cable is proportional to the square of its radius, assuming similar construction. Not only does that mean less feed-line weight to carry, it also means that you won't need larger gauge antenna wire and ropes to support the heavier feed line.

RG-8X, sometimes also called *mini-8*, is a little bit bigger and heavier than RG-58 and has a little bit less loss. The difference is, in most cases, insignificant. On a side note, the differences in loss do become quite a factor for the VHF and UHF bands. If you plan to use the same feed line for 2 meters and above, you'll find that the smaller cable can really eat up a lot of your power, even with a very well matched antenna system.

Heads up, purists—It's shuddering time again… But this is the last time, I promise. Contrary to popular opinion, RG-174 and its Teflon-dielectric cousin RG-316 can make a good feed line. Granted, the loss is considerably higher than the other lines—about 6.5 dB per 100 feet at 30 MHz. But again using our 33-foot feed-line length, the difference between RG-316/RG-174 and RG-58 is only about 1.3 dB at 30 MHz. That's just barely enough to be noticeable to the receiving station. And the loss, and difference in the loss, becomes even less as we move down in frequency.

The biggest advantage to this small coax is that it's about half the diameter (and only a quarter of the weight) of RG-58. The biggest drawback is that connectors are not as ubiquitous as are those for the larger cables. Most of the available connectors require a crimping tool, which can be a significant dollar investment and require learning some new skills. I prefer to use the RG-316 with Teflon dielectric. Many of the connectors require you to solder on a center pin, and the Teflon won't melt and deform if your technique (like mine) is not exactly "ISO-9000 certified." Quicksilver Radio Products, **www.qsradio.com**, among others, offers made-up RG-316 cables in varying lengths with a variety of connectors. Especially if you're backpacking, you might want to give some serious consideration to using this lightweight cable.

Ladder line is far and away the low-loss champion in our group, with about 0.15 dB per 100 feet at 30 MHz. Many antenna tuners support a balanced-line output as well as a coax output. They typically incorporate a 4:1 balun to make the transformation between the balanced line and the unbalanced coax going to the rig. Ladderline is not as easy to deploy or roll up as coax, although it's reasonably lightweight. Unlike coax, you need to avoid sharp bends when routing it, and you should make sure to keep it clear of nearby metallic objects. Nevertheless, if lowest loss is your paramount consideration, ladder line and an antenna tuner is the way to go.

WHAT'S THE FREQUENCY?

Looks like we're set to go. Rig, power, antenna and feed line, and are all hooked up. It's time to get on the air! But where?

A detailed discussion of propagation is beyond the scope of this chapter, but the short answer is that all of the HF bands are suitable for Unplugged operation. During daylight hours at sunspot peaks, 10, 12 and 15 meters can be open to all parts of the world using a very modest antenna. Since it's relatively easy to erect full-sized antennas on these bands, when conditions permit you can have a great time here.

Even at sunspot lows, band openings can pop up and be a lot of fun. My favorite band, at least during the day, is 17 meters. You'll often find other Unplugged operators around 18.157 MHZ and interesting DX can pop up at any time, anywhere on the band. You can almost always count on 20 meters being open to someplace, virtually around the clock. With low power and a compromise antenna, busting those big pileups will be a challenge, to say the least. But tune around and you're likely to find a nice QSO.

For CW and digital aficionados, it's hard to beat 30 meters. The 200-W power limit means a more level playing field on this relatively uncrowded band. If you're looking for regional contacts during the day, 40 meters is a great choice. A relatively low, in terms of wavelength, dipole can provide great coverage within a 300 to 500-mile radius. This is known as NVIS (Near Vertical Incident Skywave) propagation. There's a wealth of information on the Internet covering this topic. As the sun sets and the shortwave broadcasters come booming in on 40, you can find pleasant conversation on 80 meters. With our Unplugged antennas becoming increasingly small fractions of wavelengths, efficiency suffers on the lower bands. But if you tune around or call CQ, you'll often be able to scare up a QSO. To be honest, I've never had much success on 160 meters while operating Unplugged, but by all means give it a try. You may be pleasantly surprised at

what you can work.

INFORMATION, PLEASE

The Internet has become a wonderful resource for all types of information, and Unplugged operation is no exception. Visit **www.hfpack.com** and sign up for their e-mail reflector. There's a large number of folks with a whole lot of Unplugged experience and they're generally happy to share their expertise. This site is a must-see before you head out. Most of the manufacturers' Web sites listed above contain a good degree of technical information and encourage questions from potential customers. If you don't already own a copy of *The ARRL Antenna Book*, order one today.

And by all means, join your local radio club. You may well find an expert who has already traveled the same road you're starting on. By dint of experience and experimentation, you may even become an expert!

Put your thinking cap on, scrounge around, dig out that mobile antenna, hit the hardware store, throw some wire in a tree and get out and on the air Unplugged. See you on the bands. **Fig 5-11** is a photo of my portable radio, set up on a picnic bench ready to go.

Fig 5-11—A picnic table out in the bright, warm sun makes a great place to operate in the field!

CHAPTER 6

RV Mobile

Author Al Brogdon, W1AB, has been operating mobile since 1978 and has enjoyed antenna experimentation since he was first licensed (in 1952). In an article about his "Five-Way Antenna Coupler" in the November 1958 issue of *QST*, Al (then W4UWA/2 and living in New Jersey, literally across the street from the Atlantic Ocean) reported that he had used his homebrew antenna tuner to contact W6WNI—using a wet kite twine as an antenna. In 1990, Al (then K3KMO) gave demonstrations of 2-meter antennas to several Maryland radio clubs, showing how he used a gamma match to load a slide trombone on 2 meters.

He has operated HF mobile CW for over 25 years from his Toyotas (pickup truck, vans, car), his Honda Gold Wing motorcycle, his Itasca Phasar motor home, and one of the ultimate RVs—the luxury cruise steamboat *Mississippi Queen*. He operated RV mobile extensively (both in motion and while in campgrounds) when he and his wife Maggie took a nine-month RV trip around the country to celebrate Al's retirement.

One current antenna project being considered by Al is the design of a gamma match to use his BBb tuba as a 10 meter antenna. Further information on Al's ham background may be found at the end of the chapter on motorcycle operation in this book.

INTRODUCTION

The term "recreation vehicle" is used to describe motor homes, camping trailers, and truck campers (campers that fit into the beds of pickup trucks). Fifty or more years ago, the term *camping* usually referred to *roughing it* in a tent for a weekend or a week-long vacation, cooking over a campfire and often going without bathing. However, the term now means taking your 40-foot motor home—with its matched reclining easy chairs, queen-size bed, bathtub, 31-inch color TV, DirecTV receiver, and air conditioning—on the road to places both near and far.

Of course, you will want to take your hobby activities along with you, just as the Anasazi (the Ancient Ones) did, as can be seen in **Fig 6-1**, showing a special petroglyph in the canyons in the Southwest.

Operating from an RV combines many of the best features of automobile-mobile operating and home-station operating. You should carefully read the chapter of this book on mobile operation from automobiles to get a general sense of mobile operation, equipment and antenna installation, and how to get started in mobile hamming.

With an RV you usually have more space for your ham station than you will find in the typical automobile. You also have much more latitude for installing mobile antennas, thanks to the larger structure of the RV. Best of all, you can take along portable mast systems for putting up excellent antennas when you are parked for the night—or for a few days— or for the winter.

Because of the great variations from RV to RV, and also the variations in ham equipment and antennas, the discussions of this chapter will be aimed at stimulating your imagination and giving you ideas to use in figuring out the best way to install *your* ham equipment in *your* RV.

We'll look at the differences in RV operating compared with automobile mobile operation, and in most cases the differences will be to the advantage of the RV ham. We'll also take a brief look at the mutual interference problems that you might encounter in the shoulder-to-shoulder settings of RV parks. Yes, *mutual* interference problems. In addition to the strong possibility of your causing interference to your neighbors' TVs and other consumer electronic equipment, your ham receivers may pick up unintended signals from the electronic equipment in your

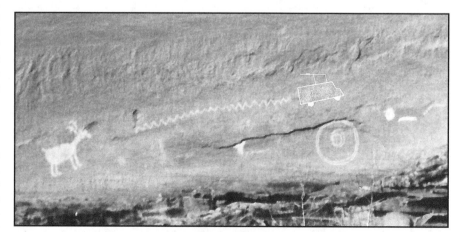

Fig 6-1—Ancient Anasazi petroglyph. *(Courtesy Larsen E. Rapp, WIOU.)*

Fig 6-2—K2TQN's RV-mounted Old Radio Museum features a 1933 ham station. *(Courtesy K2TQN.)*

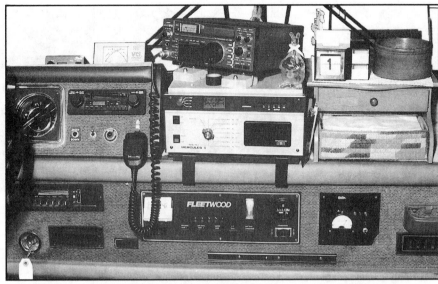

Fig 6-3—The excellent driver's-seat operating position of Charles Wilson, K4CAV. *(Courtesy K4CAV.)* [from *QST*, Oct 2003, p 52]

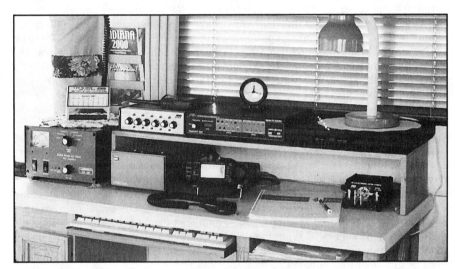

Fig 6-4—The cushy shack Dick and Ruth Stroud, W9SR and W9BRN, have in their RV (a Cruisair III) is as nice as many home stations. *(Courtesy W9SR.)* [from *QST*, Oct 2003, p 53]

neighbors' RVs.

Most RV hams will operate either SSB or CW on HF, or FM on VHF/UHF. Therefore, this chapter will address only those two types of operation. However, there are a fair number of hams who operate via the amateur satellites as well as other specialized types of operation, and we don't want them to think we are overlooking them. Indeed, we salute them for their imaginative setups and resourcefulness!

For one example of operating via the ham satellites from a truck camper, see "Satellite DXing 'To Go'," *QST*, June 2002, p 40, by David Rosenthal, N6TST. It's a fascinating and often amusing story, and it shows some good photos of how David mounted antenna masts on his camper. I suggest that you read it. Click on the first article at **www.arrl.org/members-only/tis/info/pdf/0206040.pdf** to read it on line, if you're a League member.

Before we get into the nitty-gritty of RV hamming, I want to be sure you know of John Dilks, K2TQN, and his RV, in which he houses the Old Radio Museum that he takes on the road to share his appreciation of old-time radio. In his RV, John has set up an authentic 1933 ham station. See **Fig 6-2**.

To see the 1999 article by Rick Lindquist, N1RL, about John's museum, go to **www.arrl.org/news/stories/1999/06/17/1/**. Also, visit John's home page for more travelogue stories, at **www.eht.com/oldradio/**.

You say John's call sign sounds familiar to you? And well it should, John writes the "Old Radio" column for *QST*. Now let's get on with discussing how you can get on the air with Amateur Radio from your RV.

STATION INSTALLATION

Some RV hams like to have their station installed near the driver's seat, so they can operate while on the move—true mobile

operation. With a driver's-seat installation, the ham can also sit in the driver's seat and operate when the RV is parked. (In this chapter, we'll use the shorthand expression "fixed RV" to refer to operating while the RV is parked.)

Other hams like to have a station that's more like a home station, installed in the living room or bedroom area of the RV, operating only when the vehicle is parked. Some RV hams have two separate stations: One up front for true mobile operation, and one in the coach for more leisurely hamming.

Many RV hams have VHF/UHF FM equipment installed so they can operate it from the driver's seat while in motion, plus

a more comprehensive station (HF, VHF/UHF, packet, PSK-31, etc) installed in the coach for fixed-RV operation. Thus, they can enjoy the best of both worlds.

THE "MOBILE STATION" RV APPROACH

If you install the ham equipment near the driver's seat, you will usually have more space available to install it than you will find in the typical automobile. Installing the equipment and routing the power, control, and antenna cables can be done following the methods and precautions described in the automobile mobile chapter of this book. See **Fig 6-3**.

Some hams worry about drilling holes to mount equipment and to pass cables through. I say, forget your worries and drill away! You should, of course, be absolutely certain that the locations of the holes are correct before drilling. You should also be certain that the new holes will not affect any existing structure or equipment. The holes must be water-, weather- and mouse-tight. The rule of thumb is that a mouse can squeeze through a hole about as large as your little finger. Wait a minute—maybe that should be "rule of little finger."

But don't worry about destroying the resale value of your RV by drilling a few small holes. Let's face it; the value of an RV depreciates even faster than the value of an automobile. There's an old saying, "A pleasure boat is a hole in the water into which the owner continually pours money." Well, folks, an RV is the land equivalent of the pleasure boat.

THE "HOME STATION" RV APPROACH

The main advantages of the home station approach is that you have more space for installing the equipment, and you can erect larger, more efficient antennas when operating from a fixed location. **See Fig 6-4** for a neat and functional setup.

If you install your ham station in a location other than at the driver's seat, keep in mind that it must be securely mounted, either by clamping or bolting it to the operating table. You do not want to see a couple of kilobucks worth of good radios slide off the table and crash onto the floor if you have to make a panic stop or swerve sharply to avoid a problem on the highway. See **Fig 6-5**.

Frank Clements, W6GZI, set his fixed-RV operating position up behind the front passenger seat of his Winnebago Sightseer motor home. Frank swivels the seat around 180° so that it faces the rear of the RV when he operates. Frank custom-built the cabinet to fit the ham equipment, removing a second-row swivel chair on the right-hand side to make room for the cabinet. The rear wall of the cabinet then became one side of the entrance stairwell. **See Fig 6-6**.

The IC-706 Mk II is mounted at the bottom, with an SCS PTC-II multi-mode controller above it. Frank uses the PTC-II with a Sony laptop computer for Pactor and WinLink e-mail. Frank's fixed-RV antenna is a 23-foot Shakespeare model 390 marine whip that is tuned for all HF bands with an SGC model SG-230 autotuner. The Shakespeare whip has a motor-driven mount to fold down onto the

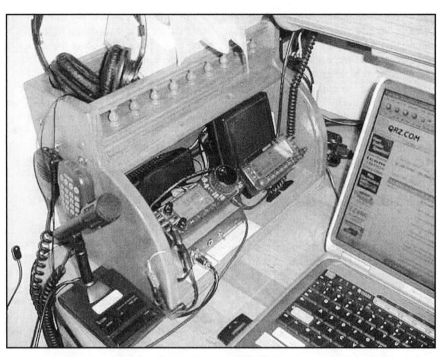

Fig 6-5—W4NFR's RV station is mounted in a small desk. *(Courtesy W4NFR.)*

roof during travel. Frank has a 144 and 440 MHz dual-band rig on the dash, in addition to the fixed-RV station shown here.

Another nice motor home installation is featured on the Web pages of Jim Cook, AL7RV, at **www.al7rv.net/radiodesk.html**, with several photos of Jim's nice fixed-RV operating position and antennas. For HF, Jim also uses the popular Shakespeare 23-foot marine whip.

Jim Thompson, W4THU, took an interesting approach to installing a station for fixed-RV operation. He built a potent station into a 27-inch wide closet that formerly held an apartment-sized washer and clothes dryer! The station featured a Kenwood TS-930 transceiver, a Collins 30L-1 amplifier and a Heathkit SB-640 antenna tuner. Accessories include a Heathkit SB-614 oscilloscope for monitoring the transmitted signal and an MFJ-490 memory keyer. Check out Jim's Web site at **www.radioworks.com/ Hamshack/ShackSB.html** for interesting photos of his fixed-RV station. Jim has moved on to another RV at this point, and is installing another fixed-RV station in the new travel trailer.

PROPER RF GROUNDING

Here is a tip about RF grounding for an RV ham station I would like to share with you, one that I learned the hard way. The concept applies to both the mobile-station approach and the home-station approach.

When I first installed my TS-50 and

Fig 6-6—W6GZI at his fixed-RV operating position. *(Courtesy W6GZI.)*

MFJ-484 memory keyer near the driver's seat in my motor home, I had problems with loading the antenna on some bands. Since I was using a mag mount, my thought was that I might need to run a ground strap from the mag mount frame to the vehicle body (which is often necessary). I installed a strap from the mag mount to vehicle ground, but the loading was still screwy.

I tried a few other things, but the antenna *still* wouldn't load properly. Finally, I realized that I had not run a ground strap (an RF ground) from the rig itself to a nearby vehicle ground point. Even though there were several dc ground paths from the rig to vehicle ground (via the various power, control and antenna cables), all of the ground paths were apparently high-impedance paths on the bands where I was having the loading problems.

I installed a ground strap from the rig to

the vehicle ground. Following that, the antenna loaded fine on all bands. This shows that you should install a low-impedance grounding strap from the transceiver to the vehicle ground, and a ground strap from the antenna mount (if mag-mounted) to the vehicle ground. Then it's likely that your antenna will load well on all bands for which it is designed.

POWERING THE RV MOBILE STATION

General Considerations

Because most modern amateur equipment requires 12 V (actually 13.8 V) primary power, you can operate the ham station directly from the coach battery. However, if you have enough space to include the transceiver's 120 V power supply in the station, it's convenient to power the ham station from the campground's ac connection, via the RV's ac distribution system.

The most flexible way to power the RV ham station is to use a high-current DPDT switch to switch the station equipment's 12 V input between the 120 V power supply's output and the 12 V coach battery. You can then use the power source that best suits you at any given time, as well as being able to get on the air if there is a failure in the power mains or when dry camping.

That scheme, incidentally, is the power system I use for my home station. If there is a power failure, I flip one heavy-duty DPDT toggle switch, and I'm on the air using the 12 V deep-cycle marine/RV battery that's stashed beneath the shack desk, and which is continuously trickle charged (using a 12 V, 300 mA "wall wart" power supply).

The ham station location should be chosen to be as near the coach battery as possible, to minimize the voltage drop along the power cable's run. Always fuse both the positive and the negative leads at both ends of the power cable run for maximum safety. And ensure that the conductor size is large enough to carry the maximum current your station will draw without excessive voltage drop. A good source of suitable 12 V power cable is copper automotive battery jumper cables. Those cables usually have some sort of coding molded into the rubber jacket to identify the positive lead, a help in keeping up with the polarity of the cable. Sometimes (but not always) they will have different colors for the insulation on the two conductors, which is a further help.

When running the power cable in the RV, be sure that it will not be subject to chafing or repeated flexing at any point in its run. Also, leave at least three feet of cable slack at the station end of the cable, in case you have to relocate the station at some later time.

Alternative Power Sources

The continuous sound of a gas-driven generator is, at best, annoying—at worst, fatiguing. Therefore, if an RV ham wants go on long periods of dry camping, he might want to consider alternative power sources. The two alternative power sources most commonly used are solar power and wind power. Both are relatively expensive but are widely available.

Here are some thinking points for you along the lines of alternative power sources, together with some examples of what's available nowadays. If you want to look further into these possibilities, you can find plenty of information on the subject, both on line and in the Amateur Radio and alternative power magazines and books.

Solar Power for the RV Ham Station

Although solar cell arrays have been around for some time now, their efficiencies are still low (typically no more than 15%) and they are pretty pricey. Smaller arrays can be used to keep 12 V batteries topped off. Larger (and more expensive) ones will provide enough current for you to operate the ham station from a 12 V battery without draining the battery's charge.

One example of a 32 W solar panel is the UniSolar FLX-32 flexible solar array (flexibility is a plus), available from Affordable Solar (**store.yahoo.com/ affordablesolar/1125.html**). The regular price is $275. You can check the price when you read this and see what has happened to it.

The Affordable Solar Web page at **store.yahoo.com/affordablesolar/ solarpanels—by-manufacturer.html** lists solar panels by manufacturer and by wattage rating. As you will see from that list, you can buy some high-powered solar panels that will provide 175 W or so, but those will set you back nearly $1000.

The prices of solar panels vary widely, and change over time. You can do a Froogle search to find out which way the sun is shining if you consider buying a panel for your RV. You can lay out your solar array on the RV roof, which is convenient and also a good location to keep the panel from being damaged.

Wind-Powered Generators

Another possibility for the RV ham is the wind-powered generators that are available from marine supply houses.

These are, for the most part, low-wattage devices that can keep your 12 V battery charge topped off.

Here's a link that will take you to a Web page describing the Solardyne Air X Wind Turbine: **store.yahoo.com/solardyne/ air403wingen.html**. The Air X will provide 100-W output with a wind speed of about 18 mph, at a cost of $639 (at the time this is being written). The Web page also has several accessories for use with the Air X turbine.

You can find information on building your own wind-power systems in the available literature. If you read about this subject extensively you will realize that, although the idea of wind power is simple, its execution is often complex and there are many potholes in the road to success.

By the way, just in case the thought crosses your mind that you could drive along the Interstate with your wind turbine running, don't!

DC-to-AC Inverters for the RV Ham Station

An ac-to-dc inverter (which connects to a 12 V battery and provides 120 V ac output) is a quiet alternative to your RV's auxiliary generator for powering a ham station or other electrical devices over short periods of time. During the last few years, inverters have come into more common use in all kinds of applications. Because of the larger quantities being manufactured today, the prices for the inverters have come down quite a bit.

Early inverters (such as the good old Heathkit MP-14 inverter that came on the market about 1961) had a square-wave AC output, which would put an annoying buzzing sound in your receiver's audio output. (A square wave is rich in odd-order harmonics of the output frequency.) For the most part, the inverters that are available nowadays have what the manufacturers call "modified sine-wave output," which approximates a sine wave. Your ham gear and audio/video equipment will work happily with those inverters.

Inverters are available at a lot of automotive supply stores and at Wal-Mart stores. Here's a link to a Wal-Mart Web page that shows you what you might expect to find in your local store: **www.walmart.com/ catalog/search-ng.gsp?search_ constraint=0&search_query=inverter& ics=20&ico=0&Continue.x=27& Continue.y=9**. Several years ago, I bought one of the MaxxSST 350 W model VEC061 Inverters at my local Wal-Mart store, and it has worked very well for me in various applications. Dang—I just noticed that Smiley Zorro has cut the price below what I paid for it! As I told you, inverter prices keep edging downward.

A quick Froogle search on "power inverter" brought forth some interesting results, including a 180 W inverter for $21 and an 800 W inverter for $62. One item that piqued my interested was a StatPower 1350 W inverter complete with a 60 Ah battery on its own little wheeled dolly for only $296 (**www.pricegrabber.com/ u s e r _ s a l e s _ g e t p r o d . p h p / masterid=4972614/lot_id=1134064/ mode=googlegfr/**). That would be a very useful unit to have around the home for general utility and emergency use, in addition to being a good way to power a ham station for reasonably long periods of time.

The prices quoted in the preceding paragraphs are current at the time this is being written. Do your own Froogle search to find out what's currently available.

RV ANTENNAS FOR HF MOBILE USE

This section will discuss HF mobile antenna types, HF mobile antenna mounting, HF fixed-RV antennas, antennas used in the W1AB HF RV station (both mobile and fixed-RV), VHF/UHF antennas for RV mobile use and VHF/UHF antennas for fixed-RV use.

As I've suggested before, search the current and recent Amateur Radio literature for further ideas that you might incorporate in your RV station. Although some of the basic ideas are 50 or more years old, some bright ham may have come up with a slick new idea for either a different antenna configuration or a different antenna mounting system. Keep yourself plugged into the network and keep your eyes and ears open for these new (or old) ideas.

HF Mobile Antenna Types

For HF mobile operating, there are several antenna types to choose from:

1. A center-loaded antenna, such as the *Hustler.*
2. A helically wound antenna, such as the *Hamstick.*
3. A bug-catcher antenna.
4. A screwdriver antenna.
5. A whip antenna with either a manual or an automatic antenna tuner at its base.

The center-loaded antenna and the helically wound antenna provide about the same efficiency. They also share the disadvantage or having a relatively narrow usable bandwidth without retuning the antenna.

For multiband use with Hustler and Hamstick types of mobile antennas, you can put a small metal plate (a *spider*) at the top of the solid mast and mount up to four or five single-band resonators on the plate. This way, you can switch bands without having to change or adjust the antenna.

Although quite a few hams attach the maximum number of resonators to their spider fixture, doing so tends to make the antenna a bit top-heavy. It's best to determine which bands you are most likely to use and install *only* those resonators on your spider.

When you have mounted the resonators on your spider, you will find that the tuning process takes a bit of time. Start with the highest-frequency resonator and tune it to resonance at the desired operating frequency. Then move to the resonator for the next-lower frequency band and tune it. And so on, until you have tuned all the resonators.

When you go back to the highest-frequency resonator, you will find that it's center frequency has moved, because of the mutual coupling among the coils of the different resonators. Retune it, then move through the rest of the resonators and retune them, from highest to lowest frequency. After three passes through all the resonators, you will likely have them tuned to resonance at the desired center frequencies.

Here's an interesting bit of speculation: Most mobile hams know about the spider approach, and many have used it with Hustler or other resonators. I've never heard of anyone mounting two resonators for the same band (say, 40 meters or 75/80 meters) on a spider and tuning one to resonance in the CW portion of the band and the other to resonance in the phone part of the band. It might work. If you try it and it works, write it up for *QST!*

Getting back to the other three kinds of antennas (the bug-catcher, the screwdriver and the whip with an antenna tuner at the base), they all have the advantage of being tunable to any frequency in the entire HF spectrum—a very useful feature if you operate on different frequency segments within a given ham band (for example, 40 meter SSB and 40 meter CW). The bug-catcher is manually tuned by moving a tap along the antenna's large-diameter center loading coil. The screwdriver is remotely tuned via an electric motor that changes the value of inductance in the antenna. The whip antenna is tuned on frequency with the antenna tuner (often an autotuner) at its base.

Anther interesting idea that can be used with mobile RV hamming is the Connecticut Longhorn antenna, designed and built by Andrew Pfeiffer, K1KLO, and described in the August 1967 issue of *QST.* To see the article on line in pdf format, go to **www.arrl.org/tis/info/Mobile-H.html/6708011[1].pdf**, scroll about a third of the way down the page, and click on "The Connecticut Longhorn." Andy's antenna produced excellent results without becoming too tall to use. The Connecticut Longhorn, by the way, is the antenna that gave me the idea of using the fold-over Hustler mast for my own RV mobile antenna.

If I were to start mobiling all over again, I would use a screwdriver antenna with an autotuning memory controller. With it, you can cover all frequencies in any of the HF ham bands, and you can forget about adjusting the antenna—the system automatically tunes as you change frequencies and change bands. Furthermore, the screwdriver controller "remembers" the matching settings for each frequency, so returning to a frequency that you have previously used is quick and easy.

A caution: As with the RV station's power cable, always leave at least three feet of slack in the coaxial cable feeding the antenna, in case you later relocate the station.

HF MOBILE ANTENNA MOUNTING

You usually have a bit more flexibility in mounting a mobile antenna on an RV than you do on an automobile. Many RVs have ladders for roof access, to which mobile antennas (and the larger fixed-portable antennas) can be easily mounted. And you can consider using any of the antenna mounting schemes for automobiles that are described in Chapter 1 of this book.

Harvey Tetmeyer, K5LJM, has a RoadTrek motor home that he wanted to outfit with HF and VHF/UHF antennas without drilling (m)any holes in the vehicle. His Web page at **www.members. cox.net/azharvey/vara/vara.htm** shows how he mounted the antennas, including a screwdriver antenna for HF mobile operation.

For fixed-RV HF operation, Harvey replaces the 5 foot whip on the top of the screwdriver with a helical radiator wound on an 18 foot fishing pole. Harvey had an antenna idea that is really slick. As most mobile hams learn sooner or later, a matching inductor is often required from the base of the mobile antenna to ground for proper impedance matching of the antenna to the coaxial cable. Harvey installed a coaxial tee connector at the base of the antenna, as shown in the accompanying photo in **Fig 6-7**, with a toroidal inductor mounted on a PL-259 plug that screws onto the outward-facing side of the coaxial tee connector.

The RV that my wife Maggie and I used for many years is an Itasca Phasar. The cab is constructed of steel, while the coach is made of fiberglass. I used a mag mount on top of the cab for my HF mobile antenna. The mag mount was made using an H-shaped frame of rectangular aluminum stock, with magnets at each of the four ends of the arms of the H and a standard connector for a mobile antenna at the center of the H.

For my mobile antenna, I made an antenna using Hustler parts. The antenna was made up of an MO-2 fold-over mast, a VP-1 spider mount and resonators for either two, three or four amateur bands. Most of the time, I used only two resonators, for 40 and 20 meters. See **Fig 6-8**.

As you see in the photo, I left the MO-2's fold-over joint unlocked and leaned the upper part of the mast back toward the rear of the vehicle, at an angle of about 15° above the horizontal. I fabricated a rear support for the antenna from PVC pipe and attached its base to the hardware around the roof vent, so that I didn't have to drill any holes for it. To stabilize the antenna, I added two side guys, made of lightweight nylon twine, from the top of the PVC support to each side of the RV.

The advantage of mounting the antenna

Fig 6-7—K5LJM's mounts his antenna-matching toroid on a coaxial T connector. (Courtesy K5LJM.)

Fig 6-8—W1AB's RV Hustler mobile antenna is on the roof of the RV; a 20-foot mast is set up beside the RV to hold up a wire antenna.

like this is that the MO-2 mast (which does virtually all of the radiating) is up and in the clear, but the highest point of the antenna is still only about 10 feet above the ground, so it will pass underneath all standard-height underpasses. The antenna works well, and I've worked long-haul DX often and easily as I motor along the highways.

HF FIXED-RV ANTENNAS

When your RV is parked (fixed-RV operation), your antenna possibilities become much more interesting! In the typical campground situation, you won't have enough room for long horizontal HF antennas, but you can use short horizontal antennas or you can erect a vertical that will outperform your mobile antenna. Another possibility you can consider is an inverted-L antenna, with a vertical leg that turns horizontal at the top of the mast for the second leg.

Several fixed-RV hams have had great success on the lower-frequency HF bands while using longer wires strung along two tall supports—one at each end of the RV—from the bottom of one support to the top of that support, then continuing with a flattop across to the top of the other support, and coming down the second support to its bottom end. A wire like that can be fed either in its center (with coax or ladder line) or at one end (with ladder line or a single-wire feed).

One simple vertical antenna can be made using a telescoping mast, made of either fiberglass or carbon fiber, with a wire running along its length. Telescoping masts made of those materials and which extend to 30 feet are available from amateur suppliers, but they are slightly expensive, in the $100 range.

For considerably less money, you can buy a 20 foot telescoping B'nM Black Widow Crappie Rod from Bass Pro Shops for only $21. (Go to **www .basspro.com,** click on Freshwater Fishing, then under Online Shopping click on Rods, then on Crappie. From there, click on B'nM Black Widow Crappie. On that final page you will find the 20 foot Model BW6 pole.) You can use the pole as is for an antenna support, or you can improvise a lower mast extension for the crappie pole (from PVC pipe or other material) to increase the mast length to 25 or 30 feet. It's a less expensive way to make up a mast of suitable height to support a wire to use as an HF vertical... even though some people might opine that you have a crappie antenna support.

In the October 1980 issue of *QST*, Charles Schechter, W8UCG, described adapting a Hustler 4BTV (a home station

vertical antenna) to mount on his Airstream trailer for fixed-RV operation. You can find the article via a clickable link at **www.arrl.org/tis/ info/Mobile-H.html**. You will also find on that page clickable links to articles about other mobile antennas that you might consider for your RV. One of those articles, by Phil Rand, W1DBM, from the March 1969 issue of *QST*, describes several RV antenna schemes, including an interesting loop antenna that Phil built to use on his Airstream trailer.

Another good antenna and antenna support for fixed-RV operation was described by John Portune, W6NBC, in the January 1994 issue of *WorldRadio*. You can find information on his antenna on his Web site (**www.w6nbc.com/rvantenna.html**). John's antenna support system was made from commonly available materials, which is a bonus. See **Fig 6-9**.

Per Moberg, N2BFH, uses a Hex-Beam antenna. The flexible arms are used to support two W-shaped wire elements that make up a Yagi-Uda array. Per comments that "We could put up our Hex-Beam antenna in half the time if we didn't have to explain what we were doing. The antenna looks like an upside-down umbrella with no fabric and has lots of pieces, so assembling it attracts lots of spectators." See **www.hexbeam.com/index.shtml** for information on the Hex-Beam antennas and also see **Fig 6-10**.

Another fixed-RV antenna possibility is the use of a pair of Hamstick mobile antennas to make a dipole. Several manufactures make an adapter for that purpose, including Hamstick's model 901 Rotatable Dipole Mount (go to **hamstick.com**, click on Mounts, then click on Dipole Mount to see it. Another dipole mount can be seen at **search.cartserver.com/ search/search.cgi?cartid=s-3018&category=shop&keywords= hamstick+dipole&keywords_1=& keywords_2=&maxhits=10&go=Go** (the third item on the page). Just connect a pair of Hamsticks to these fixtures and

Fig 6-9—W6NBC's fixed-RV antenna. (Courtesy W6NBC.)

Fig 6-10—The Hex-Beam antenna used by N2BFH for fixed-RV operation. (Courtesy N2BFH.)

Fig 6-12—The tip-up mount used by K6EAS on his RV for his 23-foot Shakespeare marine whip. (Courtesy K6EAS.)

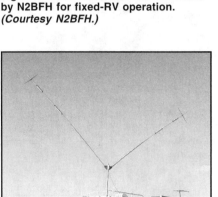

Fig 6-11—A fixed-RV dipole made from two trap vertical antennas. (Courtesy W6TEE.)

mount them on a mast to make a short dipole.

A further idea for antennas made from Hamsticks is to use them as elements for a Yagi-Uda array, which quite a few hams have done. As an example of the Hamstick beam idea, Russ Wilson, VE6VK, describes his interesting three-element mini-beam made from Hamsticks on his Web page at **www.telusplanet.net/public/ve6vk/minibeam/minibeam.html**. Russ credits Rick Dorsch, NE8Z, with the idea for the mini-beam. On that same Web page, you will find information on a very compact V-shaped beam Russ built from Hamsticks that would be good for fixed-RV use. Also on that Web page, you will find photos and a description of a 3-element beam built from Hamsticks by Bernie Farthing, NP2CB, and one built by Dietmar Fichter, VE3CG.

If you do a Google search on "Hamstick beam" you will find quite a few ham antennas built using Hamsticks. I suggest that you read all the material on the subject you can find and use the ideas, mounting system and mounting system that best suit your situation.

Along the same lines as the Hamstick dipole, the accompanying photo shows an antenna used by an RV ham at the January 27, 2005, Quartzfest in Quartzite, Arizona. Unfortunately, we can't identify the ham who made this antenna, but Les Cobb, W6TEE (who furnished the photo) recalls that it was a Canadian ham who had come down to catch some warm weather. See **Fig 6-11**, where the antenna uses two of small multiband trap vertical antennas to make a dipole. The antenna would be somewhat more effective if the center was elevated another 10 or 15 feet. If you decide to copy this idea, you could use a taller mast and mount the two trap verticals either in the horizontal plane or sloping downward instead of upward. This type of antenna would be a bit more efficient and would have a wider usable bandwidth, as compared to a Hamstick dipole.

The annual Quartzfest is held every year during the same week that the Quartzite January RV Show is held. RV hams dry-camp six miles south of town (near mile marker 99 on US-95), some distance off the highway in the desert. Seminars, demonstrations, VE sessions and a swap meet are scheduled early enough each day to allow time later in the day to explore the RV Show, to shop and to poke around in the desert. Over 90 RV hams were at the 2005 Quartzfest. The January 2005 issue of *CQ* magazine published an article written about the Quartzfest by Gordon West, WB6NOA. (Gordon had attended the 2004 and 2005 Quartzfests.)

Ed Schnelbach, K6EAS, uses a tip-up mount with a Shakespeare 23 foot marine whip. The mount is made from a welded framework of square steel tubing, anchored to the trailer bumper at its supports, to mount the Shakespeare whip. Ed's Alinco autotuner is mounted to this frame near the antenna feed point. Like all such RV mounts, the antenna is placed horizontal on the roof for travel, with a suitable homemade cradle to hold the antenna in place when traveling. See **Fig 6-12**.

Frank Clements, W6GZI, also uses a 23 foot Shakespeare marine whip on his motor home. The photo in **Fig 6-13** of Frank's

Fig 6-13—W6GZI's 23-foot Shakespeare marine whip on his RV. (Courtesy W6GZI.)

antenna was taken in an RV park near Lone Pine, California, during a Sam's Radio Hams campout. Frank's antenna is similar to the K6EAS installation, except that Frank uses a motor to raise and lower the whip onto the roof, while Ed uses a manual lanyard system to raise and lower the whip.

For his fixed-RV operation, Doug Troxel, WB7BVT, uses a screwdriver antenna mounted on a vertical garage-door-opener track running up the back of the RV. The base of the screwdriver can be run up level with the roof or lowered to the bumper by the screw in the door-opener track. See **Fig 6-14**.

Another simple possibility—if your mobile antenna is a good performer—is to use your mobile antenna for fixed-RV operation. As mentioned earlier, Harvey Tetmeyer, K5LJM, replaces the top section of his screwdriver antenna with a longer one when operating fixed-RV, for better antenna efficiency. However, some hams with good mobile antennas simply connect a run of coax from the antenna to the RV station to operate fixed-RV. See **Fig 6-15**.

Whitey Doherty, K1VV, spends many weekends operating from lighthouses to give contacts to hams around the USA and throughout the world. Whitey has developed a very efficient antenna for 40 meters, the primary lighthouse band. He uses two sections of electrical conduit (one

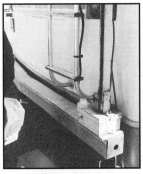

Fig 6-14—WB7BVT's fixed-RV antenna—a screwdriver antenna mounted on a vertical garage door track running up the back of the RV. *(Courtesy WB7BVT.)* [Please place the two photos side by side, with a on the left, and no "A" and "B" callouts]

Fig 6-16—K1VV's fixed-mobile antenna for excellent 40-meter performance can be used for fixed-RV operation. *(Courtesy K1VV.)*

Fig 6-15— If you have a good mobile antenna, you can park your tow vehicle near your trailer and run a piece of coax from the antenna to the RV ham station for fixed-RV operation. *(Courtesy K5LJM.)*

Fig 6-18—W1AB attaches the antenna mast to the side-door hinges with compression clamps for fixed-RV operation (one of several methods he used to attach the mast to the RV).

section 1 inch in diameter and the other 3/4-inch in diameter) with a Hustler 40-meter resonator on top of the conduit. It gives Whitey better antenna efficiency on 40 meters compared to the shorter Hustler MO-3 mast. See **Fig 6-16**.

Art Bell, W6OBB, reports that he could never get the RV antenna results he wanted using screwdriver antennas. He installed a field of short masts on top of his RV and strung a loop antenna around the RV roof, as shown in the accompanying photo. Art feeds the loop using an antenna tuner with balanced output and 450-Ω ladder line. He reports that the loop works well on all HF bands, and it also works on 160 meters if he opens the loop. Art says that "My signal is so strong that most hams I work refuse to believe I am mobile." Art took the idea of his RV loop antenna from the massive double loop antenna he uses for his home station. See **Fig 6-17**.

MOUNTING A MAST ON YOUR RV

There are many possibilities for attach-

Fig 6-17—W6OBB's roof-mounted loop antenna for HF and 160 meters. *(Courtesy W6OBB.)*

ing a fixed-location VHF/UHF antenna to your RV. The ladder on the rear of many RVs provides an excellent structure to which your antenna mast can be attached. If your RV is a trailer, the tongue of the trailer is a very solid structure for mounting antennas. With my Itasca Phasar, I attached the mast I used (which is discussed elsewhere in this section) to the two hinges of the RV's side door, resting the bottom of the mast on the ground. If you look over your RV carefully, you will be

able to find lots of places to mount antennas.

Just remember that the attachment points for your mast must be strong enough to withstand the loading caused by strong winds. When operating fixed-RV from W1AB, I sometimes attached my mast to the side-door hinge, and sometimes to the passenger's door handle, resting the base of the mast on the ground. See **Fig 6-18**.

Another way to mount a mast—*if* you are not going to move the RV from its parking place very often—is the old trick of building a plate that you can place under a vehicle tire, with a mast attachment point on the plate. You can fabricate this type of plate from steel, aluminum or wood. In use, you place the plate in front of (or behind) the vehicle wheel and gently drive the RV onto the plate to place the wheel on the plate to stabilize it. A homebrew plate of this type is shown in the photo in **Fig 6-19**, under the wheel of the pickup truck used by Whitey Doherty, K1VV, for his lighthouse operating.

Another utilitarian way to mount a mast or vertical antenna is to use a base plate such as the MFJ-1904, as shown in **Fig 6-20**. That ground-mounted base has a vertical piece with U-bolts for securing the vertical antenna or mast, four corner posts to press into the ground to stabilize the plate, and four attachment points for radi-

Fig 6-19—The homebrew mast base used by K1VV, with the vehicle holding the mast in place. (Courtesy K1VV.)

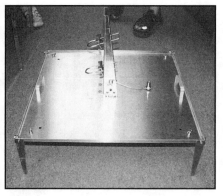

Fig 6-20—The MFJ-1904 base plate for antenna masts. (Courtesy MFJ.)

Fig 6-21—An example of a "Remove Before Flight" tag. This makes a highly visible warning to make sure your antennas are down and stored before driving off in your RV! A Google search will show you a variety of commercial items with this legend, including nightshirts, tee shirts and men's boxer shorts.

als. The MFJ-1904 is made of stainless steel, so it will last a long time in ham service.

With *any* fixed-RV antenna, you should securely attach a large note to the RV's steering wheel that says something in the order of "**STOW ANTENNAS AND MASTS BEFORE MOVING THIS VE-HICLE**," to avoid damage to your antenna and RV. Of course, you could always buy a "remove before flight" cloth tag used by pilots for a slick look! See **Fig 6-21**. You can find sources of these tags through a Froogle search.

Fig 6-22—The W1AB RV at Organ Pipe Cactus National Monument, Arizona.

THE W1AB RV HF ANTENNAS, MOBILE AND FIXED

Soon after my wife Maggie and I retired, we left on a meandering trip around the USA in our motor home, intending to be gone for three months. However, we soon discovered that because we had no schedule to keep, we took life easier and we traveled at a more leisurely pace. We took time to smell the roses, and the carnations, and the lilacs too. During that trip, we spent seven months on the road. We traveled all over the country and concentrated on sightseeing in the Southwest and along the coastal area of the West Coast states. See **Fig 6-22**.

The mobile antenna I used (which is described elsewhere in this chapter) was made of Hustler components, and it worked very well when we were in motion. I worked lots of long-haul DX (both when in motion and when sitting in camp-

grounds) with that antenna and the TS-50 at 100 watts. But hams are always thinking about *better antennas*, and so I considered possibilities for better antennas during those times when we would be in a campground for several days.

When preparing for the trip, I found two-piece telescoping aluminum tent poles at a local camping supply store. The brand name was, appropriately, *Eureka*. The longest of these telescoping poles was 4 feet, 8 inches when collapsed, and it extended to a maximum length of 8 feet, 6 inches. If you want to buy these and can't find them locally, you can order them from camping stores such as Campmor (**www.campmor.com/** and then enter 21541 (the item number) in the search window Search by Keyword/Item #).

Let me digress to note that you can buy aluminum tubing in telescoping sizes from aluminum-stock suppliers. However, it's often difficult to find the telescoping tubing locally, and it can be fairly expensive

to have the tubing shipped to you. If you can get telescoping sizes of aluminum tubing, they would be better than revamping the aluminum tent poles as I did to get aluminum tubing. That said, let's go back to the way I modified and used the collapsible tent poles.

The collapsible tent poles have a quick-release fixture that works well in their intended usage. However, I wanted a more positive way to secure the aluminum sections together. I drilled out the rivets that held the quick-release fixture onto the tubing and removed the fixture. Then I used a hacksaw to cut four longitudinal slots in each end of the larger-diameter tubing. I cut the four slots (equally spaced around the circumference of the tubing) about two inches deep. Then I bought a pair of compression clamps and placed one clamp on each end of the larger tubing. The photo in **Fig 6-23** shows my bundle of poles (and the wooden dowel rod that I used as a base insulator), held together by a Velcro strap.

I repeated this process with three more tent poles. Now I had four sections of the smaller-diameter tubing and four sections of the larger-diameter tubing, each about four and a half feet long. The sections of tubing can be assembled easily to make a long mast, alternating between the smaller-diameter tubing and the larger-diameter tubing and allowed a six-inch overlap at each joint.

My compact bundle of eight pieces of tubing stored easily and out of the way in a corner of the RV. I could assemble them all into one 33-foot length of tubing to use as a 40-meter vertical, or make up two 16-foot antenna supports for a horizontal wire.

The total cost for the four poles and the eight compression clamps was about $70. That's not a bad price for a package that can be assembled into a quarter-wave vertical antenna for 40 meters, or into two 16-foot support poles for holding up wire antennas—and which is a fairly small package that can be stowed easily, even in a small RV.

To guy the poles, I used the small-diameter braided nylon twine that you can find in hardware stores everywhere. The twine is strong (150 pounds test), lightweight and inexpensive. It even comes in two colors, white or yellow, so you can use the color that suits your fancy. I also used the nylon twine to hold up horizontal wire antennas on the RV trip. I've kept rolls of that nylon twine around for years and have used it to support lightweight antennas and for light-duty guying in various temporary and permanent antenna installations. I recommend that you consider keeping a roll of the nylon twine around, both at home and in your RV. Once you start using it you'll continually find new uses for it. It is so inexpensive that you can use lengths of it once or twice and then discard them. Oh, yes—burn the nylon twine in two rather than cutting it, to keep it from unraveling.

Another useful thing to have around both your home and your RV ham shack is the short Velcro straps made to secure cable bundles. These short straps have Velcro loops on one side and hooks on the other. They are very convenient for temporary or permanent use in bundling cables, attaching cables to supports, holding bundles of aluminum tubing together when in storage, etc

There are several variations on the Velcro straps. The best kind are those that have (1) Velcro along the entire length of both sides of the strap, and (2) a wide part on one end with a slot across its width. **Fig 6-24** shows two examples of the packages that are offered for sale, and how two of the Velcro straps can be daisy-chained together into longer lengths, such as the

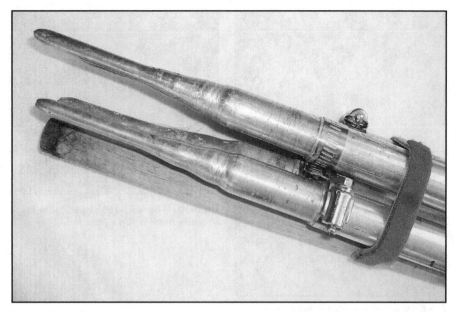

Fig 6-23—A bundle of telescoping tent poles used by W1AB for antenna masts.

straps I used to secure the bundle of aluminum tubing shown earlier.

You can find Velcro straps at most office-supply stores and sometimes in dollar stores. If you find any in dollar stores, you should lay in a large supply of them, because they are considerably less expensive there than in office supply stores.

Meanwhile, getting back to the use of the aluminum tubing: When I assembled the tubing into a 33-foot length to use as a 40-meter vertical, it was pretty wobbly and fairly difficult for two people (my wife Maggie and me) to pull into the vertical position. But we could do it. When we put up the 33-foot vertical, I placed two sets of three guy twines on it, at the 16- and 30-foot levels. After using the guy twines to raise the vertical, I tied them off on whatever supports were available.

When using the tubing as a vertical antenna, I inserted a 3-foot length of wooden dowel rod into the bottom piece of aluminum tubing as an insulator. I used the bottom compression clamp to clamp the inner conductor of the coaxial feed line against the tubing at the feed point, as well as to secure the tubing to the wooden base-insulator rod. Then I would run out one or two 33-foot lengths of wire connected to the coaxial cable's shield as an elevated counterpoise for the antenna. I would attach the elevated radials to the cable shield and clamp them off to the wooden base mast using another compression clamp. Obviously, you can't install radials in a busy campground with a lot of pedestrian traffic, but Maggie and I stayed in less heavily frequented campgrounds, so the elevated radials weren't a problem.

Another possibility for a base insulator

Fig 6-24—Examples of Velcro strap packages, and two straps daisy-chained together.

is a glass bottle. You can set the bottom of the aluminum mast on the neck of the bottle, with the bottle sitting in a small depression in the ground.

The wire I used for making my portable antennas on the big RV trip came from a hamfest many years ago, a spool of several thousand feet of #26 stranded insulated wire. I got it at a very good price because no one else thought the tiny wire was good for anything.

At this point, I can almost hear the readers gasping at the idea of using such extremely small wire—*insulated*, at that—to make antennas. After all, they have learned from reading the Amateur Radio books and magazines that you *must* use large-diameter wire to make antennas; the larger diameter, the better. I've used various sorts of random-length antennas made with the #26 wire and fed with ladder line on vacation trips (not necessarily all in our RV) and have worked stations all over the world, receiving good reports on all bands from 160 to 10 meters. As with the small nylon twine, the small wire can be used once and then thrown away, because its value is so low.

Here's another possibility for an RV fixed-location antenna, and is an extension of the idea behind the W1AB "Simple Killer Antenna" described in June 1999 *QST*. The original W1AB Killer Antenna is a 40-meter elevated ground plane antenna with a three-wire counterpoise, fed with ladder line via a balanced antenna tuner. The vertical wire and the three radials were all measured and cut exactly to length (33 feet, 3 inches). After seeing how well that antenna worked at my home station (and it still is in service), I started experimenting with variations on the idea when putting up portable antennas here and there during our travels. The technique I've used can be summarized pretty easily. Let me first warn you that the technique I use in making these portable Killer Antennas flies in the face of much of the accepted antenna theory. Some antenna theorists will insist that it is *Antenna Heresy*.

All I can say in defense of my technique is that it works very well, and that I've worked lots of good DX with 100 W (even in huge pileups) using these antennas, receiving good signal reports. Here's my technique, in six simple steps:

1. Run out one length of wire—50 feet is okay; 100 feet is better; 150 feet is dandy—and get it up in the air as high as possible. Bring the end that will be fed to a window in the ham shack. This Number One wire can be either horizontal, vertical, semi-vertical or inverted-L shaped. I've used all of those variations with good results. Note that it is not necessary to measure the wire's length; just make it long enough to fit between the supports that are available to you.

2. Run out a second wire, perhaps 40 or 50 feet long. Wire Number Two can either continue in or near the same line as wire Number One (but in the opposite direction), or it can be horizontal if Number One is vertical (or vice versa). Again, it's not necessary to measure the length; size it to fit between your available supports.

3. Run a third wire in a third direction and make wire Number Three a different length than wire Number Two (perhaps two-thirds as long). You're right; you don't need to measure its length. You're

catching on!

4. Bring one end of each wire to the outside of the RV, near the ham shack. Run a piece of ladder line from the ham station to the wires. Connect wire Number One to one side of the ladder line, and connect wires Number Two and Number Three to the other side of the ladder line.

5. Attach the ham-shack end of the ladder line to an antenna tuner. The antenna tuner must cover a wide range of impedances (up to 2000 Ω or greater), and it must have two-wire (balanced) output (or you must have an external 4:1 balun to convert an unbalanced output to balanced output). (Of course, antenna purists will point out that I'm connecting a balanced line to an unbalanced antenna, so who knows or cares whether the antenna tuner's output is balanced?)

6. Set your transceiver to low-power output (10 W or so) and adjust the antenna tuner for minimum reflected power on each band you intend to use. As you make these tuner adjustments, make a "cheat sheet" of the settings of the tuner's capacitance and inductance control settings, so you can retune to a given frequency quickly.

That's it. Now choose your band, adjust the antenna for that band, set the transceiver back to its normal output and operate.

What? You're wondering why there is a third wire and why it is a different length from the second wire? I found, when making two-wire antennas, that I would sometimes have trouble matching the antenna

with the antenna tuner to get a low SWR on a particular band (or bands). The feed point impedance was apparently a value that was outside the impedance tuning range of the antenna tuner.

I thought that if I added a third wire connected to the same side of the ladder line as the second wire, then the combination of wire Number One and wire Number Three might then fall within the tuner's impedance-matching range. I started using the third wire routinely and so far I've always been able to match my portable antennas on all bands.

You may ask, "How can that simplistic and nonscientific approach *possibly* work?!" In reply, here are some examples of my results (using my trusty TS-50 at 100 W output):

Operating from Nantucket Island, Massachusetts

I operated from Nantucket Island during a DX Test, and I snagged BY1PK through a huge pileup on 20 CW. Wire Number One was vertical and about 20 feet long, wire Number Two (horizontal, about four feet above the ground) was about 30 feet long, and wire Number Three (horizontal, also about four feet above the ground) was about 18 feet long. I worked a lot of other good DX during that contest weekend, much of it through pileups.

Operating from Hilton Head Island, South Carolina

Operating from Hilton Head, I tossed two wires out windows on opposite walls on the second-floor ham shack, and into the trees. The wires were about 70 and 50

"My Other RV Is a Motorcycle..."

More than once I've seen the bumper sticker on the back of an RV that said just that. There's a lot of truth in calling a motorcycle a recreational vehicle. The photo in **Fig 6-A** came from a Marconi Web site (**www.marconicalling.com/**), and I'm including it here for your edification and amusement. This photo—from 1914, in the very early days of radio—shows a completely portable Marconi 1/4-kW wireless telegraphy set mounted on two Douglas motorcycles for transport. The station was a miracle of miniaturization for that era. You can see the main radio set on the left-hand cycle, and the antenna masts and other antenna components on the right-hand cycle.

The technical description of the

station on the Web site explains, "They used demountable umbrella-type aerials, syntonic transmitters, and magnetic detector reception."

Fig 6-A—This 1914 Marconi 1/4-kW portable wireless telegraphy station could be carried by two Douglas motorcycles. Note the aluminum poles used to make an extension mast for the antenna. (Photo courtesy Marconi Corporation plc.)

feet and they ran sort of horizontal. I ran the third wire, about 30 feet long, around the inside the house. The feed point of the antenna was inside the second-floor ham shack. Although conditions were only fair during the two weeks we were there, I worked European and South American stations regularly on 40 and 80 meter CW, and I worked Europe and Central America on 160 meter CW.

Operating from Sint Maarten

In Sint Maarten (no, I didn't drive my RV there), our timeshare condo is right on the beach at Half Moon Bay in Philipsburg and our apartment is on the fourth floor. There was a short vertical pole about 140 feet from the condo. A wire from our balcony to that mast became wire Number One—parallel with the water's edge (about 50 feet from the water), and about 40 feet above the ground at the fed end.

Wire Number Two ran from the balcony downward into a palm tree below the balcony. It was nearly vertical and about 30 feet long. The third wire ran along the condo's side balcony (wooden) railing, around the corner of the building and along the front side of the front balcony railing. I used a piece of ladder line about 10 feet long to reach from the antenna tuner to the balcony, where I connected the three wires to it.

The results: I worked stations on most continents on all HF bands, with ease and with good reports. I worked up and down the East Coast of the USA with good signal reports on 160 CW and SSB. I was heard by two European stations (that I know of) on 160 CW, but couldn't hear European replies through the heavy static. I ran schedules often with my long-time friend Jack Colson, W3TMZ, in Florida. We usually had "pipeline" conditions, with strong signals and no interference.

My homemade QSL card, shown in **Fig 6-25**, shows where the three wires were run. The antenna wasn't anything fancy, but it sure worked well. These and my other portable Killer Antennas were made with #26 wire from my hamfest purchase. I used two or three large steel washers as a weight and tossed the wire into convenient trees. Often, after the wire was in the trees, I couldn't see it—unintentionally it became a stealth antenna.

I could give you other examples, but you get the idea. Before you pooh-pooh the concept, you should try it. You might be pleasantly surprised at how well such simple and unscientific antennas work!

OTHER RESOURCES
VHF/UHF RV ANTENNAS

VHF/UHF antennas for mobile RVing

will be essentially the same as those used for mobile automobile operating. Therefore, consult Chapter 1 in this book on automobile operation for ideas. VHF/UHF antennas for fixed-RV operation are another matter. Because of the relatively small wavelengths, you can have high-gain antennas and electrically tall masts. Here are some ideas for you to consider using with your RV.

RV Antennas for VHF/UHF Mobile Use

RV mobile operation on VHF/UHF is, for the most part, similar to automobile mobile operation. Please read the chapter in this book on VHF/UHF mobile antennas for automobiles, and you will find that most of those antennas will also work for RV mobile operation. There are a few differences, mostly in the possibilities of larger and better equipment that can be used in the mobile station, and in the possibilities of larger, high-gain antennas for FM use, such as $^5/_8$-wave verticals or two-element vertical collinear arrays.

The vertically polarized mobile VHF/UHF antennas used for FM operation are fairly short, so it's easy to mount them on

Fig 6-25—The PJ7/W1AB QSL card, with the antenna drawn in.

top of the RV, high and in the clear. If you're mounting a VHF/UHF mobile antenna on top of a fiberglass coach, you can use small-diameter aluminum rods or an aluminum sheet to make a counterpoise.

Fixed-RV Antennas for VHF/UHF

The fixed-RVer who operates VHF/UHF can enjoy the luxury of electrically tall antenna supports and/or high-gain directional antennas. Light- or medium-duty masts can support the smaller antennas. Some RVers use lightweight tower sections or aluminum or steel mast sections to mount more substantial, heavier VHF/UHF antennas, such as Yagi-Uda or log periodic[1]-dipole arrays. The towers are heavier and take up more space, but if you expect to winter over in the same location, they might be worth your while.

While taking a vacation from the Connecticut snow one recent winter, Mike Gruber, W1MG, ran across the trailer-mounted crank-up tower and tri-band beam of the Fort Meyers (Florida) Amateur Radio Club. The setup lends itself well to Field Day and emergency communication use. The accompanying photo in **Fig 6-26**, taken by Mike, shows the trailer. It's easy to see that the ideas used by the Fort Meyers ARC could be adapted for a fixed-RV antenna.

Here's an idea for you—a collapsible and easily stowable beam antenna for FM use. Back in the 1970s I planned to use a Cushcraft 10-element 2-meter beam for portable FM operation, but I wanted to modify it so it could be dismantled and stowed in a small package for transport, then reassembled quickly and easily, without needing hand tools. I won't go into the gory details, but here's the gist of what I did.

I cut the boom into two pieces and used a short piece of aluminum tubing that was

Fig 6-26—The W4LX trailer-mounted crank-up tower and tri-band beam could be adapted for fixed-RV use. (Courtesy W1MG.)

a snug fit inside the boom to support a butt splice. I drilling connecting holes through the pieces of tubing at the splice section, and used bolts and wing nuts to reconnect the boom at the splice.

Then I removed all the antenna elements from the boom and remounted them using bolts and wing nuts. With only about two hour's modification work, the beam was reconfigured so it could be taken apart and stowed in a small package. It could be re-assembled from its stowed package in about 10 minutes.

Don't forget that if you mount a VHF beam for vertical polarization for FM operation, you should *not* mount it directly onto a conductive mast. You can use a conductive mast as the lower support mast, but the top mast section should be made of nonconductive material. You can buy a three or four foot wooden dowel or a re-placement broomstick at your local hardware store that will fit inside the aluminum mast for use as the top section of the mast. Use two bolts to secure the wooden mast section to the top of the aluminum mast.

While it's more convenient to have an electrical rotator for your VHF/UHF beam antennas, it's possible to use the *Armstrong Rotator* system, quite popular with hams back in the 1950s. To use the Armstrong Rotator system, the antenna mast should be directly outside a window. When you want to rotate the beam, open the window and stick your strong arm out the window. Grasp the mast and turn the beam by hand to the desired direction. (You can have the mast out of arm's reach, but it becomes a lot less convenient to use.) You can run a long bolt through the mast to point in the same direction as the beam, as a directional indicator.

Lest you think I am pulling your leg, that system was widely used 50 years ago, when the world was a simpler place and radio technology wasn't yet so sophisti-cated. I remember seeing a photo in a *QST* of that era that showed a ham's home station with a lightweight 10 meter three-element beam mounted on the roof above the ham shack. The antenna mast came through the roof into the ham shack, ending right beside the operating position. The ham had connected an automobile steering wheel to the bottom of the mast, at about head level when he was seated at the operating position, which gave him a good handhold for rotating the mast and antenna. He had added a directional arrow to the steering wheel, so he could tell which direction he was beaming.

The Armstrong Rotator system will likely sound crude to the technical sophisti-cates in the current ham community. The only things I can think of to recommend it

for fixed-RV use is that it's inexpensive, it works well and it is not prone to mechani-cal or electrical failures.

One caution to the fixed-RV ham: As discussed in the following paragraphs, there is the strong possibility of TVI be-cause of the susceptibility of RV TV an-tenna systems and their broadband preamplifiers to RF. Therefore, if you plan to run higher power or high-gain antennas on VHF/UHF, don't plan to stay in a crowded RV park. Or be prepared to time-share the ether with your TV-viewer neighbor RVers.

MUTUAL INTERFERENCE AND THE FIXED-RV HAM

Because of the close quarters in most campgrounds, there is the strong possibil-ity of interference to other campers' TV sets (and other consumer electronic equip-ment) from your RV ham station. This is true for both HF and VHF/UHF operation.

Almost all standard-equipment TV an-tenna systems in RVs consist of shortened dipoles with built-in, high-gain, broad-band RF preamplifiers—a recipe for ham disaster. The TV antenna/preamp combi-nations make it a virtually certainty that the ham will cause TVI. Although there is a convenient switch for turning the preamp on and off, most RVers seldom leave it off, even in strong signal areas.

Of course, if you are staying in one of the upscale campgrounds that furnished cable TV at each RV campsite, you are far less likely to have TVI problems. If you do have it, one approach to the TVI problem is to introduce yourself to your camp-

ground neighbors and be completely forth-right about the problem. Explain to them that your ham station might cause TVI, because of inherent problems in the design of RV TV antenna systems. Offer to work with them on a time-sharing plan so you can operate part of the time and they can watch TV part of the time.

On the other hand, here's a sneaky trick that worked for me at a time I wanted to be on the air for long operating periods during the 1997 ARRL Phone Sweepstakes. I was operating in the contest from the San Joaquin Valley Section, using my TS-50 and wire antennas held up by my portable mast sys-tem. I was staying in an almost deserted campground in a California state park. See **Fig 6-27**. Midway through the contest, a big motor home pulled into the camping space next to mine (*Why* didn't he choose one of the 30 or so other spaces in the camp-ground?!).

Anticipating TVI problems with my new neighbor, I took a break from operating and turned on my own TV set, with my TV antenna's preamp turned off. I tuned the TV to a local station that had a strong signal and a perfect picture—with the preamp switched off, of course. I checked to be sure that I was not causing TVI and saw that the TV signal was clean on my TV. Then I turned the TV off and went back to the contest.

Sure enough, about an hour later, the campground attendant knocked on my door and said that my radio was causing interference to my neighbor's TV set. (It went without saying that the neighbor had his preamp switched on.)

I casually went over to my TV and flipped it on, showing the attendant its

Fig 6-27—The W1AB setup for the 1997 Phone Sweepstakes contest. The Hustler mobile antenna was used for the higher bands, and an inverted-V trap dipole, with its center supported by the 20-foot mast, for 40 and 75 meters. The photo on the right shows the operating position at the driver's seat.

good reception. Then I went over to the ham station and made a "Hello test" call. The attendant could hear the radio and he could see that I had no TVI on my set. At that, I commented, "That other guy must have some kind of problem with his TV setup."

The attendant left and went to the big motor home next door to tell them about the demonstration I had just given him. About ten minutes later, the big motor home pulled out of its space and took up residence at the other end of the campground. Problem solved!

Sometimes the interference shoe is on the other foot. There have been a number of cases where unintended oscillation in the TV preamps radiated signals that caused interference to both Amateur Radio and Public Safety communication in the 400 to 500 MHz range. One such case was reported in *The ARRL Letter*, Vol. 21, No. 20, May 17, 2002 (**www.arrl.org/arrlletter/02/0517/** near the bottom of the page), and another in an item in the ARRL Web "In Brief" column (**www.arrl.org/news/stories/2003/03/06/3/** about a quarter of the way down the page).

RVS FOR CONTESTING AND EMERGENCY WORK

RVs are a natural for both contesting and (real or simulated) emergency hamming. In addition to providing housing for the ham equipment, they have sleeping, cooking, and toilet facilities, and they often have on-board power sources. As you outfit your RV for hamming, you might keep these two possible uses in mind.

You can find photos and comments on contest use of RV stations and RV-mounted ARES emergency support vehicles in the Soapbox section of many of the ARRL contests. The following links will point out some comments of particular interest to the RV ham; you can browse through more of the Soapbox comments to find ideas for your own RV ham station and antennas, both mobile and fixed-RV.

- **www.arrl.org/contests/soapbox/index.html?con_id=68&ofst=40**— Soapbox comments and photos from Field Day 2004. In the first item, the KR7Q gang used several RVs for their multi-position station. The second Soapbox comment, from the W7DK group, shows some very interesting devices to get lines into tall trees for raising wire antennas (which is helpful if your RV is parked in a forested area). About two-thirds of the way down the page, you will see a photo of the portable tower arrangement used by Mac Mackey, KF4GLE, for his RV-mounted HF CW position.

- **www.arrl.org/contests/soapbox/index.html?con_id=68&bycall=**— RVer W1HP operated in the 2004 Field Day, way out in the desert, near Phoenix.
- **www.arrl.org/contests/soapbox/index.html?con_id=49&ofst=60**— Soapbox, 2003 Field Day. Ed Kazmarek, KG4TUL (now K4EAK), operated from his RV. I enjoyed reading about the new-ham enthusiasm Ed had! Ed was one of the many RV hams who used a G5RV antenna!
- **www.arrl.org/contests/soapbox/?con_id=64&call=K0MHC/4**—Jim Froemkhe, K0MHC/4, had a good mast and antenna setup for his 6-meter RV station.
- **www.arrl.org/contests/soapbox/index.html?con_id=78&ofst=30**— Alas, not all RV contesting efforts result in happy campers. Murray Lycan, VE7HA, found this to be true in the 2004 10-Meter Contest, where he operated from his 1968 Ultravan RV and enjoyed (?) excruciatingly bad conditions.

GENERAL INFORMATION ON RV HAMMING

There are quite a few World-Wide Web pages with information on and discussions of RV hamming. As always, these links can change from time to time, so by the time you're reading this, some of the following may be dead links. Here are some of the good links as we go to press:

- **www.wr6wr.com/newSite/articles/features/olderfeatures/aronthervtrail.html**—"Amateur Radio on the RV Trail," by J.M. Huckabee, AA5BU, from a *WorldRadio* article.
- **www.srhams.org/**—Sam's Radio Hams.
- **www.fmca.com/chapters/spotlight/2003/0203_arc.asp**—Family Motor Coach Association.
- **www.rvweb.net/club/fmcarc/**—the Amateur Radio Chapter of the Family Motor Coach Association.
- **www.rvweb.net/club/gsrvrnet/**—the RV Radio Network.
- **www.rvweb.net/club/wbcciarc/index.html**—the Wally Byam Caravan Club.
- **home.pacbell.net/lcobb/**—W6TEE's page, with lots of good links to other pertinent Web pages about RV hamming.
- **www.qsl.net/k4mg/HamRadioBookmarks.htm#Links/HamRadio/RV**—K4MG's ham radio links, including RV hamming.
- **www.rversonline.org/ArtHamRadio.html**—WA0JOG's comments on RV hamming.

- **www.murrah.com/hamrv/**—KY8T's comments on RV hamming.

For more information, do a Google search on "RV ham Amateur Radio" and you'll find plenty of interesting reading. Also search the ARRL Web page (**www.arrl.org**) for "RV" to find recent information.

There have also been articles published in the Amateur Radio magazines about RV hamming. One excellent published article that comes immediately to mind is the article written by Ruth (W9BRN) and Dick (W9SR) Stroud, published in October 1993 *QST*. It's a good overview of RV hamming, with Web links, net information and good photos of the Strouds' motor home ham shack. Get a copy of that issue and read the article, especially if you're just getting into RV hamming.

One last Amateur Radio site with lots of good information and useful links is the Web site of Rod Dinkins, AC6V. Of particular interest to the RV ham who is looking around for antenna ideas is Rod's page at **www.ac6v.com/antprojects.htm**.

HAM RADIO NETS FOR THE RV HAM

Many hams enjoy operating in nets on a routine basis. Others like to know how to find nets if they need some sort of help, especially in emergency situations. The following nets are of long standing, and their operating frequencies and schedules are not likely to change. Nevertheless, it is a good idea for you to make an up-to-date list of nets that you might want to check into, confirming the nets' schedules and frequencies periodically, to be sure that your list is current.

Note that, in the following list, the times are listed in UTC, for the Standard Time part of the year. During the Daylight Time period, some of the nets meet at the same local time (which makes them one hour earlier in UTC), while others adjust their time one hour earlier (to maintain the same time in UTC). You will need to check out these nets individually to make your own list of nets, current times, and frequencies.

- **Wally Byam Caravan Club ARC RV Net** (central USA), 3918 kHz, Mon-Fri, 2300 UTC.
- **New England RV Net**, 3963 kHz, daily, 0100 UTC.
- **Wally Byam CCARC RV Net** (Northeastern USA), 3963 kHz, Sun, 1230 UTC. Firebird Amateur Radio Club Net, 3977 kHz, Sun, 2000 UTC; Thurs, 2300 UTC.
- **RV Service Net**, 7233.3 kHz, daily, 1200-1400 UTC.
- **Wally Byam CCARC RV Net** (eastern and central USA), 7233.3 kHz, daily, 1200 UTC.
- **Wally Byam CCARC RV Net**(Moun-

tain Time zone), 7263 kHz (alternate 7268 kHz), Mon-Fri, 1500 UTC.
- **Wally Byam CCARC RV Net** (Pacific Coast), 7263 kHz (alternate 7268 kHz), Mon-Fri, 1600 UTC.
- **Firebird ARC Net**, 7277 kHz, Mon-Sat, 1700 UTC summer and 1800 UTC winter; also 2300 UTC Mon winter.
- **Good Sam Radio Ham Net**, 7281.5 kHz, Tue-Sat, 0130 UTC; ragchews, Sun-Mon.
- **Good Sam RV Radio Network** (eastern and central USA), 7284 kHz (alternate 7238 kHz), Tue-Sat, 0130 UTC.
- **Escapees Club** (operating sporadically), 7284 kHz, Sun, 0100 UTC, following the Good Sam net; 7268 kHz, Thurs, about 0830 ET (following the Waterways Net).
- **Family Motor Coach Assn AR Net** (USA and Canada), 14263 kHz, daily, 1900 UTC.
- **Family Motor Coach Assn AMTOR Net** (USA and Canada), 14090 kHz, daily, 1730 UTC.
- **Family Motor Coach Assn YL Net** (USA and Canada), 14263 kHz, daily, 1830 UTC.
- **Firebird ARC**, 14277 kHz, Sun, 1200 and 2100 UTC.
- **Good Sam RV Radio Network** (USA and Canada), 14240 kHz, Sun at 1900 UTC.
- **RV Service Net**, 14.307.5 kHz, Mon-Fri, 1700 and 2200 UTC.
- **Wally Byam CCARC RV Net** (USA, Canada, and Mexico), 14307.5 kHz, Mon-Fri, 1700 and 2200 UTC.
- **Avion Net**, 14307.5 kHz, Mon, Wed, Fri, 2000 UTC.
- **Firebird and Winnebago Combined Net**, 21377 kHz, Wed, 2230 UTC.
- **Firebird ARC Net**, 28377 kHz, Sat, 0000 UTC; 28343 kHz, Tue, 0000 UTC.

WHO WAS WALLY BYAM?

If you are a newbie to RVing, you may be wondering who Wally Byam (1896-1962) was. Wally Byam was the visionary who founded Airstream trailers, some of the prettiest travel trailers ever built. As an adolescent, Wally was a shepherd, living in a two-wheeled donkey cart outfitted with food and water, a kerosene cook stove and a sleeping bag. That experience apparently gave Wally a mindset toward

travel trailers that carried over into his adult years and later shaped his career.

In the 1920s, Wally became a magazine publisher. An article in one of his magazines described how to build a travel trailer. Readers soon complained that the plans, as published, did *not* work. Wally tried the plans himself and confirmed that his readers were correct.

He then designed a travel trailer and published an article about how to build the trailer for $100, selling plans for $5. Eventually, he started building travel trailers, following those plans, in his back yard and selling them.

Later on, Wally refined the design and started using construction methods developed for modern all-metal airplanes. He first used the "Airstream" name in 1934. See **Fig 6-28**.

You can find an excellent thumbnail history of Wally's RV work on the *Out West* newspaper Web site (**http://www.outwestnewspaper.com/airstream.html**). I recommend that you read it.

In surfing the Web, I found a good site that has some excellent photos of a lot of the older American travel trailers at **http://www.airstream.dk/hist.htm**. Because it's a Danish site, the captions are mostly in Danish but the photos alone are worth a visit.

Another article that provides even more information on Wally and his adventures appears at **http://www.wbcci-denco.org/Wally/Wally.html**, on the Web site of the Denver Unit of the Wally Byam Caravan Club International, a caravanning, rallying and social club for owners of Airstream RVs.

Wally wrote a book, *Trailer Travel: Here and Abroad,* that was published in 1960 by the David McKay Co. Although now out of print, a search for the book on **www.amazon.com** showed that one was available at the time this is being written, in "acceptable" condition, for $55. If you're

into Airstream trailers, you should try to find a copy of this book. You can find a review of the book written by Matt Bergstrom and published in the *Explorer Rag* at **www.geocities.com/ marmotamonax/Xrag/XRagTrailer.html**.

The Vintage Airstream Club's Web page at **www.airstream.net/wallybyam.html** has a nice photo of Wally, as well as a quote that shows Wally's desire to share the feeling of joy that came from going on the road in one's own RV. I can't think of a better way to close this chapter than to share a small part of "Wally Byam's Creed," as quoted on that page:

"... To strive endlessly to stir the venturesome spirit that moves you to follow a rainbow to its end ... and thus make your travel dreams come true."

ACKNOWLEDGMENTS

Sincere thanks to the several experienced and knowledgeable hams who helped me with information, ideas, and photographs for this chapter on RV mobile and fixed-RV hamming, especially the following:
- Les Cobb, W6TEE (or, as his CW friends call him, "Slow D"), who furnished many of the photographs in this chapter showing fixed-RV antenna systems, together with descriptions of them. Les is also to be commended for his excellent Web page about RV hamming.
- Whitey Doherty, K1VV, for information on and photos of his various mobile and portable HF antennas.
- Art Bell, W6OBB, for photos and information on his RV-mounted loop antenna that puts out such a big signal on all HF bands and 160 meters.
- Mike Gruber, W1MG, for several suggestions, and for the photo of the W4LX trailer-mounted tower and tri-band beam antenna.

Fig 6-28—Wally Byam's baby— a 1936 Airstream travel trailer. *(Courtesy Airstream.)*

APPENDIX A

Antennas and Projects

HF Mobile Antennas

This section is by Jack Kuecken, KE2QJ. Jack is an antenna engineer who has written a number of articles for ARRL publications.

An ideal HF mobile antenna is:

1. Sturdy. Stays upright at highway speeds.

2. Mechanically stable. Sudden stops or sharp turns do not cause it to whip about, endangering other vehicles.

3. Flexibly mounted. Permits springing around branches and obstacles at slow speeds.

4. Weatherproof. Handles the impact of wind, rain, snow and ice at high speed.

5. Tunable to all of the HF bands without stopping the vehicle.

6. Mountable without altering the vehicle in ways which lower the resale value.

7. Efficient as possible.

8. Easily removed for sending the car through a car wash, etc.

For HF mobile operation, the ham must use an electrically small antenna. The possibility that the antenna might strike a fixed object places a limitation on its height. On Interstate highways, an antenna tip at 11.5 feet above the pavement is usually no problem. However, on other roads you may encounter clearances of 9.5 or 10 feet. You should be able to easily *tie down* the antenna for a maximum height of about 7 feet to permit passage through low-clearance areas. The antenna should be usable while in the tied-down position.

If the base of an antenna is 1 ft above the pavement and the tip is at 11.5 ft, the length is 10.5 ft which is 0.1 λ at 9.37 MHz, and 0.25 λ at 23.4 MHz. That means that the antenna will require a matching network for all of the HF bands except 10 and 12 m.

The power radiated by the antenna is equal to the radiation resistance times the square of the antenna current. The radiation resistance of an electrically small antenna is given by:

$Rr = 395 \times (h/\lambda)^2$

where

h = radiator height in meters
λ = wavelength in meters = 300/Freq in MHz

The capacitance in pF of an electrically small antenna is given approximately by:

$$C = \frac{55.78 \times h}{((den1) \times (den2))}$$

where

$(den1) = (\ln(h/r) - 1)$

$(den2) = (1 - (f \times h/75)^2)$
ln = natural logarithm
r = conductor radius in meters
f = frequency in MHz

Characteristics of a 10.5-ft (3.2 m) whip with a 0.003 m radius and, assuming a base loading coil with a Q of 200 and coil stray capacitance of 2 pF, are given in **Table 1**.

Radiation resistance rises in a nonlinear fashion and the capacitance drops just as dramatically with increase in the ratio h/λ. **Fig 1** shows the relationship of capacitance to height. This can be used for esti-

---

Table 1

Characteristics of a 10.5-foot whip antenna

F (MHz)	C (pF)	Rr	Impedance	Efficiency	L (μH)
1.8	30.1	0.146	13.72 −j2716	0.01064	240
3.5	30.6	0.55	7.43 −j1375	0.074	62.5
7	32.8	2.2	7.04 −j644	0.312	14.6
10	36.5	4.5	6.5 −j408	0.692	6.49
14	46.5	8.8	10 −j232	0.88	2.64

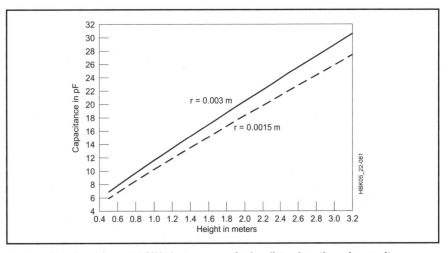

Fig 1 — Relationship at 3.5 MHz between vertical radiator length and capacitance. The two curves show that the capacitance is not very sensitive to radiator diameter.

mating antenna capacitance for other heights.

Fig 2 shows that capacitance is not very sensitive to frequency for h/λ less than 0.075, 8 MHz in this case. However, the sensitivity increases rapidly thereafter.

Table 1 shows that at 3.5 MHz an inductance of 62.5 μH will cancel the capacitive reactance. This results in an impedance of 7.43 Ω which means that additional matching is required. In this case the radiation efficiency of the system is only 0.074 or 7.4%. In other words, nearly 93% of energy at the terminals is wasted in heating the matching coil.

System Q is controlled by the Q of the coil. The bandwidth between 2:1 SWR points of the system = 0.36 × f/Q. In this case, bandwidth = 0.36 × 3.5/200 = 6.3 kHz.

If we could double the Q of the coil, the efficiency would double and the bandwidth would be halved. The converse is also true. In the interest of efficiency, the highest possible Q should be used!

Another significant factor arises from the high Q. Let's assume that we deliver 100 watts to the 7.43 Ω at the antenna terminals. The current is 3.67 A and flows through the 1375-Ω reactance of the coil giving rise to 1375 × 3.67= 5046 VRMS (7137 Vpeak) across the coil.

With only 30.6 pF of antenna capacitance, the presence of significant stray capacitance at the antenna base shunts currents away from the antenna. RG-58 has about 21 pF/foot. A 1.5-foot length would halve the radiation efficiency of our example antenna. For cases like the whip at 3.5 MHz, the matching network has to be right at the antenna!

Base, Center or Distributed Loading

There is no clear-cut advantage in terms of radiation performance for either base or center-loaded antennas for HF mobile. Antennas with distributed (or continuous) loading have appeared in recent use. How do they compare?

Base Loading

In the design procedure, one estimates the capacitance, capacitive reactance and radiation resistance as shown previously. One then calculates the expected loss resistance of the loading coil required to resonate the antenna. There is generally additional resistance amounting to about half of the coil loss which must be added in. As a practical matter, it is usually not possible to achieve a coil Q in excess of 200 for such applications.

Using the radiation resistance plus 1.5 times the coil loss and the power rating desired for the antenna, one may select the

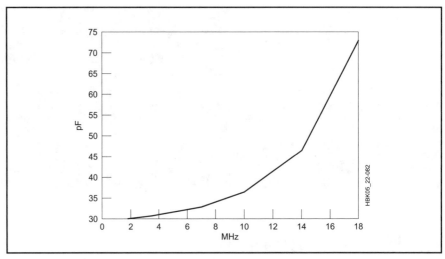

Fig 2 — Relationship between frequency and capacitance for a 3.2-meter vertical whip.

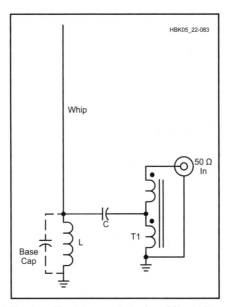

Fig 3 — The base-matched mobile whip antenna.

wire size. For high efficiency coils, a current density of 1000 A/inch^2 is a good compromise. For the 3.67 A of the example we need a wire 0.068-inch diameter, which roughly corresponds to #14 AWG. Higher current densities can lead to a melted coil.

Design the coil with a pitch equal to twice the wire diameter and the coil diameter approximately equal to the coil length. These proportions lead to the highest Q in air core coils.

The circuit of **Fig 3** will match essentially all practical HF antennas on a car or truck. The circuit actually matches the antenna to 12.5 Ω and the transformer boosts it up to 50 Ω. Actual losses alter the required values of both the shunt inductor and the series capacitor. At a frequency of 3.5 MHz with an antenna impedance of 0.55 −j1375 Ω and a

Table 22.22
Values of L and C for the Circuit of Fig 22.83 on 3.5 MHz

Coil Q	L (μH)	C (pF)	System Efficiency
300	44	11.9	0.083
200	29.14	35	0.0372
100	22.2	58.1	0.014

base capacitance of 2 pF results in the values shown in **Table 2.** Inductor and capacitor values are highly sensitive to coil Q. Furthermore, the inductor values are considerably below the 62.5 μH required to resonate the antenna.

This circuit has the advantage that the tuning elements are all at the base of the antenna. The whip radiator itself has minimal mass and wind resistance. In addition, the rig is protected by the fact that there is a dc ground on the radiator so any accidental discharge or electrical contact is kept out of the cable and rig. Variable tuning elements allow the antenna to be tuned to other frequencies.

Connect the antenna, L and C. Start with less inductor than required to resonate the antenna. Tune the capacitor to minimum SWR. Increase the inductance and tune for minimum SWR. When the values of L and C are right, the SWR will be 1:1.

For remote or automatic tuning the drive motors for the coil and capacitor and the limit switches can be operated at RF ground potential. Mechanical connections to the RF components should be through insulated couplings.

Center Loading

Center loading increases the current in

the lower half of the whip as shown in **Fig 4**. One can start by calculating the capacitance for the section above the coil just as done for the base loaded antenna. This permits the calculation of the loading inductance. The center loaded antenna is often operated without any base matching in which case the resistive component can be assumed to be 50 Ω for purposes of calculating the current rating and selecting wire size for the inductor.

The reduced size top section results in reduced capacitance which requires a much larger loading inductor. Center loading requires twice as much inductive reactance as base loading. For equal coil Qs, loss resistance is twice as great for center loading. If the coil is above the center, the inductance must be even larger, and the loss resistance increases accordingly. These factors tend to negate the advantage of the improved current distribution.

Because of the high value of inductance required, optimum Q coils are very large. One manufacturer of this type coil does not recommend their use in rain or inclement weather. The large wind resistance necessitates a very sturdy mount for operation at highway speed. Owing to the Q of these large coils the use of a base matching element in the form of either a tapped inductor or a shunt capacitor is usually needed to match to 50 Ω.

Another manufacturer places the coil above the center and uses a small extendable *wand* for tuning. To minimize wind resistance, the coil lengths are several times their diameters. These antenna coils are usually close wound with enameled wire. The coils are covered with a heat-shrink sleeve. If used in heavy rain or snow for extended periods water may get under the sleeving and seriously detune and lower the Q of the coils. These antennas usually do not require a base matching element. The resistance seems to come out close enough to 50 Ω.

It is possible to make a center-loaded antenna that is remotely tunable across the HF bands; however, this requires a certain amount of mechanical sophistication. The drive motor, limit switches and position sensor can be located in a box at the antenna base and drive the coil tuning mechanism through an electrically isolated shaft. Alternatively, the equipment could be placed adjacent to the loading coil requiring all of the electrical leads to be choked off to permit RF feeding of the base. The latter choice is probably the most difficult to realize.

Continuously Loaded Antennas

Antennas consisting of a fiberglass sleeve with the radiator wound in a continuous spiral to shorten a CB antenna from 8.65 feet to 5 or 6 feet have been on the market for many years. This modest shortening has little impact on the efficiency but does narrow the bandwidth.

One line of mobile antennas uses periodic loading on a relatively small diameter tube. A series of taps along the length are used to select among the HF bands. An adjustable tip allows one to move about a single band. Because the length to diameter ratio is so large the loading coil Q is relatively low. The antenna is most effective above 20 meters.

The Screwdriver Antenna

The screwdriver antenna consists of a top whip attached to a long slender coil about 1.5 inches in diameter. The coil screws itself out of a base tube which has a set of contact fingers at the top. For lower frequencies more of the coil is screwed out of the base tube and at maximum frequency the coil is entirely *swallowed* by the base tube.

The antenna is tunable over a wide range of frequencies by remote control. It has the advantage that the drive mechanism is operated at ground potential with RF isolation in the mechanical drive shaft. On the other hand, the antenna is not easily extended to 10.5 foot length for maximum efficiency on 80 and 40 meters. Because of its shape, coil Q will not be very high.

Digital Versus Analog Couplers

Digital HF antenna couplers were first used by the military about 1960 for radios with Automatic Link Establishment. In this mode, the military radio has a list of frequencies ranging 2 to 30 MHz. It will try these in some sequence and will *lock* on the frequency giving the best reception. During the search, frequencies change much too fast to permit the use of conventional roller coils and motor driven vacuum capacitors. By comparison the digital coupler can jump from one memory setting to another in milliseconds.

For matching a mobile whip, the circuit shown in **Fig 5** will suffice. The inductor and capacitor can each be made up of about 8 binary sequenced steps. For example, at 3.5 MHz, the 10.5-foot antenna has an impedance of about 0.55 −j1684 Ω. From Table 2 we see that we could use a series inductance sequence of 20, 10, 5, 2.5, 1.25, 0.625, 0.32 and 0.16 μH. We can use a relay to short unwanted elements. In this way we could theoretically produce any value of inductance between 0 and 39.84 μH in steps of 0.16 μH. In reality you will never reach a zero inductance. With all of the relays shorted, the wiring inductance and contact inductance of 8 relays appear in series. Also, each of the coils will have the open circuit capacitance of a relay contact across it in addition to the normal stray capacitance.

With most relays it does not make sense to switch less than 2 pF. For that reason, the capacitance chain would consist of 2, 4, 8, 16, 32, 64, 128 and possibly 256 pF. This would give a maximum of 510 pF and

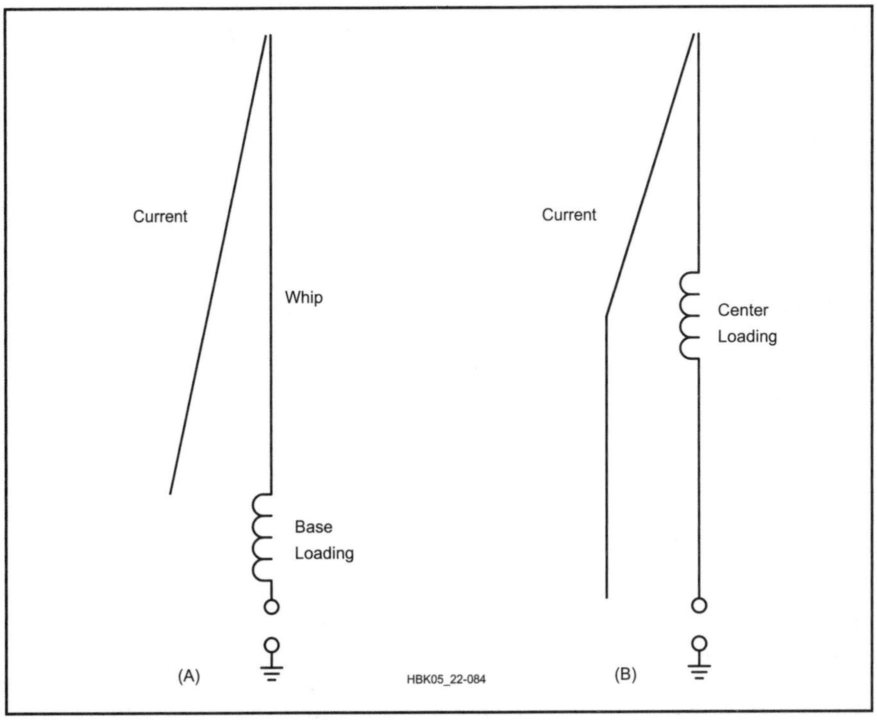

Fig 4 — Relative current distribution on a base-loaded antenna is shown at A and for a center-loaded antenna at B.

HBK05_22-084

a step size of 2 pF. Each relay has an open circuit capacitance of about 1 pF, and that gives a minimum capacitance of 8 to 9 pF. As a practical matter, there is also the stray capacitance between the relay contacts and the coil windings.

In a high Q matching circuit that handles 100 W, the individual relays must handle 4 or 5 kV with the contacts open and several amperes of RF with the contacts closed. If we can unkey the transmitter so that the coupler will not have to switch under power, we'll still need some sizeable relays. If the inductors have lower Q, the voltages and currents will be correspondingly lower. Some military couplers use Jennings vacuum latching relays. This is expensive, as each of the 16 or 17 relays costs more than $100.

If coil and antenna Qs are kept or forced low, the voltages and currents become more reasonable. However, if the antenna size is restricted this reduction comes only at the cost of decreased efficiency. A commercially available ham/marine digital coupler employs RF reed relays rated for 5 kV and 1 A, and restricts the power at low frequencies if the antenna is small. Another ham/marine unit uses small relays in series where voltage requirements are great and in parallel where current requirements are great—not good engineering practice. A third offering is not too specific about the power rating with very high Q loads.

There are no successful examples of 100 W plus couplers that use PIN diode switching. Their use is highly problematic given the high-Q loads they would handle.

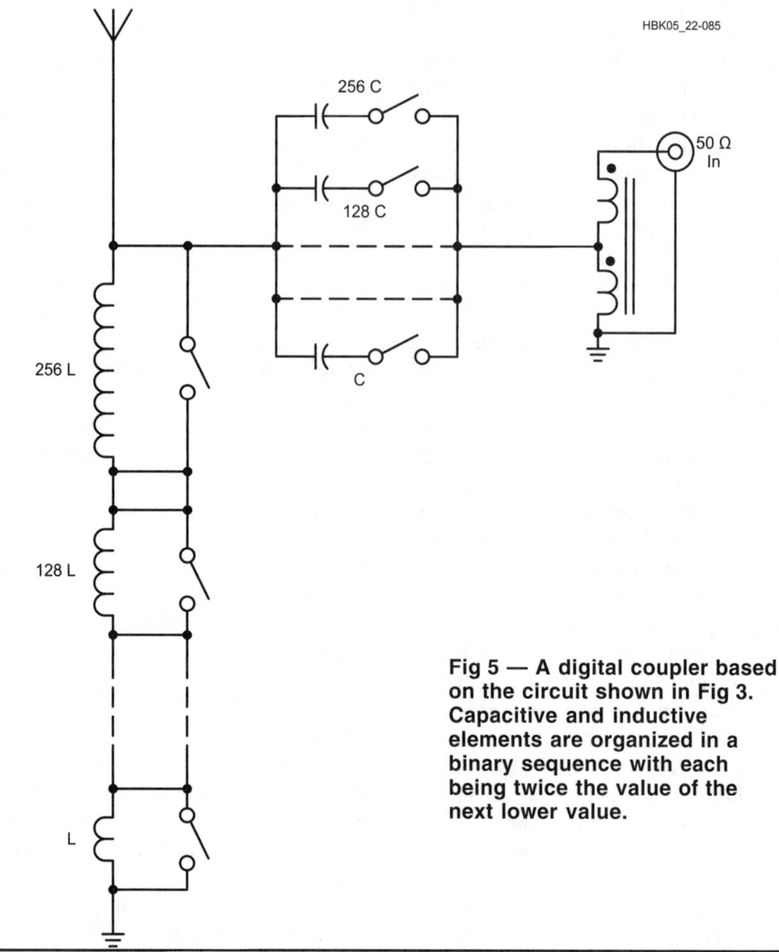

A Remotely Tuned Analog Antenna Coupler

KE2QJ built an antenna coupler designed for 100-W continuous-duty operation that will tune an antenna 10 feet or longer to any frequency from 3.5 to 30 MHz. With longer antennas, the power rating is higher and the lowest frequency is lower. The design requires only hand tools to build; however, access to a drill press and a lathe could save labor.

The roller coils and air variable capacitors to be used are not widely manufactured these days. Tube-type linear amplifiers still use air variable capacitors but these are generally built on order for the manufacturer and are not readily available to consumers in small quantity.

Until the 1970s, E. F. Johnson manufactured roller coils and air variable capacitors that were suitable for kilowatt amplifier finals and high power antenna couplers. On occasion one or more of these may be found in the original box, but they tend to be expensive. Ten Tec and MFJ

Fig 5 — A digital coupler based on the circuit shown in Fig 3. Capacitive and inductive elements are organized in a binary sequence with each being twice the value of the next lower value.

both manufacture antenna couplers and offer some components in small quantities.

The following data refer to generic motors, capacitors and inductors. The descriptions are intended to aid the builder in selecting items from surplus, hamfest flea market offerings or salvage of old equipment.

The Motors

Two motors are required, one to drive the inductor and one to drive the capacitor. The design employs permanent magnet dc gearhead motors with a nominal 12-V rating. A permanent magnet (PM) motor can be reversed by simply reversing the polarity of the drive voltage, and its speed can be controlled over a wide range by pulsing the power on and off with a variable duty cycle. The motor should have an output shaft speed on the order of 60 to 180 r/min (1 to 3 r/s) although this is not critical. New, such motors, can cost as much as $65 to $150 in small quantities. However, they can be found surplus and in repair shops for a few dollars.

The motor you are looking for is 1 to 1.5 inches in diameter and perhaps 2.5 inches long. It might be rated 12 or 24 V and have a $1/_4$-inch diameter output shaft. At 12 V it should have enough torque to make it hard to stop the shaft with your fingers. Tape recorders, fax machines, film projectors, windshield wipers and copiers often use this type of motor.

Limit Switches

On a remotely operated unit it is usually necessary to have limit switches to prevent the device from *crashing* into the ends. On an external roller coil these can be microswitches with paddles mounted on each end of the coil. As the coil is wound to one end, the roller operates the paddle and opens the limit switch which stops the motor.

Fig 6A illustrates a simple motor control circuit. Relay K1 is arranged as a DPDT polarity reversing switch. If switch CCW is pressed, the motor rotates CCW and the steering diode D2 prevents the relay from operating. If CW is pressed, relay K1 operates reversing the polarity at the motor. The motor is energized through the steering diode.

Fig 6B shows how to add limit switches. The diodes across the switches are called anti-jam diodes. When a switch opens, the diode permits current to flow in the reverse direction and the motor to move the roller away from the open switch.

The photograph of **Fig 7** shows the mounting of the switches on the coil. The diode should be a power rectifier type rated for several times the motor current and at least 60 V.

Position Reader

While not necessary, it is worthwhile to have a way to determine inductor position. An easy way to do this is to couple a 10-turn potentiometer to the coil shaft or drive gears. Make sure that the potentiometer turns less than 10 turns between limits. Don't try to make it come out exact.

Because the potentiometer is a light mechanical load, a belt drive reduction works well and won't slip if properly tensioned. Fig 7 shows the potentiometer and the gear drive.

You may be able to find suitable gears. However, a belt drive requires less precise shaft positioning than fine tooth gears. With a lathe, pulleys can be made in almost any ratio. Vacuum cleaner belts and O rings make handy belts.

Couplings

In this coupler circuit, both ends of the capacitor are *hot* with RF although the end adjacent to the transformer is at relatively low voltage. Nevertheless, the capacitor shaft must be insulated from the motor shaft. The coil can be driven from the grounded end. Insulation is not necessary, but use a coupler between the motor shaft and the coil to compensate for any misalignment. Universal joints and insulating couplings are available from most electronics supply houses. You can make a coupling from a length of flexible plastic tubing which fits snugly over the shafts. Clamp the tubing to the shafts to avoid slippage.

The Capacitor

The easiest capacitor to use is an air variable. It should have a range of approximately 10 to 250 pF. The plate spacing should be 2 mm ($^1/_{16}$ inch) or more, and the

Fig 6 — At A, motor control circuit used by KE2QJ. This circuit uses pulse modulation for speed control with good starting torque. Direction of rotation is controlled by the relay. At B, how to add limit switches to the circuit. See text.

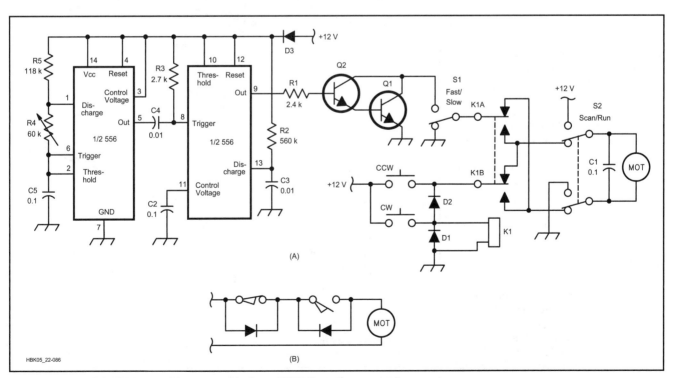

Fig 7 — Photo of the inductor drive assembly from KE2QJ's antenna coupler.

plate edges should be smooth and rounded. The capacitor should be capable of continuous 360° rotation, and it would be nice if it had ball bearings. The straight-line capacitance design is best for this application. Several capacitors of this type are available in military surplus ARC-5 series transmitters. These are approximately 2 × 2 × 3 inches.

The capacitor must be mounted on stand-off insulators although high voltage will not be present on the frame. A cam that briefly operates a microswitch when the capacitor goes through minimum can be used to flash an LED on the remote control panel. This provides an indication that the capacitor is turning.

The Inductor

As calculated earlier, and assuming an inductor Q a bit under 300 is attainable, the roller inductor for this coupler should have a maximum inductance on the order of 40 µH. The wire should be at least #14 AWG wound about 8 t/inch.

You can use **Table 3** as a guide to buy a roller coil at a hamfest. The seller may not know the inductance of the coil. The antenna loading coil from an ARC-5 transmitter will work, but the wire is a bit small.

You could make the loading coil by threading 2, 2.5 or 3-inch diameter white, thick-wall PVC pipe with 8 t/inch. If the pipe is threaded in a lathe, the wire can be wound into the threads under considerable tension. This helps to prevent the wires from coming loose with wear or temperature.

The Transformer

The transformer consists of a bifilar winding on an Amidon FT-114-61 core. Start with two 2-foot lengths of #18 insulated wire; Teflon insulation is preferable. Twist the wire with a hand drill until there are about 5 t/inch (not critical). Wind 12 turns onto the core. This should about fill it up. Attach the starting end of one wire to the finish end of the other. This is the 12.5-Ω tap. One of the free ends is grounded and the other is the 50-Ω tap. Mount the coil on a plastic or wooden post through the center of the coil. A metal screw can be used as long as it does not make a complete turn around the core.

Construction

For ease of service, mount the inductor, its drive motor, position sensing potentiometer and limit switch assembly on a single aluminum plate. A plug and socket assembly permits rapid disconnection and removal. Make a similar assembly for the capacitor, its drive motor, transformer and the interrupter. Both assemblies should be

Table 3
Data for 40 µH Coils

Diameter	Length	Turns
2.3 inch (58 mm)	5.625 inch (143 mm)	45
2.8 inch (71 mm)	4.25 inch (108 mm)	34
3.3 inch (84 mm)	3.375 inch (86 mm)	27

made on $^1/_{16}$ to $^1/_4$-inch thick aluminum. These individual assemblies make it easier to fix problems.

The chassis shown in **Fig 8A** is made of a single piece $^1/_{16}$ to $^3/_{32}$-inch aluminum bent in an L shape. Two chassis-stiffening braces are riveted in place. Alternatively, the chassis can be made of flat sheets with aluminum angles riveted around the edge.

Mount the coil and capacitor assemblies parallel to the long leg of the L. Punch a 1-inch hole in the center of the short end of

the L. Cover the hole with an insulator made of PVC, Teflon or other suitable material.

The rest of the case is a 4-sided wooden assembly as shown in Fig 8B. The back wall of the box is drilled to accept the two pivot pins. The box is slid over the chassis and the pivot pins engaged. The tie-down screw secures the box. For service, remove the tie-down screw and slide off the cover. The works of the coupler are very easy to get at!

The box is made of $^1/_4$-inch exterior grade plywood except for the back plate, which is $^3/_8$ or $^1/_2$-inch plywood. The sides, top and back should overlap the flanges on the chassis by $^1/_2$ inch. The inside corner seams of the box should be reinforced with $^1/_2$ or $^3/_4$-inch square strips. Assembly can be with any water resistant glue.

Finish the box, inside and out, with several coats of clear urethane varnish, sanding lightly between coats. This leaves a smooth plastic finish. This can be sprayed

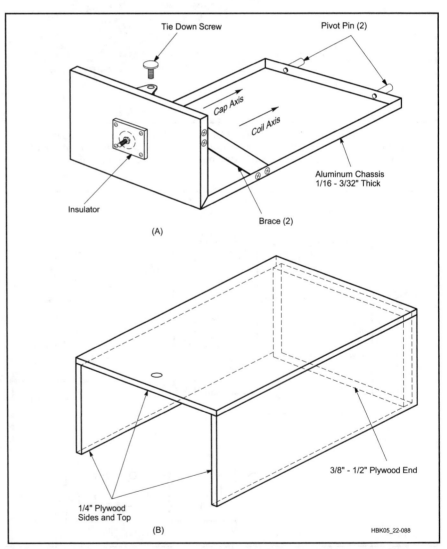

Fig 8 — At A, the chassis for the coupler mounting box. At B, the box cover.

with an exterior paint that matches your car's color.

If the sides of the box fit closely over the flanges, no fastening beside the tie-down screw is required. A nearly perfect seal will leak out hot air when the sun shines on it and will draw in cold damp air in the evening, trapping moisture inside. A moderate fit will keep rain and snow out and permit the box to *breathe* freely, thereby keeping the inside dry.

Mounting the Whip and the Box

Plastics in bumpers and bodies makes the mounting of a mobile whip antenna problematic. Modern bumpers are covered with plastic and the bumper is attached to the car unibody through a *5-MPH* shock absorber. The latter item is an unreliable ground.

The arrangement of **Fig 9** solves many of these problems. It uses a $^1/_4$-inch aluminum plate 6 to 8 inches wide and long enough to fit between a reasonably strong place on the unibody and the place behind the bumper where the antenna wants to be. This plate is fitted with an angle bracket for the lower bolt on the shock/bumper mounting. This plate is stiffened with a length of $1 \times 1 \times \frac{1}{4}$ inch aluminum angle bolted in several places.

Near the forward edge of the plate, two $^1/_2$-inch diameter aluminum shear posts are fitted. The bottom of each is tapped 10-32 and bolted through the mount plate with a stainless 10-32 screw. At the top of these shear posts another piece of $1 \times 1 \times$ $^1/_2$-inch angle is attached which is screwed to the unibody with three or four #10 stainless sheet metal screws. A bracket attaches the mount plate to the bumper's shock absorber. The angle bracket may either be welded to the plate or bolted with angle stock. In the event that the car is hit from behind or backs into an obstacle, the two 10-32 screws will shear off, thereby preventing the mount from defeating the 5 MPH crushable shock absorber. The part protruding behind the bumper may be cut down in width to 3 inches and rounded for appearance and safety.

Any type of base insulator may be used, but try to bring the base of the antenna to the height of the coupler output terminal. You can make a good base insulator from thickwall white PVC $1^1/_2$-inch pipe. Reinforce each end. Start with a $1^1/_2$-inch length of pipe. Remove a $^5/_8$-inch wide strip so the remaining portion can be rolled and pressed into the open end of the insulator. Apply PVC pipe glue just before pressing in the piece; this gives the insulator a double wall thickness at each end. Aluminum plugs can be turned for a snug fit and tapped for $^3/_8$-24 hardware. The plugs can be held in place with 8-32 stainless screws.

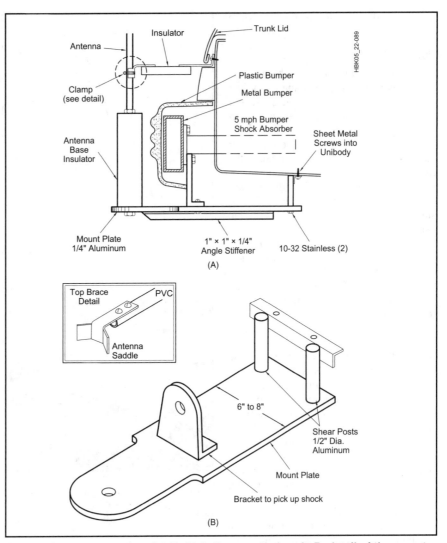

Fig 9 — Antenna mounting detail. At A, the overall plan. At B, detail of the mount plate.

The upper antenna brace has an aluminum plate at one end that goes under the trunk lid (see Fig 9). A length of $^1/_2$-inch diameter heavy-wall white PVC pipe, which serves as an insulator, is screwed to this. At the other end of the insulator, another aluminum piece is bent to form a saddle for the antenna which is clamped to the saddle. This clamp should be as high as convenient above the mount plate, preferably not less than a foot. The mount plate should be sturdy enough for you to stand on and with the brace will easily hold a whip upright at 70 MPH or more.

The coupler box mounting is shown in **Fig 10**. Brackets can be made of $^1/_8 \times$ 2-inch aluminum with a brace going perhaps 2 inches from the corner. The brackets bolt or rivet to the chassis. The bracket reaches through the gap between the trunk lid and the plastic top of the bumper. For reinforcement, a pair of reinforcement plates $1.5 \times 2 \times ^1/_4$-inch thick are bolted to

the plastic on the under side of the bumper. Ground the reinforcement plates to the unibody with some $^3/_4$ or 1 inch ground braid.

Two 10-32 stainless screws hold each reinforcement plate to the plastic bumper and a central $^1/_4$-20 tapped hole holds down the box bracket. One need only remove two screws to get the box off the car for car wash, etc. You have to open the trunk to remove the antenna coupler, and this provides a measure of security.

The Spring and Whip

A section of 1-inch diameter aluminum tubing extends from the top of the insulator to the base of the spring. It's usually best to have the spring about 4 ft above the pavement. The type used for CB whips works well. A 7-ft whip brings the top to about 11.5 ft above the pavement. The 7-ft whip can be a cut down CB unit. Don't use the type with helical winding. When the antenna is tied down, the bow of the whip should be about 7 ft above the pavement.

Tuning

It is best to initially tune the antenna using low power. A power attenuator just after the transceiver will limit SWR, but your SWR indicator must be on the antenna side of the attenuator.

For a first tune-up, set the capacitor control to SCAN and slowly advance the inductor from minimum inductance toward maximum. As the inductor approaches the correct value the SWR will start to kick down. At this point take the capacitor off of SCAN and JOG it to a best tune. Next, JOG the inductor and repeat; the SWR should go down. Continue until a 1:1 SWR is obtained. Record the potentiometer setting. The next time you want to use this frequency run the coil directly to the logged setting.

In the SCAN position the capacitor motor runs at full voltage. When you JOG the capacitor for low SWR you will find the speed far too fast for sharp tuning. The slow speed tuning is provided by using duty-factor modulation of the motor current. The circuit of Fig 6A supplies fixed width pulses with variable timing. At the slowest speed, the unit will supply about one pulse per second and the motor shaft will rotate one degree or so per pulse. The full voltage pulse provides good starting torque.

If the SWR cannot be brought to 1:1, examine the coil and capacitor to see whether either is at maximum or minimum. At high frequencies above 24 MHz it may be necessary to place a capacitor between the coupler and the antenna base.

Radiation Patterns

At the lower frequencies the pattern tends to be essentially round in azimuth. At 20 meters the pattern tends to become more and more directive. The patterns in **Fig 11** were calculated using *EZNEC*. The frequency is 18.13 MHz and the antenna is mounted at the left rear corner of a mid-size sedan. It may be seen that the pattern has more than 10 dB maximum-to-minimum ratio with the

broad maximum along the diagonal of the vehicle occupied by the antenna. If the antenna were mounted in the center of the vehicle, the omnidirectional characteristics would be improved. However, the antenna would have to be much shorter to stay under 11.5 feet. The shorter antenna would likely be weaker in its best direction than the taller antenna is in its worst.

References

J. S. Belrose, VE3BLW, "Short Antennas for Mobile Operation," *QST*, Sept 1953, pp 30-35, 109.

J. Kuecken, KE2QJ, *Antennas and Transmission Lines*, MFJ Publishing, Ch 25.

J. Kuecken, KE2QJ, "A High Efficiency Mobile Antenna Coupler," *The ARRL Antenna Compendium, Vol 5*, pp 182-188.

J. Kuecken, KE2QJ, "Easy Homebrew Remote Controls," *The ARRL Antenna Compendium, Vol 5*, pp 189-193.

J. Kuecken, KE2QJ, "A Remote Tunable Center Loaded Antenna," *The ARRL Antenna Compendium, Vol 6*.

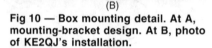

Fig 10 — Box mounting detail. At A, mounting-bracket design. At B, photo of KE2QJ's installation.

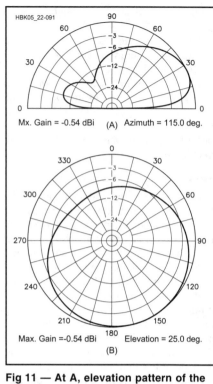

Fig 11 — At A, elevation pattern of the KE2QJ mobile antenna. The pattern is in the plane that runs diagonal through the car. At B, azimuth pattern at 25° elevation for the same antenna. The operating frequency is 18.130 MHz.

By Robin Rumbolt, WA4TEM

From *QST*, March 1993

An Easy, On-Glass Antenna with Multiband Capability

Are you mobile on one VHF/UHF band, or two?
Either way, this on-glass antenna design is for you!

W ith every new car purchase comes the agonizing decision of where to punch the hole for my 2-meter antenna. Recently I purchased a dual-band transceiver, and the problem became where to punch two holes. I'd rather punch no holes at all!

An on-glass antenna seemed like the ideal solution. Such antennas couple RF through the windshield without the need to drill holes for cables and mounting hardware. Being a builder at heart, I designed an on-glass antenna to suit my needs. Not only does it feature the ability to disconnect the radiating element quickly (for car washes, etc), it has multiband capability, too!

Construction

I built the base of my antenna out of heatsink material (see Figure 1). I happened to find a piece of bare aluminum heat-sink stock with long, straight fins. Each fin was spaced about 1/4 inch apart. You can find similar heat-sink material at your local hamfest flea market. It's cheap and relatively easy to machine. You can also use aluminum channel stock, which is available from a variety of sources.

The first step was to cut out a piece roughly 1 5/8 inches square and remove all but the two middle fins. The fins were a bit too tall, so I carefully trimmed them to 1/2 inch in height. I used a grinding wheel to round the corners and drilled 9/64-inch holes in the centers of both fins.

The antenna coupling plate is cut from a piece of sheet steel. Its dimensions equal those of the antenna mount. (Avoid using aluminum for the coupling plate, since it's very difficult to solder.)

The quick-disconnect assembly is made from two hexagonal brass standoffs just wide enough to fit snugly between the fins. One standoff has a hole threaded through its entire length. The other standoff has a threaded stub on one end and a threaded hole in the other. I carefully drilled a 9/64-inch hole through the open end of the sec-

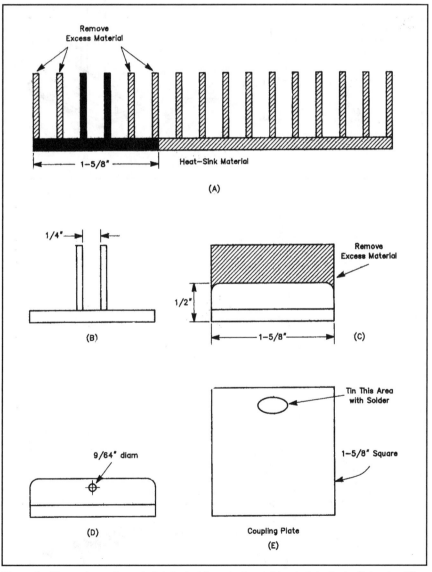

Figure 1—The antenna base is fabricated from a piece of heat-sink stock. Cut out a 1 5/8-inch section and remove all but the two center fins (A and B). Round off the sharp corners of the fins and trim for a 1/2-inch height (C). Drill a 9/64-inch hole through the centers of both fins (D). The coupling plate is cut from a 1 5/8-inch section of sheet steel. Tin a small area as shown at E.

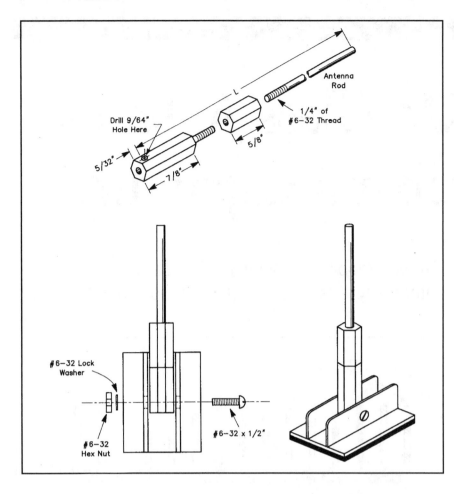

Figure 2—The on-glass antenna is made of brass welding rod attached to two threaded standoffs (see text). Drill a #6-32 hole in the bottom standoff as shown. Using a #6-32 × 1/2-inch screw, nut and lock washer, secure the antenna to the base. The total length of the antenna (L) is measured from the tip of the welding rod to the mounting screw. Use the lengths listed in Table 1 and then trim as necessary to obtain a low SWR.

ond standoff. Using a #6-32 × 1/2-inch screw, I assembled the standoff to the base as shown (see Figure 2).

The radiating element is made of 3/32-inch brass welding rod. I cut a #6-32 thread about 1/4 inch up one end. This end is screwed tightly onto the first standoff. If you lack the tools to thread the rod yourself, use 1/16-inch welding rod and solder it to the standoff.

The total antenna length depends on the band you wish to use. See Table 1 for approximate lengths for various bands. As you can see in Figure 2, the finished section screws onto the stub of the base-mounted standoff. Whenever I need to remove it, a few twists is all it takes!

The coupling plate and the antenna base are attached to the windshield with double-sided foam tape (Radio Shack 64-2361). One tape strip isn't wide enough to cover the base and the plate, so I applied two strips side-by-side. It was a simple matter to cut the strips, peel off the backing and apply the tape to each piece. Any excess is easily trimmed away. The important thing to remember is not to peel the paper backing from the tape until just before you're ready to install the antenna.

Mounting

As you search for just the right spot to mount your antenna, bear in mind that you must ground the coaxial cable shield to the car body near the mount. In most cars, the top center of the front or rear windshields is best. Older cars usually have screws to attach the molding in these areas. These screws can often be used for grounding. If you own a newer car without strategically located screws, you'll have to install one yourself. In my Dodge Caravan, I drilled a small hole in the roof support (not the roof itself!) and used a small sheet-metal screw to fasten a solder lug in place. Whichever approach you use, check the screw with your VOM and ensure that it really makes contact with the chassis of your car. Many screws anchor in metal, but the metal isn't always grounded!

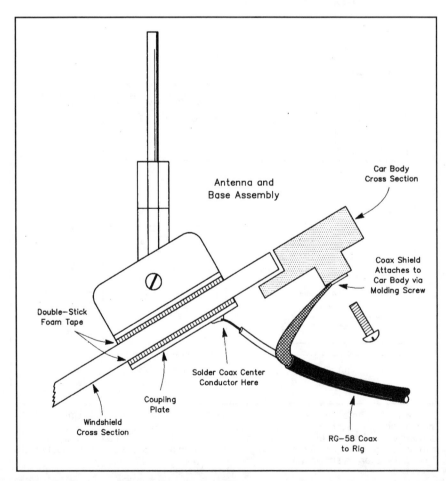

Figure 3—Use strong foam tape to hold the base and the coupling plate to the windshield. The coaxial cable center conductor is soldered to the coupling plate. The braid is grounded to the car body via a nearby molding screw. The braid must be grounded at the antenna for proper performance.

Appendix A.10   Antennas and Projects

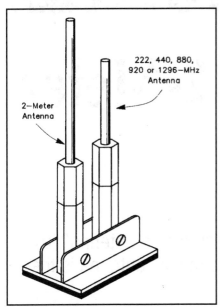

Figure 4—The multiband option. You can mount two antennas in the same base. This is ideal for today's dual-band, VHF/UHF transceivers.

Table 1
Antenna Lengths for Various Bands

Frequency (MHz)	Length (inches)
145	23¹/₄
223	15¹/₁₆
444	7¹⁷/₃₂
880 (cellular telephone)	11³/₈ (³/₄ λ)
920	11 (³/₄ λ)
1296	7³/₄ (³/₄ λ)

Hold the base to your windshield in the area where you intend to install it. Adjust the antenna until it is vertical, then tighten the nut. Remove the mount and spray paint the entire assembly black—or whatever color looks best with your car!

When the paint is dry, clean the glass thoroughly (inside and outside). Check your chosen antenna location one more time. Is it in the path of windshield wipers? If you open the trunk or hatch, will the antenna be crushed?

If everything looks safe, peel the paper from the foam tape and attach the base to the outside glass. Press firmly to ensure that the tape sticks to the surface. Attach the coupling plate to the inside glass directly opposite the base. Solder the center coax conductor to the coupling plate and connect the coax shield to the ground screw or lug.

Tuning

With an accurate SWR/power meter, make SWR measurements and begin prun-ing the antenna for the lowest SWR. In my 2-meter installation, with 50 watts forward power, the needle doesn't even wiggle in the reverse-power position (a 1:1 SWR). If you can't get the SWR below 3:1, check your coax ground at the antenna. This is often the culprit.

Multibanding

I saved the best for last! You can create a dual-band antenna by simply mounting a second antenna and quick-disconnect assembly on the same mounting base (see Figure 4). For example, here's a fancy system for hams who own cellular phones: Install 2-meter and 880-MHz antennas in the same mount. The 2-meter whip will do double duty on 2 meters and 70 cm, while the 880-MHz antenna is perfect for your cellular telephone. An antenna farm on glass! (This configuration must be fed with a single feed line and an appropriate diplexer must be purchased or homebrewed.[1])

Conclusion

I am extremely happy with the antennas I have made using this on-glass method. No external holes were necessary and the antennas disconnect easily. SWR is low on every band and the antenna's radiation efficiency seems to rival any hole-mounted antenna I've used in the past!

[1]D. Jenkins, "A Simple VHF/UHF Diplexer," QST, October 1991, pp 18-25.

By Bill English, N6TIW

From *QST*, April 1991

A Glass-Mounted 2-Meter Mobile Antenna

Want a no-holes, no-paint-scratching antenna? This easy-to-build glass-mounted mobile antenna is the answer!

To me, mag-mount antennas are a pain in the neck. Yet, I couldn't bear to drill a hole in my car to install a permanent mobile antenna. So, I found an alternative. In this article, I'll tell you how to build an attractive glass-mounted antenna that looks like those used for cellular-telephone service.

The antenna is easy to build. No special tools or skills are required, yet the antenna's appearance is compact and high tech. Instead of having an untidy mess on your car roof, people will think you are one of the nouveau riche. If you're not up to building one of the antennas, you can get one ready to install.[1]

The System

Figure 1 shows a schematic of a glass-mounted antenna system. At A is the electrical equivalent of a $1/4$-λ antenna. Why the need for L and C, and how does the signal get from the coax to the antenna without a connecting wire? The reason—and the way—is capacitance. The mounting plates on either side of the glass act as capacitor plates separated by a glass dielectric—your windshield or window. This capacitor exhibits a negative reactance. To cancel the negative reactance, inductance (L) is used. Because C and L have equal but opposite reactances at the frequency of interest, the feed line sees just the $1/4$-λ antenna. This antenna can be a simple $1/4$-λ whip, or it can be physically shortened with a loading coil to give it the cellular-phone look. More on this later.

Note that the feed-line shield is grounded near the capacitor. Because we're using a $1/4$-λ antenna, this is an important part of the system. Obviously glass makes a bad ground plane, but there is still plenty of metal around to do the job.

Antenna Design

The size of the antenna mount was

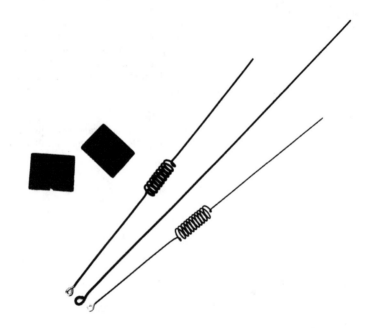

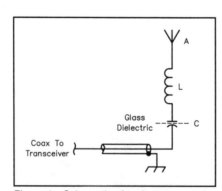

Figure 1—Schematic of a glass-mounted antenna system. The signal passes through the glass via the plates of the antenna mount, which form capacitor C. This capacitance is canceled by inductance L. Therefore, the feed line just sees the $1/4$-λ antenna A.

determined by structural and appearance considerations. I made the mount as big as my taste would allow. I calculated the mount's capacitance to be 10 to 20 pF. (I can't calculate it any closer because I'm uncertain about the dielectric constant and thickness of the glass and adhesive.) This capacitance range equates to a reactance of about 60 to 120 ohms at 2 meters. Therefore, the inductive reactance required to cancel the mount's capacitive reactance is 60 to 120 ohms.

How do you build the right amount of inductance into the antenna? Look at Figure 2. This graph is from *The ARRL Antenna Book* (Fig 36, p 2-36). When an antenna is shorter than 90 degrees, it looks capacitive to the feed line. When it's longer than 90 degrees, it looks inductive. Armed with this information, I went on two design paths:

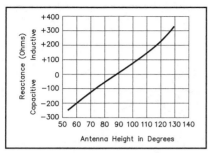

Figure 2—Approximate reactance of a vertical antenna over perfectly conducting ground. The wavelength/diameter ratio is about 2000. Actual reactance values vary considerably with wavelength/diameter ratio.

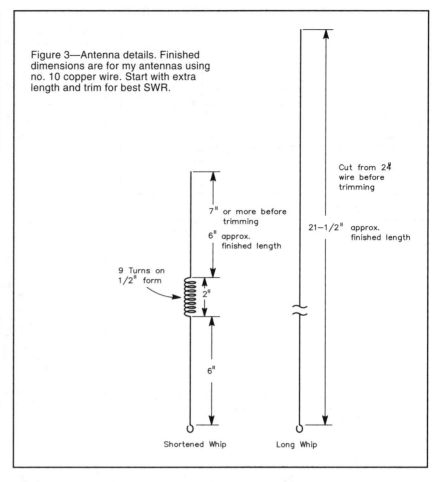

Figure 3—Antenna details. Finished dimensions are for my antennas using no. 10 copper wire. Start with extra length and trim for best SWR.

• Make a whip antenna longer than 90 degrees to provide the inductance needed to cancel the base mounting capacitance. From Figure 2, this is about 96-106 degrees, which calls for an antenna length of 20 to 22½ inches. Using no. 10 copper wire, my antenna ended up being 21½ inches long.

• Make a shortened antenna with a loading coil large enough to cancel the capacitive reactances of the base mounting capacitance and the shortened (less than 90 degrees) antenna length. I picked a 60-degree antenna length, which requires about 200 ohms of inductive reactance.[2] Adding an appropriate amount of inductance to cancel the base capacitance—and allowing some extra—I decided on a ½-inch diameter, 9-turn coil 2 inches long. This is all very approximate, but we don't have to worry too much about precision here. All sins will be forgiven when the antenna is trimmed.

Antenna Elements

I tried several materials for making antenna elements. Stainless steel is by far the most durable. It is hard to work in sizes over 1/16 inch, but 1/16-inch welding rod is workable with ordinary tools. Bare copper wire is by far the easiest to work, but it is too flexible unless first stresshardened by stretching.[3] (To ensure proper mounting-flange spacing, be sure to stretch the wire before making the antenna mount described in the next section.) Stress hardened copper wire stands up well to freeway driving. Feel free to experiment with other materials.

Figure 3 gives finished dimensions for antennas I've made. The length for best SWR may be slightly different when the antenna is mounted on your car, but that shouldn't matter because you'll be starting with a longer element length and trimming to size.

Making a long whip is easy. Cut a 24-inch length of straight wire. Bend one end into a small loop with needle-nose pliers. Be sure the loop will pass a no. 6 screw. Except for painting, you're done. Making the shortened antenna is only a little more complicated (see Figure 4). Sand the antenna wire before forming the coil. To make the coil, clamp one end of a 3-ft length of straight wire in a vise. Bend the free end into a small loop for the long whip. Then place the center of a ½-inch wooden dowel six inches from the center of the loop and roll the wire up on the dowel for 9 turns, moving toward the vise. Remove the wire from the vise and cut off the bent end. Hold one end of the coil with pliers and bend the free end so it cuts through the axis of the coil perpendicular to it. Then hold the bent portion with pliers near the coil edge and bend the wire so it is on the axis of the coil. Bend the free end on the other side of the coil along the coil's axis in the same way. Cut the end that was in the vise 7 inches or more from the coil. Space the turns so that the finished coil is about 2 inches long.

The Antenna Mount

The antenna mount (see Figs 5 and 6) is made from brass strips that are readily available at hobby stores. The details are shown in Figure 5. The inside plate is a 1½ × 2-inch piece of 0.016-inch brass with a 3/16-inch-wide tab cut from it. Brass of this thickness can be cut easily with metal shears. Just make two cuts ¼ inch long and bend the tab down. The tab is shown at the top of the plate in Figs 5 and 6, but it could be placed on the side if that better suits you. (The feed line will eventually be soldered to this tab.)

Now assemble the outside mount. Note that the right-angle flanges that hold the antenna are made from 1-inch-wide by 0.025-inch-thick brass strip. You'll need heavy-duty metal shears or a hacksaw to cut this brass. Start with 1 × 1-inch pieces and cut the corners off as shown in Figure 5. Then bend them by clamping them in a vise and hammering them over. Drill the hole in one flange before soldering to the mounting plate. To solder the flanges to the outside mounting plate, first apply solder to both parts. Then, press the flange onto the plate with a soldering iron to melt the solder. Before removing the iron, apply pressure with a screwdriver to keep the flange in place until the solder hardens. An iron with a reasonable amount of power is needed. (I used my 140-W gun.) After mounting the first flange, space the second flange from the first using the antenna wire. After both flanges are soldered in place, drill the second hole using the first hole as a guide. Be sure to file off all burrs, and for a neat appearance, round the corners and file off all solder blobs and flows. Set the mount aside for now.

Painting

Although it's not necessary to do so, I painted my mount and antenna. (*Don't paint the sides of the mounts that will be glued to the glass.*) Flat black paint provides a clean, finished look. Before painting, clean off all flux and sand surfaces

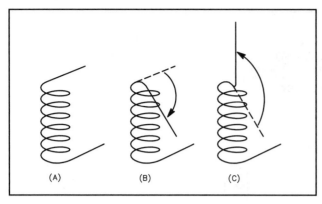

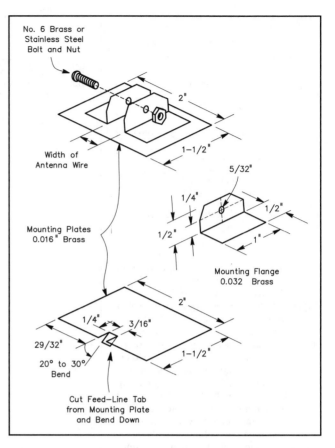

Figure 4—Bending the shortened antenna. After forming the coil (A), bond one free end so it crosses the coil axis perpendicular to it (B). Then bend up along the coil axis (C). Repeat this procedure on the other end so both ends are along the coil axis.

Figure 5—Antenna-mount details. Solder mounting flanges centered on the top mounting plate. The mounting flanges are cut from brass strips available at hobby stores.

with fine sandpaper until they are uniformly clean and bright. To paint, bolt the antenna in the outside mount and stand it upright. This makes it easier to paint and ensures that the electrical contact surfaces are not coated.

Brass doesn't take paint well. I primed my first prototype with auto primer. Although the paint looked fine, it chipped easily. An etching primer works much better. A primer designed for brass is best, but primers designed for aluminum will work too. Flecto Ferrothane Surfabond no. 52 is one of these. (Marine supply stores carry brass primer.) Once the metal is primed, apply a finish coat of a good quality flat black paint intended for metal. I used Rust-Oleum Bar-B-Q Black. It is a high-temperature paint that stands up to soldering.

Assembly

The most popular position for mounting on-the-glass antennas seems to be the top center of the rear window.[4] I have a station wagon, so this wasn't convenient for me. I located my antenna at the top center of the windshield. I like it in this position because the inside mount is neatly behind the rear view mirror and the coax route to the rig is short.

Before sticking the mount to the glass, solder the center conductor of the feed line to the inside plate's tab. Solder the coax braid to a ring lug, leaving enough braid exposed to reach a nearby screw in the headliner trim or mirror mount for grounding to the vehicle body. The grounding screw should be close to the mount, near the glass. If you don't have a screw conveniently close, put one there. The antenna won't work right without a properly grounded shield. If needed, drill a small hole in the headliner trim into the mounting metal, and insert a self-tapping or sheet-metal screw. After installing the mount, remove the screw, place the ring lug over it and replace the screw.

I tried several adhesives for the mount, but the easiest to use is double-sided foam tape. Radio Shack tape (RS 64-2361) ap-

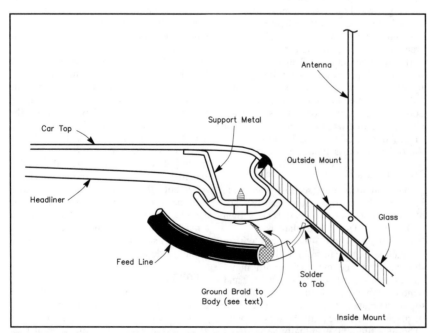

Figure 6—Antenna-mount installation. Mounting plates are applied to opposite sides of the glass. The antenna is bolted to the outside mount. The center conductor of the feed line is soldered to the inside mount. Ground the coax braid to the body using a nearby screw.

pears to be Scotch no. Y-4950, which is described as having "high peel strength and excellent weatherability." The tape comes in 1-inch-wide strips so use two strips side by side on each plate. Clean the glass and the mounting plates with rubbing alcohol before application. Apply tape to the mounting plate first, then stick it to the glass. Once it's applied, don't move it. If you make a mistake, remove the mounting plate, clean it up and put on a new adhesive strip. Although this tape should weather well, it's a good idea to seal around the edges of the outside plate with a thin bead of a silicone sealer (such as Radio Shack 64-2314).

Use RG-58 cable for the feed line if you mount the antenna on the windshield. This small-diameter coax is easily routed to the transceiver. RG-58 has rather high loss at 2 meters. For the lengths used here, however, losses are below 1 dB, so they won't be noticeable. For line lengths over 10 or 15 feet, use RG-8X (Radio Shack RG-8M) coax. It has a slightly larger diameter than RG-58, but losses are lower.

Routing the feed line to your rig requires some ingenuity if you want the feed-line run to be invisible, or almost so. If you install the antenna on the vehicle's rear window, you may want to route the coax under the headliner. Look for screws or other headliner retainers around the trim edges. You should be able to drop some of the headliner and snake the coax through. I tucked the feed line into a gap between the headliner and the windshield trim along the top of the windshield, then used small dabs of cement to run it in an indentation along the door post to my transceiver. This worked well because my car's black trim matched the coax outer covering. Terminate the transceiver end of the feed line with a connector to match the one on your rig. If you're going to use a hand-held transceiver, cut the coax a bit long.

Pruning the Antenna

Once the antenna is installed, trim it for minimum SWR. Insert an SWR meter between the rig and the feed line. With your rig set for low power, check the SWR near the bottom and top ends of the 2-meter band. The SWR should be lower at the low end of the band. If not, and both SWR readings are over 2:1, the antenna is too short. There isn't much you can do to fix this except make a new antenna element. Most likely, however, everything will be okay. Using wire cutters, trim about 1/8 inch from the antenna tip, rechecking the SWR near the center of the band. It's best to remove only 1/8 inch at a time to avoid cutting off too much. Keep trimming the antenna length and checking the SWR until you get it near 1:1. Check the SWR near the band edges and at your usual frequencies of operation. Trim carefully if you want better SWR at the higher end of the band. I trimmed my antenna for lowest SWR at 146 MHz. At the band edges, the SWR of my antenna is 1.4:1, giving good performance over the whole band.

Summary

There you have it! With a glass-mounted antenna, there is no need to drill holes through your roof or put up with a magmount mess. A glass-mounted antenna gives you convenience and a high-tech look.

My operating experience confirms that these antennas perform about the same as a regular $1/4$-λ antenna. The long whip should have slightly more gain than the shortened one, but both should be within 1 dB of a roof-mounted $1/4$-λ antenna. This is plenty good unless you are in a fringe area. In fact, higher-gain antennas won't necessarily help you if you are in a "hole" in hilly terrain. This situation is common where I live in the San Francisco Bay area. I have been very pleased with the shortened antenna's performance and appearance.

Notes
[1]Contact William J. English, 81 Meadow View Rd, Orinda, CA, 94563. (The ARRL and *QST* in no way warrant this offer.)
[2]Actually, this coil-sizing method applies only to a base-loaded antenna with a coil small enough that its radiation can be neglected. Our antenna is center loaded (this increases the inductance required) with significant radiation (reduces inductance required). Rather than sorting these effects out mathematically, I just made the element long and trimmed for best SWR.
[3]Here's how to stretch no. 10 copper wire. Place one end of a length of wire in a vise. Wrap several turns of the other end around a crowbar about 6 inches from the fulcrum end. Leave enough free wire between the vise and bar for the antenna you are making. Place the fulcrum end of the bar against the edge of your work bench opposite the vise, or against another solid object. Pull until you feel the wire give several inches. Not only will this stiffen the wire, but it also straightens out all the kinks. Cut the wire at the crowbar end. Leave the straight wire clamped to the vise to form the coil if you are making the shortened antenna.
[4]A letter from Joseph Butcher, KE9FZ, prompted your editor to seek some information on the RF properties of tinted window glass. Mounting on glass antennas to factory-tinted (deep-tint) windows, he says, results in problems with high SWR.

Rex Greenshade of the Ford Motor Company told me that he believes OEM-specified tinted glass does not cause the problem: Aftermarket heavily tinted glass is the gremlin. Rex said that the maximum OEM glass tinting must conform to certain specifications, one reason being that police officers must be able to view the occupants of the vehicle when approaching on foot.

Ken Brown of the Ford Motor Company told me that no Ford OEM glass tints interfere with RF. However, Instaclear—not a tint—does interfere with the transmission of AF. Radar cannot penetrate it and it does interfere with the installation of glass-mount antennas. Instaclear is a Ford option that permits rapid clearing of ice and snow from a car's windshield. Instaclear glass has a conductive powder added to the layers of windshield glass. Although Instaclear is not designed as a tinting element, under certain lighting conditions and viewing angles it can appear as a pink or bronze tint that is highly reflective and virtually opaque. Ken said that Instaclear does not violate any existing state laws governing tinted window glass.

Scott Staedtler, N8ILG, of Antenna Specialists Co. told me that in his experience, the degree of tinting does indeed have an effect on glass-mounted antennas. He's aware of Instaclear, but claims that the OEM privacy glass tint (as used on Ford vans and Broncos, for instance) as well as aftermarket tints can cause problems. Rear-window defoggers/deicers do not interfere with on-glass antennas so long as you avoid mounting the antenna on the wire trace(s).—*Ed.*

Index

FEEDBACK

Please use this form to give us your comments on this book and what you'd like to see in future editions, or e-mail us at **pubsfdbk@arrl.org** (publications feedback). If you use e-mail, please include your name, call, e-mail address and the book title, edition and printing in the body of your message. Also indicate whether or not you are an ARRL member.

Where did you purchase this book?
   ☐ From ARRL directly   ☐ From an ARRL dealer

Is there a dealer who carries ARRL publications within:
   ☐ 5 miles   ☐ 15 miles   ☐ 30 miles of your location?   ☐ Not sure.

License class:
   ☐ Novice   ☐ Technician   ☐ Technician with code   ☐ General   ☐ Advanced ☐ Amateur Extra

Name _____ ARRL member? ☐ Yes ☐ No

Call Sign _____

Daytime Phone (   ) _____ Age _____

Address _____

City, State/Province, ZIP/Postal Code _____

If licensed, how long? _____ e-mail address: _____

Other hobbies _____

_____

Occupation _____

For ARRL use only	ARM
Edition	1 2 3 4 5 6 7 8 9 10 11 12
Printing	1 2 3 4 5 6 7 8 9 10 11 12

From _____

_____

_____

EDITOR, AMATEUR RADIO ON THE MOVE
ARRL—THE NATIONAL ASSOCIATION FOR AMATEUR RADIO
225 MAIN STREET
NEWINGTON CT 06111-1494

— — — — — — — — — — — — — — — please fold and tape — — — — — — — — — — — — — — — — —